Fracture Mechanics

Fracture Mechanics

Edited by **Arthur Bloomberg**

NY RESEARCH PRESS
New York

Published by NY Research Press,
23 West, 55th Street, Suite 816,
New York, NY 10019, USA
www.nyresearchpress.com

Fracture Mechanics
Edited by Arthur Bloomberg

International Standard Book Number: 978-1-63238-204-7 (Hardback)

Printed in China.

Contents

Preface

In my initial years as a student, I used to run to the library at every possible instance to grab a book and learn something new. Books were my primary source of knowledge and I would not have come such a long way without all that I learnt from them. Thus, when I was approached to edit this book; I became understandably nostalgic. It was an absolute honor to be considered worthy of guiding the current generation as well as those to come. I put all my knowledge and hard work into making this book most beneficial for its readers.

Fracture mechanics is the field of mechanics related to the study of the propagation of cracks in materials. This book provides a collection of researches regarding the functions of fracture mechanics procedures in materials science, medicine and engineering. Major topics discussed in this book include the strength of biological tissues, safety of nuclear reactor components, effects of fatigue in pipelines, environmental effects on fracture etc. The book also discusses mathematical and computational steps underlying fracture mechanics functions, and the advancements in statistical modeling of fatigue. It will be most beneficial to mechanical and civil engineers, and material scientists from industry, research and education.

I wish to thank my publisher for supporting me at every step. I would also like to thank all the authors who have contributed their researches in this book. I hope this book will be a valuable contribution to the progress of the field.

Editor

Section 1

Computational Methods of Fracture Mechanics

Chapter 1

Foundations of Measurement Fractal Theory for the Fracture Mechanics

Lucas Máximo Alves

Additional information is available at the end of the chapter

1. Introduction

A wide variety of natural objects can be described mathematically using fractal geometry as, for example, contours of clouds, coastlines , turbulence in fluids, fracture surfaces, or rugged surfaces in contact, rocks, and so on. None of them is a real fractal, fractal characteristics disappear if an object is viewed at a scale sufficiently small. However, for a wide range of scales the natural objects look very much like fractals, in which case they can be considered fractal. There are no true fractals in nature and there are no real straight lines or circles too. Clearly, fractal models are better approximations of real objects that are straight lines or circles. If the classical Euclidean geometry is considered as a first approximation to irregular lines, planes and volumes, apparently flat on natural objects the fractal geometry is a more rigorous level of approximation. Fractal geometry provides a new scientific way of thinking about natural phenomena. According to Mandelbrot [1], a fractal is a set whose fractional dimension (Hausdorff-Besicovitch dimension) is strictly greater than its topological dimension (Euclidean dimension).

In the phenomenon of fracture, by monotonic loading test or impact on a piece of metal, ceramic, or polymer, as the chemical bonds between the atoms of the material are broken, it produces two complementary fracture surfaces. Due to the irregular crystalline arrangement of these materials the fracture surfaces can also be irregular, i.e., rough and difficult geometrical description. The roughness that they have is directly related to the material microstructure that are formed. Thus, the various microstructural features of a material (metal, ceramic, or polymer) which may be, particles, inclusions, precipitates, etc. affect the topography of the fracture surface, since the different types of defects present in a material can act as stress concentrators and influence the formation of fracture surface. These various microstructural defects interact with the crack tip, while it moves within the material, forming a totally irregular relief as chemical bonds are broken, allowing the microstructure to be separated from grains (transgranular and intergranular fracture) and microvoids are joining (coalescence of microvoids, etc..) until the fracture surfaces depart. Moreover, the characteristics of macrostructures such as the size and shape of the sample and notch from which the fracture is initiated, also influence the formation of the fracture surface, due to the type of test and the stress field applied to the specimen.

After the above considerations, one can say with certainty that the information in the fracture process are partly recorded in the "story" that describes the crack, as it walks inside the material [2]. The remainder of this information is lost to the external environment in a form of dissipated energy such as sound, heat, radiation, etc. [30, 31]. The remaining part of the information is undoubtedly related to the relief of the fracture surface that somehow describes the difficulty that the crack found to grow [2]. With this, you can analyze the fracture phenomenon through the relief described by the fracture surface and try to relate it to the magnitudes of fracture mechanics [3 , 4 , 5 , 6 , 7, 8, 9 - 11, 12, 13]. This was the basic idea that brought about the development of the topographic study of the fracture surface called fractography.

In fractography anterior the fractal theory the description of geometric structures found on a fracture surface was limited to regular polyhedra-connected to each other and randomly distributed throughout fracture surface, as a way of describing the topography of the irregular surface. Moreover, the study fractographic hitherto used only techniques and statistical analysis profilometric relief without considering the geometric auto-correlation of surfaces associated with the fractal exponents that characterize the roughness of the fracture surface.

The basic concepts of fractal theory developed by Mandelbrot [1] and other scientists, have been used in the description of irregular structures, such as fracture surfaces and crack [14], in order to relate the geometrical description of these objects with the materials properties [15].

The fractal theory, from the viewpoint of physical, involves the study of irregular structures which have the property of invariance by scale transformation, this property in which the parts of a structure are similar to the whole in successive ranges of view (magnification or reduction) in all directions or at least one direction (self-similarity or self-affinity, respectively) [36]. The nature of these intriguing properties in existing structures, which extend in several scales of magnification is the subject of much research in several phenomena in nature and in materials science [16 , 17 and others]. Thus, the fractal theory has many contexts, both in physics and in mathematics such as chaos theory [18], the study of phase transitions and critical phenomena [19, 20, 21], study of particle agglomeration [22], etc.. The context that is more directly related to Fracture Mechanics, because of the physical nature of the process is with respect to fractal growth [23, 24, 25, 26]. In this subarea are studied the growth mechanisms of structures that arise in cases of instability, and dissipation of energy, such as crack [27, 28] and branching patterns [29]. In this sense, is to be sought to approach the problem of propagation of cracks.

The fractal theory becomes increasingly present in the description of phenomena that have a measurable disorder, called deterministic chaos [18, 27, 28]. The phenomenon of fracture and crack propagation, while being statistically shows that some rules or laws are obeyed, and every day become more clear or obvious, by understanding the properties of fractals [27, 28].

2. Fundamental geometric elements and measure theory on fractal geometry

In this part will be presented the development of basic concepts of fractal geometry, analogous to Euclidean geometry for the basic elements such as points, lines, surfaces and fractals volumes. It will be introduce the measurement fractal theory as a generalization of Euclidean measure geometric theory. It will be also describe what are the main mathematical conditions to obtain a measure with fractal precision.

2.1. Analogy between euclidean and fractal geometry

It is possible to draw a parallel between Euclidean and fractal geometry showing some examples of self-similar fractals projected onto Euclidean dimensions and some self-affine fractals. For, just as in Euclidean geometry, one has the elements of geometric construction, in the fractal geometry. In the fractal geometry one can find similar objects to these Euclidean elements. The different types of fractals that exist are outlined in Figure 1 to Figure 4.

2.1.1. Fractais between $0 \le D \le 1$ *(similar to point)*

An example of a fractal immersed in Euclidean dimension $I = d + 1 = 1$ with projection in $d = 0$, similar to punctiform geometry, can be exemplified by the Figure 1.

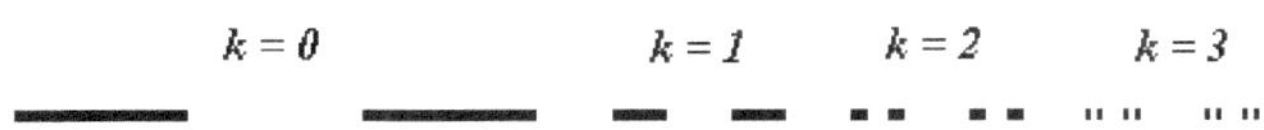

Figure 1. Fractal immersed in the one-dimensional space where $D \cong 0{,}631$.

This fractal has dimension $D \cong 0{,}631$. This is a fractal-type "stains on the floor." Other fractal of this type can be observed when a material is sprayed onto a surface. In this case the global dimension of the spots may be of some value between $0 \le D \le 1$.

2.1.2. Fractais between $1 \le D \le 2$ *(similar to straight lines)*

For a fractal immersed in a Euclidean dimension $I = d + 1 = 2$, with projectionin $d = 1$, analogous to the linear geometry is a fractal-type peaks and valleys (Figure 2). Cracks may also be described from this figure as shown in Alves [37]. Graphs of noise, are also examples of linear fractal structures whose dimension is between $1 \le D \le 2$.

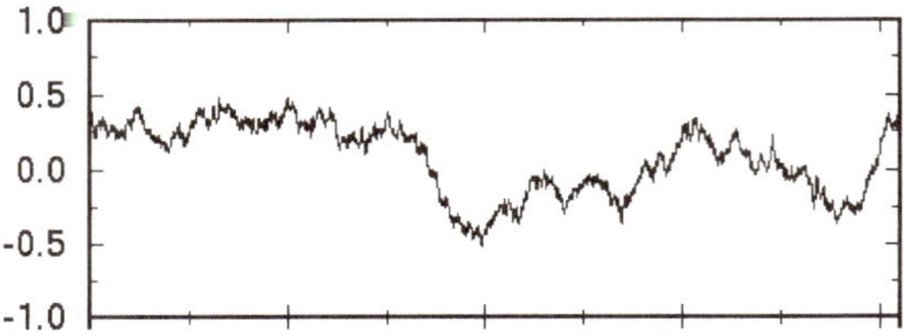

Figure 2. Fractal immersed in dimension $d = 2$. rugged fractal line.

2.1.3. Fractals between $2 \leq D \leq 3$ (similar to surfaces or porous volumes)

For a fractal immersed in a Euclidean dimension, $I = d + 1 = 3$ with projection in $d = 2$, analogous to a surface geometry is fractal-type "mountains" or "rugged surfaces" (Figure 3). The fracture surfaces can be included in this class of fractals.

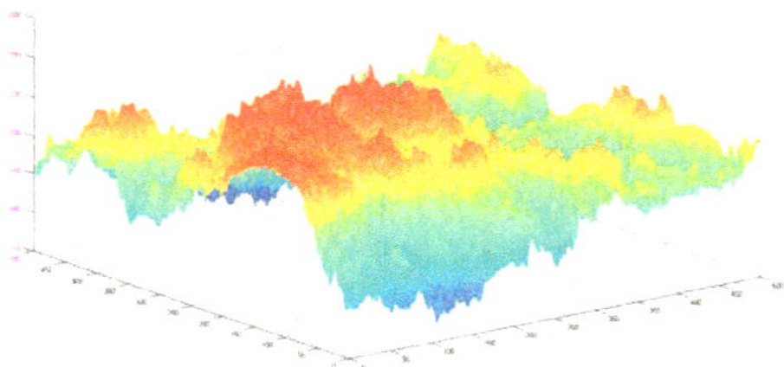

Figure 3. Irregular or rugged surface that has a fractal scaling with dimension D between $2 \leq D \leq 3$.

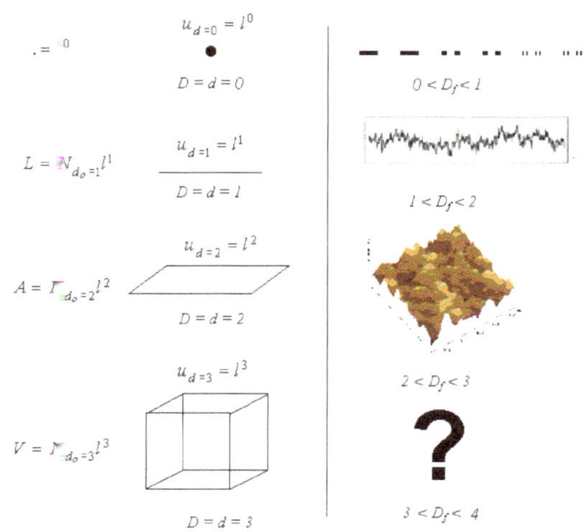

Figure 4. Comparison between Euclidean and fractal geometry. D, d and D_f represents the topological, Euclidean and fractal dimensions, of a point, line segment, flat surface, and a cube, respectively

Making a parallel comparison of different situations that has been previously described, one has (Figure 4)

2.2. Fractal dimension (non-integer)

An object has a fractal dimension, $D,(d \leq D \leq d+1=I)$, where I is the space Euclidean dimension which is immersed, when:

$$F(\varepsilon L_0)=\varepsilon^{-D}F(L_0) \tag{1}$$

where L_0 is the projected length that characterizes an apparent linear extension of the fractal ε, is the scale transformation factor between two apparent linear extension, $F(L_0)$ is a function of measurable physical properties such as length, surface area, roughness, volume, etc., which follow the scaling laws, with homogeneity exponent is not always integers, whose geometry that best describe, is closer to fractal geometry than Euclidean geometry. These functions depend on the dimensionality, I, of the space which the object is immersed. Therefore, for fractals the homogeneity degree n is the fractal dimension D (non-integer) of the object, where ε is an arbitrary scale.

Based on this definition of fractal dimension it can be calculates doing:

$$\varepsilon^{-D}=\frac{F(\varepsilon L_o)}{F(L_o)} \tag{2}$$

taking the logarithm one has

$$D=-\frac{\ln\left[\dfrac{F(\varepsilon L_o)}{F(L_o)}\right]}{\ln(\varepsilon)} \tag{3}$$

From the geometrical viewpoint, a fractal must be immersed into a integer Euclidean dimension, $I=d+1$. Its non-integer fractal dimension, D, it appears because the fill rule of the figure from the fractal seed which obeys some failure or excess rules, so that the complementary structure of the fractal seed formed by the voids of the figure, is also a fractal.

For a fractal the space fraction filled with points is also invariant by scale transformation, i.e.:

$$P(L_o)=\frac{F(\lambda L_o)}{F(L_o)}=\frac{1}{N(L_o)} \tag{4}$$

Thus,

$$\varepsilon^{D}=P(L_0) \text{ ou } N(L_0)=\varepsilon^{-D} \tag{5}$$

where $P(L_0)$ is a probability measure to find points within fractal object

Therefore, the fractal dimension can be calculated from the fllowing equation:

$$D = -\frac{\ln N(L_0)}{\ln \varepsilon} \tag{6}$$

If it is interesting to scale the holes of a fractal object (the complement of a fractal), it is observed that the fractal dimension of this new additional dimension corresponds to the Euclidean space in which it is immersed less the fractal dimension of the original.

2.3. A generalized monofractal geometric measure

Now will be described how to process a general geometric measure whose dimension is any. Similarly to the case of Euclidean measure the measurement process is generalized, using the concept of Hausdorff-Besicovitch dimension as follows.

Suppose a geometric object is recovered by α-dimensional, geometric units, u_D, with extension, δ_k and $\delta_k \leq \delta$, where δ is the maximum α-dimensional unit size and α is a positive real number. Defining the quantity:

$$M_D(\alpha, \delta, \{\delta_k\}) = \sum_k \delta_k^{\alpha} \tag{7}$$

Choosing from all the sets $\{\delta_k\}$, that reduces this summation, such that:

$$M_D(\alpha, \delta) = \inf_{\{\delta_k\}} \sum_k \delta_k^{\alpha} \tag{8}$$

The smallest possible value of the summation in (8) is calculated to obtain the adjustment with best precision of the measurement performed. Finally taking the limit of δ tending to zero, $(\delta \to 0)$, one has:

$$M_D(\alpha) = \lim_{\delta \to 0} M_D(\alpha, \delta) \tag{9}$$

The interpretation for the function $M_D(\alpha)$ is analogous to the function for a Euclidean measure of an object, i.e. it corresponds to the geometric extension (length, area, volume, etc.) of the set measured by units with dimension, α. The cases where the dimension is integer are same to the usual definition, and are easier to visualize. For example, the calculation of $M_D(\alpha)$ for a surface of finite dimension, $D = 2$, there are the cases:

- For $\alpha = 1 < D = 2$ measuring the "length" of a plan with small line segments, one gets $M_D = \infty$, because the plan has a infinity "length", or there is a infinity number of line segment inside the plane.
- For, $\alpha = 2 = D = 2$ measuring the surface area of small square, one gets $M_D = A_{d=2} = A_0$. Which is the only value of α where M_D is not zero nor infinity (see Figure 5.)
- For $\alpha = 3 > D = 2$ measuring the "volume" of the plan with small cubes, one gets $M_D = 0$, because the "volume" of the plan is zero, or there is not any volume inside the plan.

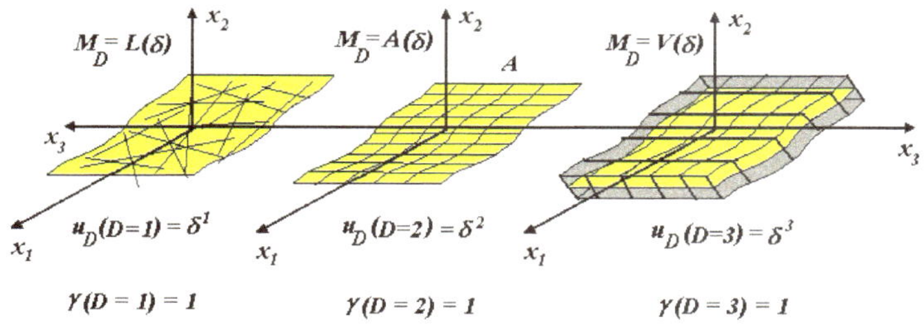

Figure 5. Measuring, $M_D(\delta)$ of an area A with a dimension, $D=2$ made with different measure units u_D for $D=1,2,3$.

Therefore, the function, M_D possess the following form

$$M_D(\alpha)=\begin{cases}0 \; para \; \alpha > D \\ M \; para \; \alpha = D \\ \infty \; para \; \alpha < D\end{cases}. \tag{10}$$

That is, the function M_D only possess a different value of 0 and ∞ at a critical point $\alpha = D$ defining a generalized measure

2.4. Invariance condition of a monofractal geometric measure

Therefore, for a generalized measurement there is a generalized dimension which the measurement unit converge to the determined value, M, of the measurement series, according to the extension of the measuring unit tends to zero, as shown in equations equações (9) and (10), namely:

$$M_D\left(\alpha,\delta,\{\delta_k\}\right)=\sum_k \delta_k^{\ \alpha} = M_{Do}\left(\delta\right)\varepsilon^{\alpha-D} \tag{11}$$

where $M_{D0}(\delta)$ is the Euclidean projected extension of the fractal object measured on α -dimensional space

Again the value of a fractal measure can be obtain as the result of a series.

One may label each of the stages of construction of the function $M_D(\delta)$ as follows:

i. the first is the *measure itself*. Because it is actually the step that evaluates the extension of the set, summing the geometrical size of the recover units. Thus, the extension of the set is being overestimated, because it is always less or equal than tthe size of its coverage.
ii. The next step is the *optimization* to select the arrangement of units which provide the smallest value measured previously, i.e. the value which best approximates the real extension of the assembly.
iii. The last step is the *limit*. Repeat the previous steps with smaller and smaller units to take into account all the details, however small, the structure of the set.

As the value of the generalized dimension is defined as a critical function, $M_{\alpha=D}(\delta)$ it can be concluded, wrongly, that the optimization step is not very important, because the fact of not having all its length measured accurately should not affect the value of critical point. The optimization step, this definition, serves to make the convergence to go faster in following step, that the mathematical point of view is a very desirable property when it comes to numerical calculation algorithms.

2.5. The monofractal measure and the Hausdorff-Besicovitch dimension

In this part we will define the dimension-Hausdorf Besicovicth and a fractal object itself. The basic properties of objects with "anomalous" dimensions (different from Euclidean) were observed and investigated at the beginning of this century, mainly by Hausdorff and Besicovitch [32,34]. The importance of fractals to physics and many other fields of knowledge has been pointed out by Mandelbrot [1]. He demonstrated the richness of fractal geometry, and also important results presented in his books on the subject [1, 35, 36].

The geometric sequence, S is given by:

$$S = \sum_k S_k \quad onde\ k = 0,1,2,.... \tag{12}$$

represented in Euclidean space, is a fractal when the measure of its geometric extension, given by the series, $M_{\alpha}(\delta_k)$ satisfies the following Hausdorf-Besicovitch condition:

$$M_d(\delta_k) = \sum_k \gamma(d)\delta_k{}^{\alpha} = N_d(\delta_k)\gamma(d)\delta_k{}^{\alpha} \underset{\delta_k \to 0}{} \begin{cases} 0; & \alpha > D \\ M_D; & \alpha = D, \\ \infty; & \alpha < D \end{cases} \tag{13}$$

where:

$\gamma(d)$ is the geometric factor of the unitary elements (or seed) of the sequence represented geometrically.

δ : is the size of unit elements (or seed), used as a measure standard unit of the extent of the spatial representation of the geometric sequence.

$N(\delta)$: is the number of elementary units (or seeds) that form the spatial representation of the sequence at a certain scale

α : the generalized dimension of unitary elements

D : is the Hausdorff-Besicovitch dimension.

2.6. Fractal mathematical definition and associated dimensions

Therefore, fractal is any object that has a non-integer dimension that exceeds the topological dimension ($D < I$, where I is the dimension of Euclidean space which is immersed) with some invariance by scale transformation (self-similarity or self-affinity), where for any continuous contour that is taken as close as possible to the object, the number of points N_D , forming the fractal not fills completely the space delimited by the contour, i.e., there is always empty, or excess regions, and also there is always a figure with integer dimension, I , at which the fractal can be inscribed and that not exactly superimposed on fractal even in the limit of scale infinitesimal. Therefore, the fraction of points that fills the fractal regarding its Euclidean coverage is different of a integer. As seen in previous sections - 2.2 - 2.5 in algebraic language, a fractal is a invariant sequence by scale transformation that has a Hausdorff-Besicovitch dimension.

According to the previous section, it is said that an object is fractal, when the respective magnitudes characterizing features as perimeter, area or volume, are homogeneous functions with non-integer. In this case, the invariance property by scaling transformation (self-similar or self-affinity) is due to a scale transformation of at least one of these functions.

The fractal concept is closely associated to the concept of Hausdorff-Besicovitch dimension, so that one of the first definitions of fractal created by Mandelbrot [36] was:

"Fractal by definition is a set to which the Haussdorf-Besicovitch dimension exceeds strictly the topological dimension".

One can therefore say that fractals are geometrical objects that have structures in all scales of magnification, commonly with some similarity between them. They are objects whose usual definition of Euclidean dimension is incomplete, requiring a more suitable to their context as they have just seen. This is exactly the Hausdorff-Besicovitch dimension.

A dimension object, D, is always immersed in a space of minimal dimension $I = d + 1$, which may present an excessive extension on the dimension d, or a lack of extension or failures in one dimension $d + 1$. For example, for a crack which the fractal dimension is the dimension in the range of $1 \le D \le 2$ the immersion dimension is the dimension $I = 2$ in the case of a fracture surface of which the fractal dimension is in the range $2 \le D \le 3$ the immersion dimension is the $I = 3$. When an object has a geometric extension such as completely fill a Euclidean dimension regular, d, and still have an excess that partially fills a superior dimension $I = d + 1$, in addition to the inferior dimension, one says that the object has a dimension in excess, d_e given by $d_e = D - d$ where D is the dimension of the object. For example, for a crack which the fractal dimension is in the range $1 \le D \le 2$ the excess dimension is $d_e = D - 1$, in the case of a fracture surface of which the fractal dimension is in the range of $2 \le D \le 3$ the excess dimension is $d_e = D - 2$. If on the other hand an object partially fills a Euclidean regular dimension, $I = d + 1$ certainly this object fills fully a Euclidean regular dimension, d, so that it is said that this object has a lack dimension $d_{fl} = I - D = d + 1 - D$, where $d_e = 1 - d_{fl}$. For example, for a crack which the fractal dimension is the range of $1 \le D \le 2$ the lack dimension is $d_{fl} = 2 - D$. In the case of a fracture surface of which the fractal dimension is the range of $2 \le D \le 3$ the lack dimension is $d_{fl} = 3 - D$.

2.7. Classes and types of fractals

One of the most fascinating aspects of the fractals is the extremely rich variety of possible realizations of such geometric objects. This fact gives rise to the question of classification, and the book of Mandelbrot [1] and in the following publications many types of fractal structures have been described. Below some important classes will be discussed with some emphasis on their relevance to the phenomenon of growth.

Fractals are classified or are divided into: mathematical and physical (or natural) fractals and uniform and non-uniform fractals. Mathematical fractals are those whose scaling relationship is exact, i.e., they are generated by exact iteration and purely geometrical rules and does not have cut-off scaling limits, not upper nor lower, because they are generated by rules with infinity interactions (Figure 6a) without taking into account none phenomenology itself, as shown in Figure 6a. Some fractals appear in a special way in the phase space of dynamical systems that are close to situations of chaotic motion according to the Theory of Nonlinear Dynamical Systems and Chaos Theory. This approach will not be made here, because it is another matter that is outside the scope of this chapter.

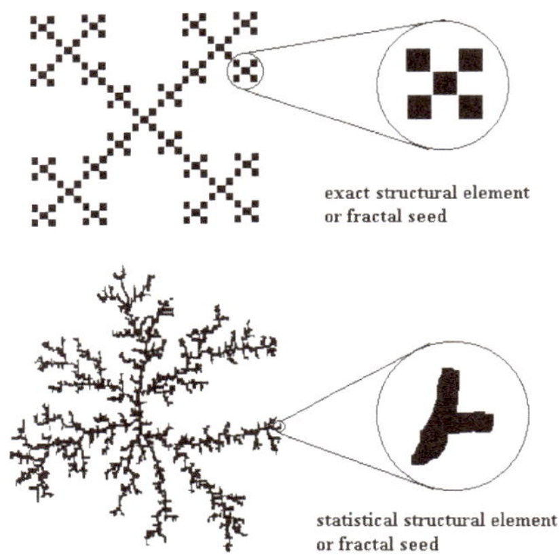

Figure 6. Example of branching fractals, showing the structural elements, or elementary geometrical units, of two fractals. a) A self-similar mathematical fractal. b) A statistically self-similar physical fractal.

Real or physical fractals (also called natural fractals) are those statistical fracals, where not only the scale but all of fractal parameters can vary randomly. Therefore, their scaling relationship is approximated or statistical, i. e., they are observed in the statistical average made throughout the fractal, since a lower cutoff scale, $\varepsilon_{\min}$, to a different upper cutoff scale $\varepsilon_{\max}$ (self-similar or self-affine fractals), as shown in Figure 6b. These fractals are those which appear in nature as a result of triggering of instabilities conditions in the natural processes [24] in any physical phenomenon, as shown Figure 6b. In these physical or natural fractals the extension scaling of the structure is made by means of a homogeneous function as follows:

$$F(\delta) \sim \delta^{d-D}, \tag{14}$$

where d is the Euclidean dimension of projection of the fractal and D is the fractal dimension of self-similar structure.

It is true that the physical or real fractals can be deterministic or random. In random or statistical fractal the properties of self-similarity changes statistically from region to region of the fractal. The dimension cannot be unique, but characterized by a mean value, similarly to the analysis of mathematical fractals. The Figure 6b shows aspects of a statistically self-similar fractal whose appearance varies from branch to branch giving us the impression that each part is similar to the whole.

The mathematical fractals (or exact) and physical (or statistical), in turn, can be subdivided into uniform and nonuniform fractal.

Uniforms fractals are those that grow uniformly with a well behaved unique scale and constant factor, λ, and present a unique fractal dimension throughout its extension.

Non-uniform fractals are those that grow with scale factors λ_i's that vary from region to region of the fractal and have different fractal dimensions along its extension.

Thus, the fractal theory can be studied under three fundamental aspects of its origin:

1. From the geometric patterns with self-similar features in different objects found in nature.
2. From the nonlinear dynamics theory in the phase space of complex systems.
3. From the geometric interpretation of the theory of critical exponents of statistical mechanics.

3. Methods for measuring length, area, volume and fractal dimension

In this section one intends to describe the main methods for measuring the fractal dimension of a structure, such as: the compass method, the Box-Counting method, the Sand-Box Method, etc.

It will be described, from now, how to obtain a measure of length, area or fractal volume. In fractal analysis of an object or structure different types of fractal dimension are obtained, all related to the type of phenomenon that has fractality and the measurement method used in obtaining the fractal measurement. These fractal dimensions can be defined as follows.

3.1. The different fractal dimensions and its definitions

A fractal dimension D_f in general is defined as being the dimension of the resulting measure of an object or structure, that has irregularities that are repeated in different scales (a invariance by scale transformation). Their values are usually noninteger and situated between two consecutive Euclidean dimensions called projection dimension d of the object and immersion dimension, $d+1$, i.e. $d \leq D_f \leq d+1$.

In the literature there is controversy concerning the relationship between different fractal dimensions and roughness exponents. The term "fractal dimension" is used generically to refer to different fractional dimensions found in different phenomenologies, which results in formation of geometric patterns or energy dissipation, which are commonly called fractals [1]. Among these patterns is the growth of aggregates by diffusion (DLA - Diffusion Limited

Aggregation), the film growth by ballistic deposition (BD), the fracture surfaces (SF), etc.. The fractal dimensions found in these phenomena are certainly not the same and depend on both the phenomenology studied as the fractal characterization method used. Therefore, to characterize such phenomena using fractal geometry, a distinction between the different dimensions found is necessary.

Among the various fractal dimensions one can emphasize the Hausdorff-Besicovitch dimension, D_{HB}, which comes from the general mathematical definition of a fractal [32, 33,34]. Other dimensions are the dimension box, D_B, the roughness dimension or exponent Hurst, H, the Lipshitz-Hölder dimension, α, etc.. Therefore, a mathematical relationship between them needs to be clearly established for each phenomenon involved. However, is observed, then that relationship is not unique and depends not only on phenomenology, but also the characterization method used.

Therefore, the phenomenological equation of the fracture phenomenon can also, in theory, provide a relationship between fractal dimension and roughness exponent of a fracture surface, as happens to other phenomenologies. In this study, there was obtained a fractal model for a fracture surface, as a generalization of the box-counting method. Thus, will be discussed the relationship between the local and global box dimension and the roughness dimension, which are involved in the characterization of a fracture surface, and any other dimension necessary to describe a fractal fracture surface.

3.1.1. Compass methods and divider dimension, D_D

The divider dimension D_D is defined from the measure of length of a roughened fractal line, for example, when using the compass method. This measure is obtained by opening a compass with an aperture δ and moving on the line fractal to obtain the value of the line length rugosa (see Figure 7). The different values of the rough line length due to the compass aperture determines the dimension divider.

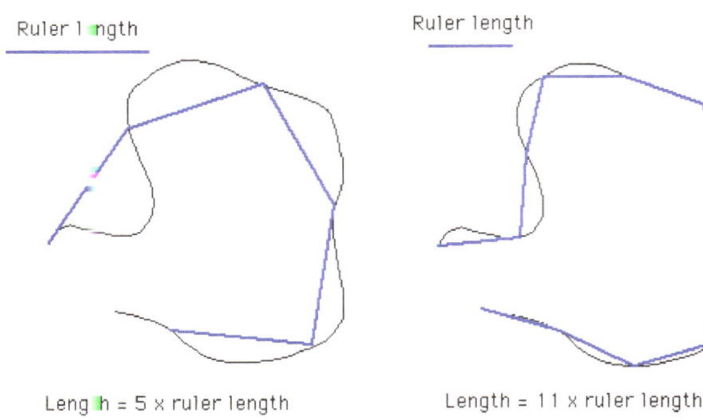

Figure 7. Compass method applied to a rugged line.

For a fractal rough line the divider dimension can be defined as:

$$D_D \equiv -\frac{\ln\left(\frac{L}{\delta}\right)}{\ln\left(\frac{\delta}{L_0}\right)}. \tag{15}$$

where L_0 is the projected length obtained from the rugged fractal length L

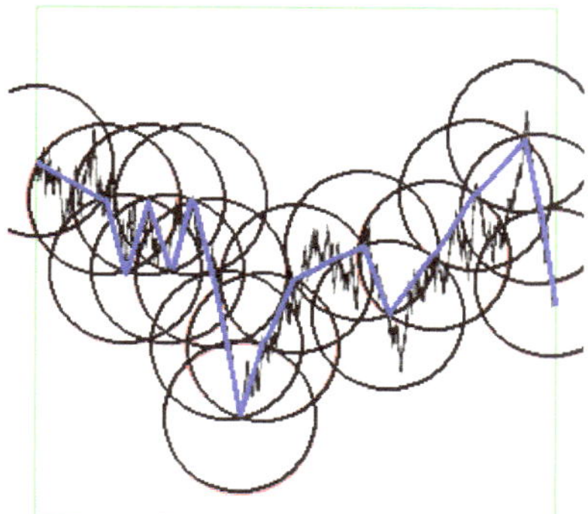

Figure 8. Compass method applied on a line noise or a rough self-affine fractal.

Several methods for determining the fractal dimension based on the compass method, among them stand out the following methods: the Coastlines Richardson Method, the Slit Island Method, etc.

3.2. Methods of measurement for determining the fractal dimension of a structure

There are basically two ways to recover an object with boxes for fractal dimension measuring. In the first method, boxes of different sizes extending from a minimum size $\delta_{\min}$ until to a maximum size $\delta_{\max}$, from a fixed origin recovering the whole object at once time. In the second case, one side of the recovering box is kept fixed, and with a minimum size ruler, $\delta_{\min}$, then recovers the figure by moving the boundary of that recovering from the minimum $\delta_{\min}$ to maximum size $\delta_{\max}$ of the object. The first method is known as a method Box-Counting exemplified in Figure 9 and the second method is known as Sand-box, shown in Figure 10. The advantage of the second over the first is that it detects the changes in dimension D with the length of the object. If the object under consideration has a local dimension for boxes with size $\delta \to 0$, unlike the global dimension, $\delta \to \infty$, it is said that the object is self-affine fractal. Otherwise the object is said self-similar. These two main methods of counts of structures which may lead to determination of the fractal dimension of an object [38].

3.2.1. Box-counting method by static scaling of the elements in a fractal structure

The Box-Counting method, comes from the theory of critical phenomena in statistical mechanics. In statistical mechanics there is an analogous mathematical method to describing phenomena which have self-similar properties, permitting scale transformations without loss of generality in the description of physical information of the phenomenon ranging from quantities such as volume up to energy. However, in the case described here, the Box-Counting method is performed filling the space occupied by a fractal object with boxes of arbitrary size δ, and count the number $N(\delta)$ of these boxes in function its size, (Figure 9 and Figure10). This number $N(\delta)$ of boxes is given as follows:

$$N(\delta) = C\delta^D \qquad (16)$$

Plotting the data in a $\log \times \log$ graph one obtains from the slope of the curve obtained, the fractal dimension of the object.

In the Box-Counting method (Figure 9), a grid that recover the object is divided into $n_k = L_0 / \delta_k$ boxes of equal side δ_k and how many of these boxes that recovering the object is counted. Then, varies the size of the boxes and the counting is retraced, and so on. Making a logarithm graph of the number N_k of boxes that recovering the object in function of the scale for each subdivision $(\varepsilon_k = \delta_k / L_0)$, one obtains the fractal dimension from the slope of this plot. Note that in this case the partition maximum is reached when, $N_\infty = L_0 / \delta_k (k \to \infty) = L_0 / l_0$, where $L_{max} = L_0$ is the projected crack length $\delta_\infty = l_0$ is the length of the shortest practicable ruler.

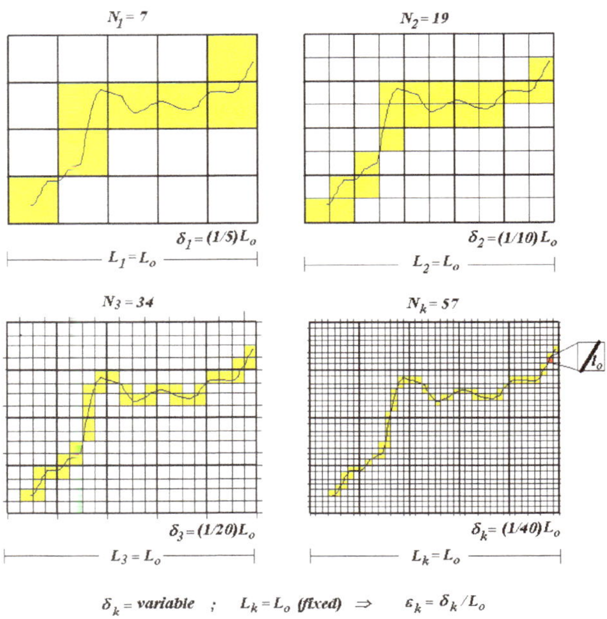

Figure 9. Fragment of a crack on a testing sample showing the variation of measurement of the crack length L with the measuring scale, $\varepsilon_k = \delta_k / L_0$ for a partition, $\delta_k = variável$ and $L_k = L_0$ (fixed), with sectioning done for counting by one-dimensional Box-Counting scaling method.

Therefore, the number $N_k(\delta_k)$ depending on the size, δ_k, of these boxes is given as follows:

$$N_k(\delta_k) = \left(\frac{\delta_k}{\delta_{max}}\right)^{-D} \tag{17}$$

In the Figure 9 is illustrated the use of this method in a fractal object. Are present different grids, or meshes, constructed to recover the entire structure, whose fractal dimension one wants to know. The grids are drawn from an original square, involving the whole space occupied by the structure. At each stage of refinement of the grid (L_0) (the number of equal parts in the side of the square is divided) are counted the number of squares $N(L_0)$ which contain part of the structure. Repeatedly from the data found, is constructed the graph of $\log L_0 \times \log N(L_0)$. If the graph thus obtained is a straight line, then the fractal behavior of the structure has self-similarity or statistical self-affinity whose dimension D is obtained by calculating the slope of the line. For more compact structure, it is recommended to make a statistical sampling, that is, the repeat the counting of the squares $N(L_0)$ for different squares constructed from the gravity center (counting center) of the in the structure. Thus, one obtains a set of values $N(L_0)$ for another set of values L_0. These data must be statistically treated to obtain the value of fractal dimension, $"D"$.

From the viewpoint of experimental measurement, one can consider using different methods of viewing the crack to obtain the fractal dimension, such as optical microscopy, electron microscopy, atomic force microscope, etc.., Which naturally have different rules δ_k and therefore different scales of measurement ε_k,.

The fractal dimension is usually calculated using the Box-Counting shown in Figure 9, i.e. by varying the size of the measuring ruler δ_k and counting the number of boxes, N_k that recover the structure. In the case of a crack the fractal dimension is obtained by the following relationship:

$$D = -\frac{\ln N}{\ln(l_o / L_o)} \tag{18}$$

The description of a crack according to the Box-Counting method follows the idea shown in Figure 9, which results in:

$$D = -\frac{\ln 57}{\ln(1/40)} = 1.096 \ . \tag{19}$$

The same result can be obtained using the Box-Sand method, as shown in Figure 10.

3.2.2. *The sand-box counting method of the elements by static scaling of a fractal structure*

The Sand-box method consists in the same way as the Box-Counting method, to count the number of boxes, $N(u)$, but with fixed length, u, as small as possible, extending gradually up the boundary count until to reach out to the border of the object under consideration. This is done initially by setting the counting origin from a fixed point on the object, as shown in Figure10. This method seems to be the most advantageous, as well as to establish a coordinate system, or a origin for calculating the fractal dimension, it also allows, in certain cases, to infer dynamic data from static scaling, as shown by Alves [47].

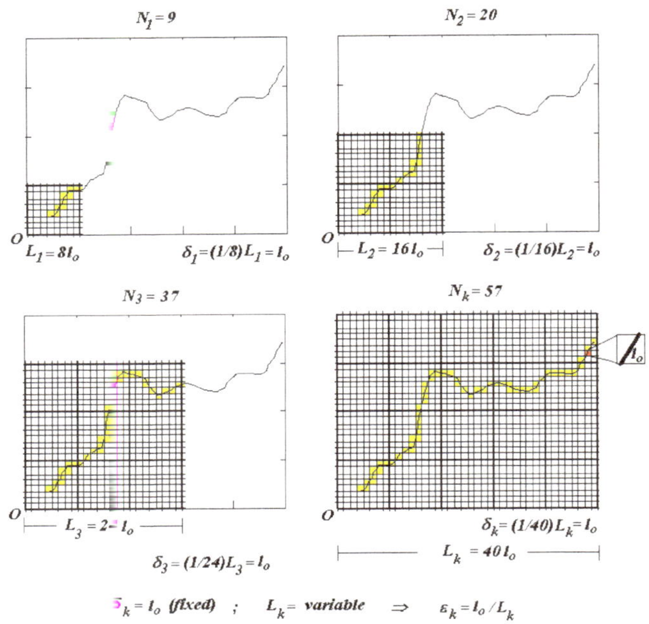

Figure 10. Fragment of a crack on test specimen showing the variation of measurement of the crack length L with the measuring scale, $\varepsilon_k = \delta_k / L_0$ for a partition $L_k = variável$, and $\delta_k = l_0$ (fixed), with sectioning done for counting by one-dimensional Sand-Box scaling method.

In the Sand-Box method (Figure10), the figure is recovered with boxes of different sizes L_k, no matter the form, which can be rectangular or spherical, however, fixed at a any point "O" on figure called origin, from which the boxes are enlarged. It is counted the number of elementary structures or seeds, which fit within each box. Plotting the graph of $\log N_k \times \log\left(\varepsilon_k = \delta_{\min}/L_k\right)$ in the same manner as in the above method the fractal dimension is obtained. Note that in this case the maximum partition is achieved when $N_\infty = L_k\left(k \to \infty\right)/\delta_{\min} = L_0/l_0$, where $L_\infty = L_0$ is the projected crack length and $\delta_{min} = l_0$ it is the length of the lower measuring ruler practicable.

3.2.3. *The global and local box dimensions*

To define the box dimension, D_B, is assumed that all the space containing the fractal is recovered with a grid (set of α -dimensional units juxtaposed in the same shape and size, δ)

with maximum size, $\delta_{\max}$, which inscribes the fractal object. Defining the relative scale, ε on the grid size, $\delta_{\max}$, as being given by:

$$\varepsilon = \frac{\delta}{\delta_{\max}} \tag{20}$$

countting the number of boxes $N(\varepsilon)$ that have at least one point of the fractal. The box dimension is therefore defined as:

$$D_B = -\lim_{\varepsilon \to 0} \frac{\ln N(\varepsilon)}{\ln \varepsilon} \tag{21}$$

At this point, there are two ways to obtain the actual value of the measure, or taking the limit when $\varepsilon \to 0$ and allows that the dimension D fits the end value of $N(\varepsilon)$, or it is considered a linear correlation in value of $\ln N(\varepsilon) \times \ln \varepsilon$, which D is the slope of the line, and this defines the measure independently of the scale.

In the case of numerical estimation, one can not solve the limit indicated in the equation (21). Then, D_B is obtained as a slope, $\ln N(\varepsilon) \times \ln \varepsilon$, when ε it is small. The value $N(\varepsilon)$ is obtained by an algorithm known as *Box-Counting*.

Self-affine fractals requiring different variations in scale length for different directions. Therefore, one can use the Box-Counting method with some care being taken, in the sense that the box dimension D_B to be obtained has a crossing region between a local and global measure of the dimensions. From which follows that for each region is used the following relationships:

$$\lim_{l_0 \to 0} N(L_0) = \left(\frac{L_0}{l_0} \right)^{D_{Bg}} \quad p/ \; L_0 << L_{0s} \tag{22}$$

for a global measurement

$$\lim_{l_0 \to 0} N(L_0) = \left(\frac{L_0}{l_0} \right)^{D_{Bl}} \quad p/ \; L_0 >> L_{0s} \tag{23}$$

where L_{0s} is the threshold saturation length which the fractal dimension changes its behavior from local to global stage.

For measurement, generally, for any self-affine fractal structure the local fractal dimension is related to the Hurst exponent, H, as follow,

$$D_{Bl} = d + 1 - H_{q=1} \tag{24}$$

At this point, one observes that for a profile the relationship $D_{Bl} = 2 - H$ commonly used, only serves for a local measurements using the box counting method. While for global measures one can not establish a relationship between D_{Bg} and H. For the global fractal dimension, $D_g = d$ and $I = d+1$ the Euclidean dimension where the fractal is embedded one has

$$d \le D_{Bg} \le d+1 \tag{25}$$

Some textbooks on the subject show an example of calculation of local and global fractal dimension of self-affine fractals, obtained by a specific algorithm [18, 22, 23, 26, 38,39].

In crossing the limit of fractal dimension local D_l to global D_g, there is a transition zone called the "crossover" and the results obtained in this region are somewhat ambiguous and difficult to interpret [39]. However, in the global fractal dimension, the structure is not considered a fractal [42, 43].

3.2.4. The Relationship between box dimension and HausdorffBesicovitch dimensions

The mathematical definition of generalized dimension of Haussdorff-Besicovitch need a method that can measure it properly to the fractal phenomenon under study. Some authors [23, 40, 44, 45] have discussed the possibility of using the Box-Counting method as one of the graphical methods which obtains a box dimension D_B, very close to generalized Haussdorff Besicovitch, D_{HB}, i.e. [44]:

$$D_B \cong D_{HB} \tag{26}$$

In this sense the box dimension, D_B is obtained for self-asimilar fractals that may be rescaled for the same variation in scales lengths in all directions by using the relationship:

$$N(L_0) = \left(\frac{L_0}{l_0}\right)^{D_B} \tag{27}$$

where l_0 is the grid size used and L_0 is the apparent size of the fractal to be characterized.

The analytical calculation of the Hausdorff dimension is only possible in some cases and it is difficult to implement by computation. In numerical calculation, is used another more appropriate definition called box dimension, D_B, which in the case of dynamic systems, has the same value of the Haussdorff dimension, D [44]. Thus, it is common to call them without distinction as *fractal dimensions*, D as will be shown below.

All the definitions related to fractal exponents that are shown here, and all numerical evaluation of these, always calculates the inclination of some amount ε against on a logarithmic scale.

The two definitions of, *Hausdorff-Besicovitch Dimension*, D_H and *Box-Dimension*, D_B are allocated the same amount, but in a way somewhat different from each other. In inaccurate way, one can think that the connection between the two is done considering that:

$$M_D(\alpha \to D) \sim N(\varepsilon)\varepsilon^d, \tag{28}$$

by analogy with equation (13), i.e. approximating to the geometric extension of the object by the number of boxes (of the same size) necessary to recover it. But, since the definition of the box dimension there is no optimization step, and its value is directly dependent on $N(\varepsilon)$ (which is not the case with the Hausdorff dimension) in practice one has often the geometric extension is overestimated, particularly for ε large, i. e. upper limit $(\varepsilon \to 1)$ and thus $D_B \le D$. However, for the lower limit, i.e. $\varepsilon \to 0$, the Hausdorff-Besicovitch dimensions, D_H and the box dimension, D_B are equal, becoming valid the measure of geometric extension process, $M_D(\delta)$ at box counting algorithm.

Considering from (28) that:

$$N(\varepsilon) \sim \varepsilon^{-D}\,(d \le D \le d+1), \tag{29}$$

and that

$$N(\varepsilon_{\max}) \sim \varepsilon_{\max}{}^{-D}\,(d \le D \le d+1) \tag{30}$$

Therefore, dividing (29) by (30) has:

$$\frac{N(\varepsilon)}{N(\varepsilon_{\max})} \sim \left(\frac{\varepsilon}{\varepsilon_{\max}}\right)^{-D}\,(d \le D \le d+1) \tag{31}$$

taking $\varepsilon_{\max}$ the total grid extension that recover the object, one has:

$$\varepsilon_{\max} \to 1 \tag{32}$$

From as early as (31)

$$N(\varepsilon) \to \varepsilon^{-D}\,(d \le D \le d+1) \tag{33}$$

Substituting (33) in (28) has:

$$M_D(\alpha \to D) \sim \varepsilon^{\alpha - D}, \tag{34}$$

This equation is analogous to the fundamental Richardson relationship for a fractal length.

4. Crack and rugged fracture surface models

The two main problematics of mathematical description of Fracture Mechanics are based on the following aspects: the surface roughness generated in the process and the field stress/strain applied to the specimen. This section deals with the fractal mathematical description of the first aspect, i.e., the roughness of cracks on Fracture Mechanics, using fractal geometry to model its irregular profile. In it will be shown basic mathematical assumptions to model and describe the geometric structures of irregular cracks and generic fracture surfaces using the fractal geometry. Subsequently, one presents also the proposal for a self-affine fractal model for rugged surfaces of fracture. The model was derived from a generalization of Voss [48] ([1]) equation and the model of Morel [49] for fractal self-affine fracture surfaces. A general analytical expression for a rugged crack length as a function of the projected length and fractal dimension is obtained. It is also derived the expression of roughness, which can be directly inserted in the analytical context of Classical Fracture Mechanics.

The objectives of this section are: (i) based geometrical concepts, extracted from the fractal theory and apply them to the CFM in order to (ii) construct a precise language for its mathematical description of the CFM, into the new vision the fractal theory. (iii) eliminate some of the questions that arise when using the fractal scaling in the formulation of physical quantities that depend on the rough area of fracture, instead of the projected area, in the manner which is commonly used in fracture mechanics. (iv) another objective is to study the way which the fractal concept can enrich and clarify various aspects of fracture mechanics. For this will be done initially in this section, a brief review of the major advances obtained by the fractal theory, in the understanding of the fractography and in the formation of fracture surfaces and their properties. Then it will be done, also, a mathematical description of our approach, aiming to unify and clarify aspects still disconnected from the classical theory and modern vision, provided by fractal geometry. This will make it possible for the reader to understand what were the major conceptual changes introduced in this work, as well as the point from which the models proposed progressed unfolding in new concepts, new equations and new interpretations of the phenomenon.

4.1. Application of fractal theory in the characterization of a fracture surface

In this section one intends to do a brief history of the fractography development as a fractal characterization methodology of a fractal fracture surface.

4.1.1. Geometric aspects and observations extracted from the quantitative fractography of irregular fracture surface

The technique used for geometric analysis of the fracture surface is called fractography. Until recently it was based only on profilometric study and statistical analysis of irregular surfaces [50]. Over the years, after repeated observations of these surfaces at various magnifications, was also revealed a variety of self-similar structures that lie between the micro and macro-structural level, characteristic of the type of fracture under observation. Since 1950 it is known that certain structures observed in fracture surfaces by microscopy, showed the phenomenon of invariance by magnification. Such structures recently started to be described in a systematic way by means of fractal geometry [51, 25]. This new approach allows the description of patterns that at first sight seem irregular, but keep an invariance by

[1] Voss present a fractal description for the noise in the Browniano mouvement

scale transformation (self-similarity or self-affinity). This means that some facts concerning the fracture have the same character independently of the magnification scale, i.e. the phenomenology that give rise to these structures is the same in different observation scales.

The Euclidean scaling of physical quantities is a common occurrence in many physical theories, but when it comes to fractality appears the possibility to describe irregular structures. The fracture for each type of material has a behavior that depends on their physical, chemical, structural, etc. properties. Looking at the topography and the different structures and geometrical patterns formed on the fracture surfaces of various materials, it is impossible to find a single pattern that can describe all these surfaces (Figure 11), since the fractal behavior of the fracture depends on the type of material [52]. However, the fracture surfaces obtained under the same mechanical testing conditions and for the same type of material, retains geometric aspects similar of its relief [53] (see Figure 12).

This similarity demonstrates that exist similar conditions in the fracture process for the same material, although also exist statistically changinga from piece to piece, constructed of the same material and under the same conditions [54; 55]. Based on this observation was born the idea to relate the surface roughness of the fracture with the mechanical properties of materials [50].

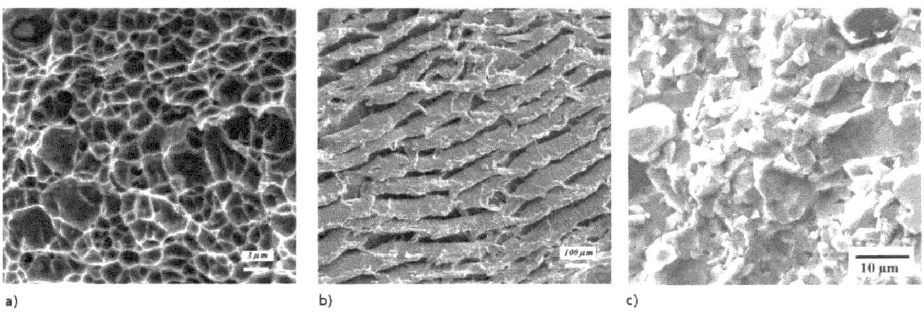

Figure 11. Various aspects of the fracture surface for different materials: (a) Metallic B2CT2 sample, (b) Polymeric, sample PU1.0, with details of the microvoids formation during stable crack propagation, (c) Ceramic [56].

4.1.2. Fractal theory applied to description of the relief of a fracture surface

Let us now identify the fractal aspects of fracture surfaces of materials in general, to be obtained an experimental basis for the fractal modeling of a generic fracture surface. The description of irregular patterns and structures, is not a trivial task. Every description is related to the identification of facts, aspects and features that may be included in a class of phenomena or structure previously established. Likewise, the mathematical description of the fracture surface must also have criteria for identifying the geometric aspects, in order to identify the irregular patterns and structures which may be subject to classification. The criteria, used until recently were provided by the fractográfico study through statistical analysis of quantities such as average grain size, roughness, etc. From geometrical view point this description of the irregular fracture surface, was based, until recently, the foundations of Euclidean geometry. However, this procedure made this description a task too complicated. With the advent of fractal geometry, it became possible to approach the problem analytically, and in more authentic way.

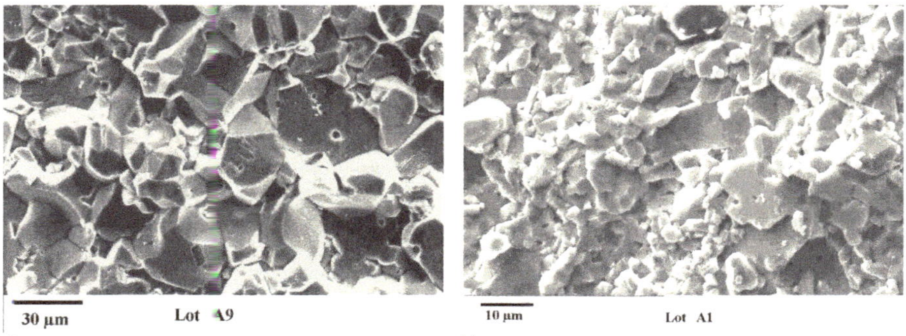

a) b)

Figure 12. Fracture surfaces of different parts mades with the same material, a) Lot A9 b) Lot A1 [56 1999].

Inside the fractography, fractal description of rugged surfaces, has emerged as a powerful tool able to describe the fracture patterns found in brittle and ductile materials. With this new characterization has become possible to complement the vision of the fracture phenomenon, summarizing the main geometric information left on the fracture surface in just a number, " D ", called fractal dimension. Therefore, assuming that there is a close relationship between the physical phenomena and fractal pattern generated as a fracture surface, for example, the physical properties of these objects have implications on their geometrical properties. Thinking about it, one can take advantage of the geometric description of fractals to extract information about the phenomenology that generated it, thereby obtaining a greater understanding of the fracture process and its physical properties. But before modeling any irregular (or rough) fracture surface, using fractal geometry, will be shown some of the difficulties existing and care should be taken in this mathematical description.

4.2. Fractal models of a rugged fracture surface

A fracture surface is a record of information left by the fracture process. But the Classical Fracture Mechanics (CFM) was developed idealizing a regular fracture surface as being smooth and flat. Thus the mathematical foundations of CFM consider an energy equivalence between the rough (actual) and projected (idealized) fracture surfaces [57]. Besides the mathematical complexity, part of this foundation is associated with the difficulties of an accurate measure of the actual area of fracture. In fact, the geometry of the crack surfaces is usually rough and can not be described in a mathematically simple by Euclidean geometry

[52]. Although there are several methods to quantify the fracture area, the results are dependent on the measure ruler size used [56]. Since the last century all the existing methods to measure a rugged surface did not contribute to its insertion into the analytical mathematical formalism of CFM until to rise the fractal geometry. Generally, the roughness of a fracture surface has fractal geometry. Therefore, it is possible to establish a relationship between its topology and the physical quantities of fracture mechanics using fractal characterization techniques. Thus, with the advent of fractal theory, it became possible to describe and quantify any structure apparently irregular in nature [1]. In fact, many theories based on Euclidean geometry are being revised. It was experimentally proved that the fracture surfaces have a fractal scaling, so the Fracture Mechanics is one of the areas included in this scientific context.

4.2.1. Importance of fracture surface modeling

The mathematical formalism of the CFM was prepared by imagining a fracture surface flat, smooth and regular. However, this is an mathematical idealization because actually the microscopic viewpoint, and in some cases up to macroscopic a fracture surface is generally a rough and irregular structure difficult to describe geometrically. This type of mathematical simplification above mentioned, exists in many other areas of exact sciences. However, to make useful the mathematical formalism developed over the years, Irwin started to consider the projected area of the fracture surface [57] as being energetically equivalent to the rugged surface area. This was adopted due to experimental difficulties to accurately measure the true area of the fracture, in addition to its highly complex mathematics. Although there are different methods to quantify the actual area of the fracture [56], its equationing within the fracture mechanics was not considered, because the values resulting from experimental measurements depended on the "ruler size" used by various methods. No mathematical theory had emerged so far, able to solve the problem until a few decades came to fractal geometry. Thus, modern fractal geometry can circumvent the problem of complicated mathematical description of the fracture surface, making it useful in mathematical modeling of the fracture.

In particular, it was shown experimentally that cracks and fracture surfaces follow a fractional scaling as expected by fractal geometry. Therefore, the fractal modeling of a irregular fracture surface is necessary to obtain the correct measurement of its true area. Therefore, fracture mechanics is included in the above context and all its classical theory takes into account only the projected surface. But with the advent of fractal geometry, is also necessary to revise it by modifying its equations, so that their mathematical description becomes more authentic and accurate. Thus, it is possible to relate the fractal geometric characterization with the physical quantities that describe the fracture, including the true area of irregular fracture surface instead of the projected surface. Thought this idea was that Mandelbrot and Passoja [58] developed the fractal analysis by the "slit island method ". Through this method, they sought to correlate the fractal dimension with the physical well-known quantities in fracture mechanics, only an empirical way. Following this pioneering work, other authors [3 , 4 , 5 , 6, 7 , 8, 11, 12, 13, 59] have made theoretical and geometrical considerations with the goal of trying to relate the geometrical parameters of the fracture surfaces with the magnitudes of fracture mechanics, such as fracture energy, surface energy, fracture toughness, etc.. However, some misconceptions were made regarding the application of fractal geometry in fracture mechanics.

Several authors have suggested different models for the fracture surfaces [60-63]. Everyone knows that when it was possible to model generically a fracture surface, independently of the fractured material, this will allow an analytical description of the phenomena resulting the roughness of these surfaces within the Fracture Mechanics. Thus the Fracture Mechanics will may incorporate fractal aspects of the fracture surfaces explaining more appropriately the material properties in general. In this section one propose a generic model, which results in different cases of fracture surfaces, seeking to portray the variety of geometric features found on these surfaces for different materials. For this a basic mathematical conceptualization is needed which will be described below. For this reason it is done in the following section a brief bibliographic review of the progress made by researchers of the fractal theory and of the Fracture Mechanics in order to obtain a mathematical description of a fracture surface sufficiently complete to be included in the analytical framework of the Mechanics Fracture.

4.2.2. Literature review - models of fractal scaling of fracture surfaces

Mosolov [64] and Borodich [3] were first to associate the deformation energy and fracture surface involved in the fracture with the exponents of surface roughness generated during the process of breaking chemical bonds, separation of the surfaces and consequently the energy dissipation . They did this relationship using the stress field. Mosolov and Borodich [64, 3] used the fractional dependence of singularity exponents of this field at the crack tip and the fractional dependence of fractal scaling exponents of fracture surfaces, postulating the equivalence between the variations in deformation and surface energy. Bouchaud [62] disagreed with the Mosolov model [64] and proposed another model in terms of fluctuations in heights of the roughness on fracture surfaces in the perpendicular direction to the line of crack growth, obtaining a relationship between the fracture critical parameters such as K_{IC} and relative variation of the height fluctuations of the rugged surface. In this scenario has been conjectured the universality of the roughness exponent of fracture surfaces because this did not depend on the material being studied [63]. This assumption has generated controversy [61] which led scientists to discover anomalies in the scaling exponents between local and global scales in fracture surfaces of brittle materials. Family and Vicsék [39, 65] and Barabasi [66] present models of fractal scaling for rugged surfaces in films formed by ballistic deposition. Based on this dynamic scaling Lopez and Schimittibuhl [67, 68] proposed an analogous model valid for fracture surfaces, where they observed in your experiments anomalies in the fractal scaling, with critical dimensions of transition for the behavior of the roughness of these surfaces in brittle materials. In this sense Lopez [67, 68] borrowed from the model of Family and Vicsék [39, 65] analogies that could be applied to the rough fracture surfaces.

4.2.3. The fractality of a crack or fracture surface

By observing a crack, in general, one notes that it presents similar geometrical aspects that reproduce itself, at least within a limited range of scales. This property called invariance by scale transformation is called also self-similarity, if not privilege any direction, or self-affinity, when it favors some direction over the other. Some authors define it as the property that have certain geometrical objects, in which its parts are similar to the whole in in successive scales transformation. In the case of fracture, this takes place from a range of minimum cutoff scale , $\varepsilon_{\min}$ until a maximum cutoff scale, $\varepsilon_{\max}$, contrary to the proposed by Borodich [3], which defines an infinite range of scales to maintain the mathematical definition fractal. In the model proposed in this section, one used the fractal theory as a form closer to reality to describe the fracture surface with respect to Euclidean description. This was done in order to have a much better approximatation to reality of the problem and to use fractal theory as a more authentic approach.

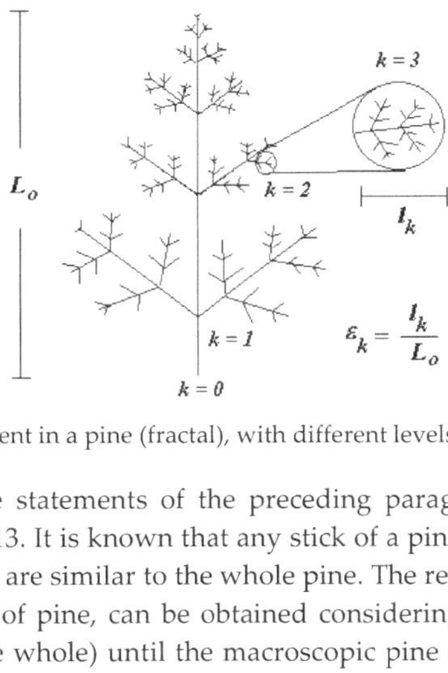

Figure 13. Self-similarity present in a pine (fractal), with different levels of scaling, *k*.

To understand clearly the statements of the preceding paragraph, one can use the pine example shown in Figure 13. It is known that any stick of a pine is similar in scale, the other branches, which in its turn are similar to the whole pine. The relationship between the scales mentioned above, in case of pine, can be obtained considering from the size of the lower branch (similar to the pine whole) until the macroscopic pine size. Calling of $\delta_{\min} = l_0$, the size of the lower branch and $\delta_{\max} = L_0$, the macroscopic size of whole pine one may be defined cutoff scales lower and upper (minimum and maximum), subdivided, therefore, the pine in discrete levels of scales as suggested the structure, as follows:

$$\varepsilon_{min} = \frac{l_o}{L_o} \leq \varepsilon_k = \frac{l_k}{L_o} \leq \varepsilon_{max} = \frac{L_o}{L_o} = 1 \begin{cases} static \quad case\ L_0 = L_{0\max} \\ dynamic\ case\ L_0 = L_0(t) \end{cases} \tag{35}$$

where an intermediate scale $\varepsilon_k \left(\varepsilon_{\min} \leq \varepsilon_k \leq \varepsilon_{\max}\right)$ can also be defined as follows:

$$\varepsilon_k = \frac{l_k}{L_o}. \tag{36}$$

The magnitude ε_k represents the scaling ratio which depicts the size of any branch with length, l_k, in relation to any pine whole. l_0 is related to the Mishnaevsky minimum size for a crack which is shown in section - 4.2.6 and $L_0 = L_{0\max}$ is the maximum leght if the fracture already been completed.

Similarly it is assumed that the cracks and fracture surfaces also have their scaling relations, like that represented in equations (35) and (36). In the continuous cutoff scale levels, lower and upper (minimum and maximum), are thus defined as follows:

$$\varepsilon_{min} = \frac{l_o}{L_o} \leq \varepsilon = \frac{l}{L_o} \leq \varepsilon_{max} = \frac{L_o}{L_o} = 1 \quad (37)$$

Note that the self-similarity of the pine so as the crack self-affinity, although statistical, is limited by a lower scale $\varepsilon_{\min}$ as determined by the minimum size, l_0, and a upper scale $\varepsilon_{\max}$, given by macroscopic crack size, L_0.

From the concepts described so far, it is verified that the measuring scale ε_k to count the structure elements is arbitrary. However, in the scaling of a fracture surface, or a crack profile, follows a question:

Which is the value of scale ε_k to be properly used in order to obtain the most accurate possible measurement of the rugged fracture surface?

There is a minimum fracture size that depends only on the type of material?

Surely the answer to this question lies in the need to define the smallest size of the fractal structure of a crack or fracture surface, so that its size can be used as a minimal calibration measuring ruler([2]).

Since an fracture surface or crack, is considered a fractal, first, it is necessary to identify in the microstructure of the material which should be the size as small as possible a of a rugged fracture, i.e. the value of $l_{\min}$. This minimal fracture size, typical of each material, must be then regarded as an elementary structure of the formation of fractal fracture, so defining a minimum cutoff scale $\varepsilon_{\min}$, for the fractal scaling, where $\varepsilon_{\min} = l_0/L_0$, where l_0 it is a planar projection of $l_{\min}$. In practice, from this value the minimum scale of measurement, $\varepsilon_{\min} = l_0/L_0$ one defines a minimum ruler size $\delta_{\min}$, for this case, equal to the value of the plane projection the smallest possible fracture size, i.e. $\delta_{\min} = l_0$. Thus, the fractal scaling of the fracture surface, or crack, may be done by obtaining the most accurate possible value of the rough length, L. However, the theoretical prediction of the minimum fracture, $l_{\min}$, must be made from the classical fracture mechanics, as will be seen below.

4.2.4. Scaling hierarchical limits

Mandelbrot [58] pointed out in his work that the fracture surfaces and objects found in nature, in general, fall into a regular hierarchy, where different sizes of the irregularities

---

[2] This must be done so that the measurement scales are not arbitrary and may depend on some property of the material.

described by fractal geometry, are limited by upper and lower sizes, in which each level is a version in scale of the levels contained below and above of these sizes. Some structures that appear in nature, as opposed to the mathematical fractals, present the property of invariance by scale transformation (self- similarity of self-affinity) only within a limited range of scale transformation ($\varepsilon_{min} \leq \varepsilon \leq \varepsilon_{max}$) . Note in Figure 13 that this minimal cutoff scale ε_{min} , one can find an elementary part of the object similar to the whole, that in iteration rules is used as a seed to construct the fractal pattern that is repeated at successive scales, and the maximum cutoff scale $D \in \mathbb{R}$ one can see the fractal object as a whole.

One must not confuse this mathematical recursive construction way, with the way in which fractals appear in nature really. In physical media, fractals appears normally in situations of local or global instability [24], giving rise to structures that can be called fractals, at least within a narrow range of scaling ($\varepsilon_{min} \leq \varepsilon \leq \varepsilon_{max}$) as is the case of trees such as pine, cauliflower, dendritic structures in solidification of materials, cracks, mountains, clouds, etc. From these examples it is observed that, in nature, the particular characteristics of the seed pattern depends on the particular system. For these structures, it is easy to see that that fractal scaling occurs from the lowest branch of a pine, for a example, which is repeated following the same appearance, until the end size of the same, and vice versa. In the case of a crack, if a portion of this crack, is enlarged by a scale, ε , one will see that it resembles the entire crack and so on, until reaching to the maximum expansion limit in a minimum scale, ε_{min} , in which one can not enlarge the portion of the crack, without losing the property of invariance by scaling transformation (self-similar or self-affinity). As the fractal growth theory deals with growing structures, due to local or global instability situations [24], such scaling interval is related to the total energy expended to form the structure. The minimum and maximum scales limit is related to the minimum and maximum scale energy expended in forming the structure, since it is proportional to the fractal mass. The number of levels scaling, k, between ε_{min} and ε_{max} depends on the rate at which the formation energy of fractal was dissipated, or also on the instability degree that gave rise to the fractal pattern.

4.2.5. The fractal geometric pattern of a fracture and its measurement scales

Considering that the fracture surface formed follows a fractal behavior necessarily also admits the existence of a geometric pattern that repeats itself, independent of the scale of observation. The existence of this pattern also shows that a certain degree of geometric information is stored in scale, during the crack growth. Thus, for each type of material can be abstract a kind of geometric pattern, apparently irregular with slight statistical variations, able to describe the fracture surface.

Moreover, for the same type of material is necessary to observe carefully the enlargement or reduction scales of the fracture surface. For, as it reduces or enlarges the scale of view, are found pattern and structures which are modified from certain ranges of these scales. This can be seen in Figure 14. In this figure is shown that in an alumina ceramic, whose ampliation of one of its grains at the microstructure reveals an underlying structure of the cleavage steps, showing that for different magnifications the material shows different morphologies of the surface of fracture.

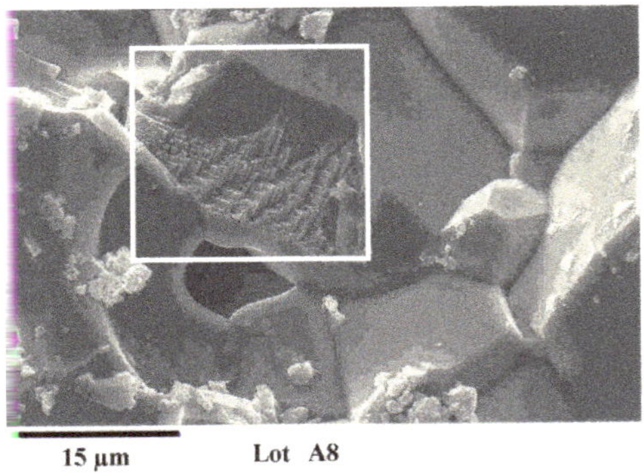

Figure 14. Changings in pattern of irregularities with the magnification scale on a ceramic alumina, Lot A8 [56].

To approach this problem one must first observe that, what is the structure for a scale becomes pattern element or structural element to another scale. For example, to study the material, the level of atomic dimensions, the atom that has its own structure (Figure 15a) is the element of another upper level, i.e., the crystalline (Figure 15b). At this level, the cleavage steps formed by the set of crystalline planes displaced, in turn, become the structural elements of microsuperfície fracture in this scale (Figure 15c). At the next level, the crystalline, is the microstructural level of the material, where each fracture microsurface becomes the structural member, although irregular, of the macroscopic rugged fracture surface, as visible to the naked eye, as is shown diagrammatically in Figure 15d. Thus, the hierarchical structural levels [69] are defined within the material (Figure 15), as already described in this section.

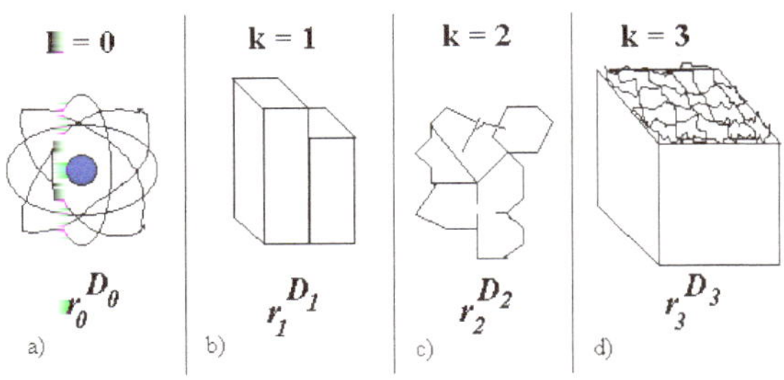

Figure 15. Different hierarchical structural levels of a fracture in function of the observation scale; a) atomic level; b) crystalline level (cleavage steps); c) microstructural level (fracture microsurfaces) and d) macrostructural level (fracture surface).

Based on the observations made in the preceding paragraph, it is observed that the fractal scaling of a fracture surface should be limited to certain ranges of scale in order to maintain the mathematical description of the same geometric pattern (atom, crystal, etc.) , which is shown in detail in section - 4.2.3. Although it is possible to find a structural element, forming a pattern, at each hierarchical level, it should be remembered that each type of structure has a characteristic fractal dimension. Therefore, it is impossible to characterize all scale levels of a fracture with only a single fractal dimension. To resolve this problem one can uses a multifractal description. However, within the purpose of this section a description monofractal provides satisfactory results. For this reason, it was considered in the first instance, that a more sophisticated would be unnecessary.

Considering the analytical problem of the fractal description, one must establish a lower and an upper observation scale, in which the mathematical considerations are kept within this range. These scales limits are established from the mechanical properties and from the sample size, as will be seen later. Obviously, a mathematical description at another level of scale, should take into account the new range of scales and measurement rules within this other level, as well as the corresponding fractal dimension.

As already mentioned, the description of the rough fracture surface can be performed at the atomic level, in cleavage steps level (crystalline) or in microstructural level (fracture microsurfaces), depending on the phenomenological degree of detail that wants to reach. This section will be fixed at the microstructural level (micrometer scale), because it reflects the morphology of the surface described by the thermodynamic view of the fracture. This means that the characteristics lengths of generated defects are large in relation to the atomic scale, thus defining a continuous means that reconciles in the same scale the mechanical properties with the thermodynamic properties. Meanwhile, the atomic level and the level of cleavage steps is treated by molecular dynamics and plasticity theory, respectively, which are part areas.

4.2.6. The calibration problem of a fracture minimum size as a "minimal ruler size" of their fractal

To answer the previous question, about the minimum fracture size, Mishnaevsky [70] proposes a minimum characteristic size, a, given by the size of the smallest possible microcrack, formed at the crack tip (or notch) as a result of stress concentration in the vicinity of a piling up dislocations in the crystalline lattice of the material, satisfying a condition of maximum constriction at the crack tip, where:

$$a \sim k_0 n b, \tag{38}$$

where k_0 is a proportionality coefficient. n is the number of dislocations piling up that can be calculated by:

$$n = \frac{\pi l \sigma (1 - \nu)}{b \mu}, \tag{39}$$

where υ is the Poisson's ratio, l is the length of the piling up of dislocations, σ is the normal or tangential stress, μ is the shear modulus and $\vec{b}$ is the Burgers vector. Substituting (39) in (38) one has;

$$a \sim \frac{k_o n\pi l\sigma(1-\nu)}{\mu}. \quad (40)$$

Mishnaevsky equates with mathematical elegance, the crack propagation as the result of a "physical reaction" of interaction of a crack size, $\langle L_0 \rangle$ with a piling up of dislocations, nb, forming a microtrinca size, a, i.e.;

$$< L_o > + < nb > \quad \rightarrow \quad < L_o + a >, \quad (41)$$

where $a << L_o$ e $nb << L_o$.

Mishnaevsky proposes a fractal scaling for the fracture process since the minimum scale, given by the size a, until the maximum scale, given by the macroscopic size crack, L_0.

As a consequence for the existence of a minimum fracture size, recently has arosen a hypothesis that the fracture process is discrete or quantized (Passoja, 1988, Taylor et al., 2005; Wnuk, 2007). Taylor et al. (2005) conducted mathematical changes in CFM to validate this hypothesis. Experimental results have confirmed that a minimum fractures length is given by:

$$l_0 \sim \frac{2}{\pi}\left(\frac{K_c}{\sigma_0}\right). \quad (42)$$

where K_c is the fracture toughness, σ_0 is the stress of the yielding strength before the material fracture.

4.2.7. Fractal scaling of a self-similar rough fracture surface or profile

A mathematical relationship between the extension of the self-similar contour and a extension of its projection is calculated as follows.

Being A the surface extension of the fractal contour, given by a self-similar homogenous function with fractional degree, D, where:

$$A(\varepsilon\delta) = \varepsilon^D A_u(\delta). \quad (43)$$

A_0 is the plane projection extension, given by a self-similar homogeneous function with integer degree, d, in accordance with the expression:

$$A_0(\varepsilon\delta) = \varepsilon^d A_u(\delta), \quad (44)$$

where, $A_u(\delta) = \delta^d$ is the unit area of measurement, whose values on the rugged and plane surface are the same. Thus the relationships (43) and (44) can be written in the same way as the equations (43) and (44). Therefore, by dividing these equations, one has:

$$A(\varepsilon\delta) = A_0(\delta)\varepsilon^{d-D}. \quad (45)$$

An illustration of the relationship (43), (44) and (45) can be seen in Figure 16.

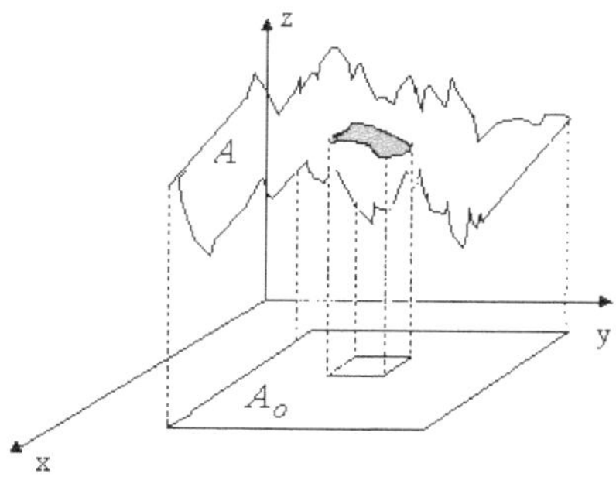

Figure 16. Rugged surface formed by a homogeneous function A, with frational degee D, whose planar projection, A_0 is a homogeneous function of integer degree d, showing the unit surface area A_u.

The rugged fracture surface, may be considered to be a homogeneous function with frational degree, D, ni.e.:

$$A = A_k \varepsilon_k^{-D}, \tag{46}$$

and its planar projection, may be considered as a homogeneous function with integer degree $d = 2$, i.e.:

$$A_0 = A_r \varepsilon_r^{-d}. \tag{47}$$

The index k was chosen to designate the irregular surface at a k-level of any magnification or reduction. The index r has been chosen to designate the smooth (or flat) surface at a r-level, and the index, 0, was chosen to designate the projected surface corresponding to rugged surface, at the k-level.

Considering that, for $k = r$ and $\varepsilon_k = \varepsilon_r$, the area unit, A_k and A_r, are necessarily of equal value and dividing relationships (46) and (47), one has:

$$A(\varepsilon_k) = A_0 \varepsilon_k^{d-D}. \tag{48}$$

The equation (48), means that the scaling performed between a smooth and another irregular surface, must be accompanied by a power term of type ε_k^{d-D}. Thus, there is the fractal scaling, which relates the two fracture surfaces in question: a rugged or irregular surface, which contains the true area of the fracture and regular surface, which contains the projected area of the fracture.

From now on will be obtained a relationship between the rugged and the projected profile of the fracture in analogous way to equation (250) for a thin flat plate (Figure 17a and Figure 17b) with thickness $e \to 0$. In this case the area of rugged surface can be written as:

$$A = Le, \tag{49}$$

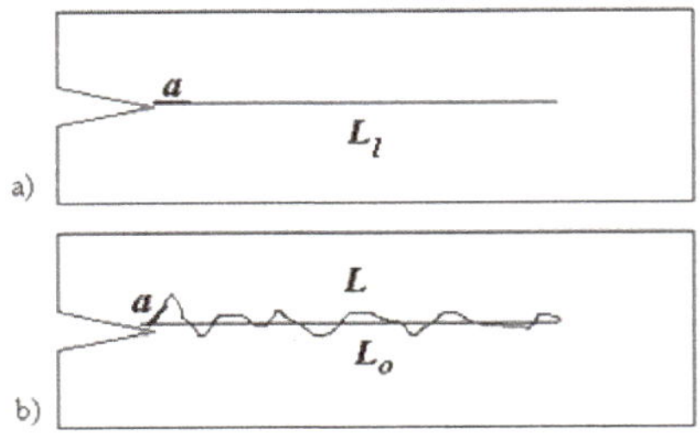

Figure 17. Scaling of a rugged profile of a fracture surface or a crack, using the Mishnaevsky minimum size as a "measuring ruler"; a) in the case of a crack is a non-fractal straight line, where $D = d = 1$; b) in the case of tortuous fractal crack, with its projected crack length, where $d \le D \le d+1$.

and the area of the projected surface as

$$A_0 = L_0 e, \tag{50}$$

According to the equation (48) the valid relationship is:

$$L(\varepsilon_k) = L_o \varepsilon_k^{\,d-D}, \tag{51}$$

where, $L(\varepsilon_k)$ is the measured crack length on the scale ε_k, L_0 is the projected crack length measured on the same scale, in a growth direction.

4.2.8. *The self-similarity relationship of a fractal crack*

The fracture is characterized from the final separation of the crystal planes. This separation has a minimum well-defined value, possibly given by theory Mishnaevsky Jr. (1994). If it is considered that below of this minimum value the fracture does not exist, and above it the crack is defined as the crystal planes moving continuously (and the formed crack tip penetrates the material), so that an increasing number of crystal planes are finally separated. One can in principle to use this minimum microscopic size as a kind of ruler (or scale) for the measurement of the crack as a whole([3]), i.e. from the start point from which the crack grows until its end characterized by instantaneous process of crack growth, for example.

The above idea can be expressed mathematically as follows:

[3] During or concurrently with its propagation, in a dynamic scaling process, or not

$$L = L_0 \varepsilon^{d-D}, \tag{52}$$

dividing the entire expression (52) above by the minimum Mishnaevsky size one has:

$$\frac{L}{a} = \left(\frac{L_0}{a}\right)\varepsilon^{d-D}, \tag{53}$$

or

$$N = N_0 \varepsilon^{d-D}, \tag{54}$$

where

$N = L/a$: is the number of crack elements a on the non-projected crack

$N_0 = L_0/a$: is the number of cracl elements a on the crack projected

and yet:

$$\varepsilon = a/L_0, \tag{55}$$

where:

ε : is the scaling factor of the fractal crack

d : is the Euclidean dimension of the crack projection

D : is the crack fractal dimension.

Within this context the number of microcracks that form the macroscopic crack is given by:

$$N = \left(\frac{a}{L_o}\right)^{-D}. \tag{56}$$

In this context (in Mishnaevsky model), the above expression is volumetric and admits cracks branching generated in the fracture process with opening and coalescence of microcracks. However, he continue equating the process in a one-dimensional way reaching an expression for the crack propagation velocity. A complete discussion of this subject, using a self-affine fractal model to be more realistic and accurate, can be done in another research paper.

The answer to the question about what should be the best scale to be used for fractal fracture scaling is then given as follows: being the limit of the crack length L_k in any scale, given by $L_k \rightarrow L$ (actual size) as well as $l_k \rightarrow l_{min}$, the value of the minimum size ruler, l_0 it must be equal to the minimum crack size, a ([4]), given by Mishnaevsky [70], through its energy balance for the fracture of a single monocrystal of the microstructure of a material. The physical reason for this choice is because the Mishnaevsky minimum size is determined by a

[4] It is possible that this minimal ruler size be very low than the scale used in fractal characterization of the fracture surfaces. However, it must to be the smallest possible size for a microcrack.

energy balance, from which the crack comes to exist, because below this size, there is no sense speak of crack length. Therefore, the scale that must be considered is given by:

$$\varepsilon_{\min} = a/L_0 \,, \tag{57}$$

where a is given by relation (40).

Therefore, the statistical self-similarity or self-affinity of a fracture surface, or a crack is limited by a cutoff lower scale $\varepsilon_{\min}$, determined by the minimum critical size, $l_0 = a$, and a cutoff upper scale $\varepsilon_{\max}$, given by the macroscopic crack length, L_0.

In two dimensions, the problem of existence of a minimum scale size (possibly given by the Mishnaevky minimum size), leads to abstraction of a microsurface with minimum area, whose shape will be investigated further, in Appendices, in terms of the number of stress concentrators nearest existing within a material.

4.3. Model of self-affine fracture surface or profiles

In this section one intend to present the development of fractal models of self-similar surfaces. From a rough fracture surface can be extracted numerous profiles also rough on the crack propagation direction. However, in this section is considered only one profile, which is representative of the entire fracture surface (Figure 18). The plane strain condition admits this assumption. Because, although the fracture toughness varies along the thickness of the material to a plastic zone reduced in relation to material thickness, it can be considered a property. This means that it is possible to obtain a statistically rough profile, equivalent to other possible profiles, which can be obtained within the thickness range considered by plane strain conditions.

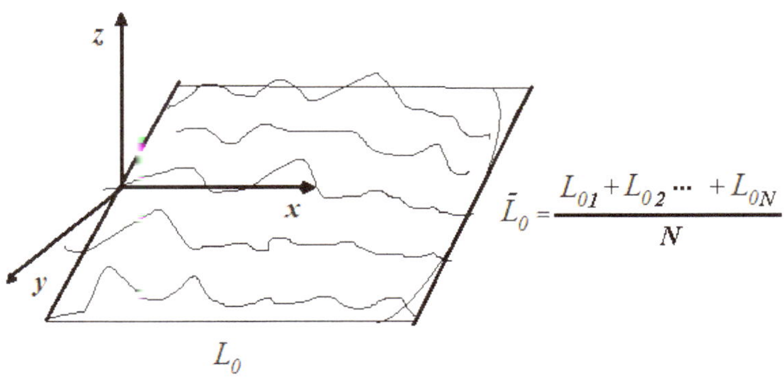

Figure 18. Statistically equivalent profiles along the thickness of the material

In order also equivalent to this, it is also possible to obtain an average projected crack length as a result of an average of the crack size along the thickness of the material thickness within the range considered by plane strain, for the purpose of calculations in CFM, it is considered this average size as if it were a single projected crack length, as recommended by the ASTM – E1737-96 [71]. Therefore, inwhat follows, is effected by reducing or lowering the dimensional degree of relationship from the two-dimensional case, shown above, for the one-dimensional case, as follows:

$$A(x,y) \to L(x). \tag{58}$$

thus, for a self-affine fractal one has:

$$L(\lambda_x x) = \lambda_x{}^H L(x), \tag{59}$$

where

$$H = 2 - D \tag{60}$$

is the Hurst exponent measuring the profile ruggedness. In one-dimensional case the fracture surface is a profile whose length L is obtained from measuring the projected length, L_0, as illustrated below in Figure 19.

4.3.1. Calculation of the rugged crack length as a function of its projected length

Considering a profile of the fracture surface as a self-affine fractal, analogous to the fractal of Figure 19, which perpendicular directions have the same physical nature the Voss [48] equation to the Brownian motion can be generalized([5]) to obtain rugged crack length L, depending on the projected crack length, L_0.

Figure 19 illustrates one of the methods for fractal measuring. This measure can be obtained by taking boxes or rectangular portions, based ΔL_0 and height ΔH_0 on the crack profile, and recovering up this profile, within these boxes, with "little boxes" (recovering units) with small sizes, l_0 and h_0, respectively (Figure 19). Instead of little boxes is also possible to use other shapes([6]) compatible with the object to be measured. Then makes the counting of the little boxes (or recovering units) needed to recover the extension of the rugged crack, centered in the box $\Delta L_0 \times \Delta H_0$. The number of these little boxes (or recovering units) of size r in function of the boxes extension (or parts), $\Delta L_0 \times \Delta H_0$, provides the fractal dimension, as shown in section 3 - Methods for Measuring Length, Area, Volume and Fractal Dimension.

Assume that the rectangular little boxes (or recovering units) of microscopic size, r, recover the entire crack length, ΔL inside the box with greater length, $\Delta L_0 \times \Delta H_0$. The number of little boxes (recovering unit) with sides of $l_0 \times h_0$ needed to recover a crack in the horizontal direction, inside the box (or stretch) of rectangular area $\Delta L_0 \times \Delta H_0$, for the self-affine fractal can be obtained by the expression:

---

[5] Voss [48], modeled the noise plot of the frational Brownian motion , where in the y-direction, he plots the amplitude, V_H, and in the x-direction, he plots the time, t.

[6] Some authors used "balls"

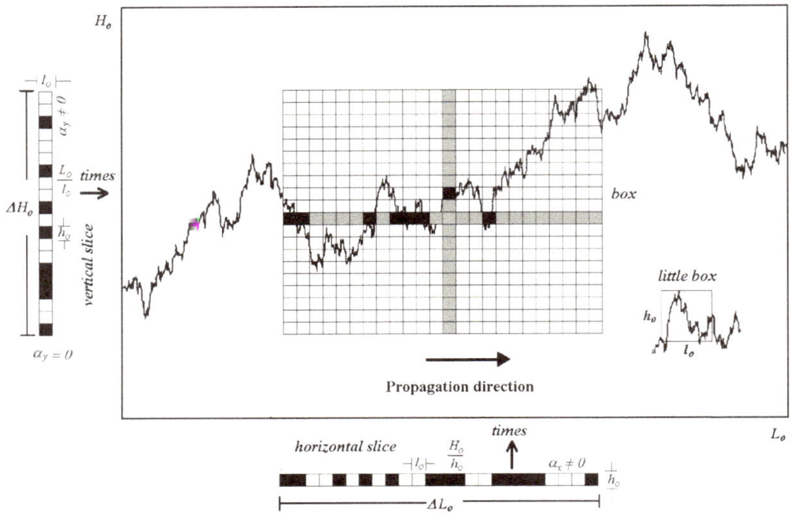

Figure 19. Self-affine fractal of Weierstrass-Mandelbrot, where $\varepsilon_k = 1/4$ and $D_x = 1.5$ and $H = 0.5$, used to represent a fracture profile (Family, Fereydoon; Vicsek, Tamas Dynamics of Fractal Surfaces, World Scientific, Singapore, 1991, p.7).

$$N_v = \frac{\Delta L_0}{l_0} \varepsilon_v{}^0 \left(\text{in vertical direction}\right) \tag{61}$$

where ΔL_0 is the crack horizontal projection and ε_v is the vertical scaling factor.

Considering that the self-affine fractal extends in the horizontal direction along L_0, and oscillates in the perpendicular direction, i.e. in the vertical direction, the number of little boxes N_h, with size, l_0 in the horizontal direction, are gathered to form the projected length L_0, while vertically the number of little boxes N_v, with size h_0, overlap each other, increasing (as power law) this number in comparison to the number of little boxes gathered horizontally. Therefore, for the vertical direction with a projection ΔH_0, the box sides $\Delta L_0 \times \Delta H_0$, an expression for the number of boxes (or units covering) can be writen as:

$$N_h = \frac{\Delta H_0}{h_0} \varepsilon_h{}^{-H} \left(\text{in horizontal direction}\right). \tag{62}$$

where H is the Hurst exponent, ΔH_0 is the total variation in height ($l_o \leq \Delta H_0 \leq \Delta L_o$) and ε_h is the scale transformation factor in the horizontal direction.

Therefore, for the corresponding rugged crack length (real) ΔL, the stretch $\Delta L_0 \times \Delta H_0$ one can writes:

$$\Delta L = N_v r \tag{63}$$

where r is equal to the rugged crack length on a microscopic scale, as a function of extension of the little boxes $l_0 \times h_0$ by:

$$r = \sqrt{l_0^{\,2} + h_0^{\,2}} \tag{64}$$

where l_0 and h_0 are the microscopic sizes of the crack length in horizontal and vertical directions, respectively. Substituting (64) in (63), one has:

$$\Delta L = N_v \sqrt{l_0^2 + h_0^2} \tag{65}$$

substituting (61) in (65), one has:

$$\Delta L = \frac{\Delta L_0}{l_0} \sqrt{l_0^{\,2} + h_0^{\,2}} \tag{66}$$

Since that in the fracture process, the scales in orthogonal directions are the same physical nature, one can choose $\varepsilon_v = \varepsilon_h = l_0 / \Delta L_0$, and one can writes from (62) that:

$$N_h = \left(\frac{\Delta H_0}{l_0}\right)\left(\frac{\Delta L_0}{l_0}\right)^{H} \tag{67}$$

being necessarily $N_h = N_v$, one has:

$$\left(\frac{\Delta L_0}{l_0}\right) = \left(\frac{\Delta H_0}{l_0}\right)\left(\frac{\Delta L_0}{l_0}\right)^{H} \tag{68}$$

rewriting the equation (66), one has:

$$\Delta L = \Delta L_0 \sqrt{1 + \left(\frac{h_0}{l_0}\right)^2} \tag{69}$$

writing h_0 from (68), as:

$$h_0 = \Delta H_0 \left(\frac{\Delta L_0}{l_0}\right)^{H-1}. \tag{70}$$

Eliminating in (69) the dependence of h_0, by substituting (70) in (69), one has:

$$\Delta L = \Delta L_0 \sqrt{1 + \left(\frac{\Delta H_0}{l_0}\right)^2 \left(\frac{\Delta L_0}{l_0}\right)^{2(H-1)}} \tag{71}$$

The curve length in the stretch, $\Delta L_0 \times \Delta H_0$ considering the Sand-Box method [38] whose counting starts from the origin of the fractal, can be written as: $\Delta L = L, \Delta L_0 = L_0$ and $\Delta H_0 = H_0$ hence the equation (71) shall be given by:

$$L = L_0 \sqrt{1 + \left(\frac{H_0}{l_0}\right)^2 \left(\frac{L_0}{l_0}\right)^{2(H-1)}}, \tag{72}$$

whose the plot is shown in Figure 20. Note that the lengths L_0 and H_0 correspond to the projected crack length in the horizontal and vertical directions, respectively.

Applying the logaritm on the both sides of equation (72) one obtains an expression that relates the fractal dimension with the projected crack length:

$$D_f \equiv \frac{\ln(L/l_0)}{\ln(L_0/l_0)} = 1 + \frac{1}{2} \frac{\ln\left\{\left(\frac{H_0}{l_0}\right)^2 \left(\frac{L_0}{l_0}\right)^{2(H-1)}\right\}}{\ln L_0} \tag{73}$$

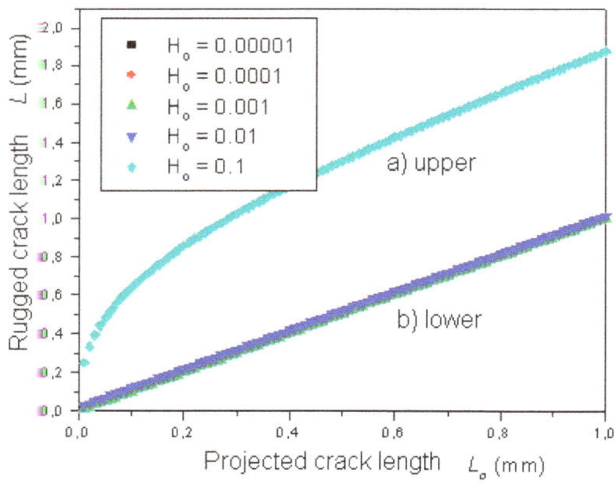

Figure 20. Graph of the rugged length L in function of the projected length L_0, showing the influence of height, H_0, of the boxes in the fractal model of fracture surface: a) in the upper curves is observed the effect of H_0 as it tends to unity ($H_0 \to 1.0$), b) in the lower curves, that appearing almost overlap, is observed the effect of H_0 as it tends to zero ($H_0 \to 0$).

The graph in Figure 20 shows the influence of the boxes height H_0 on the rugged crack length, L, as a function of the projected crack length, L_0. Note that for boxes of low height ($H_0 \to 0$), in relation to its projected length, L_0, the lower curves (for $H_0 = 0.01, 0.001, 0.0001$), denoted by the letter "b", almost overlap giving rise to a linear relation between these lengths (Figure 21). While for boxes of high height ($H_0 \to 1.0$) in relation to its projected length, L_0, the relation between the lengths become each more distinct from the linear relationship for the same exponent roughness, H.

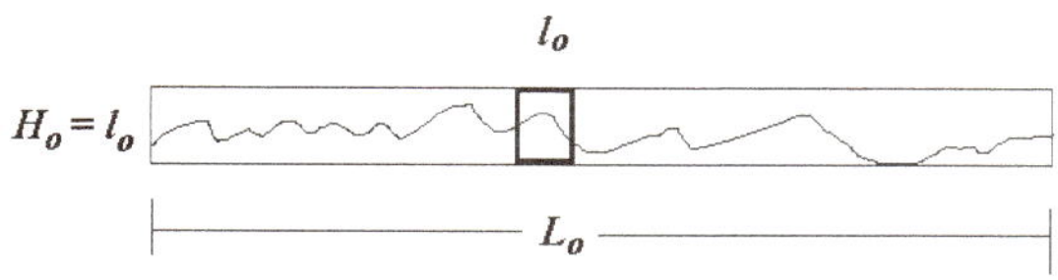

Figure 21. Counting boxes (or strechts) with rectangular sizes $L_o \; x \; H_o$ where the boxes that recovers the profile have different extensions in the horizontal and vertical directions.

Making up the counting boxes (or stretch) with rectangular sizes $L_o \; x \; H_o$ where the boxes recovering the profile have different extensions in the horizontal and vertical directions respectively, i.e., $H_o = l_o$ the equation (72). Is simplified to:

$$L = L_0 \sqrt{1 + \left(\frac{l_o}{L_o}\right)^{2H-2}} . \tag{74}$$

which plot is shown in Figure 22.

The graph in Figure 22 shows the influence of the roughness dimension on the rugged crack length, L, in function of the projected length, L_0. Note that for $H_0 \to 1.0$, corresponding to a smooth surface, the relation between the rugged and projected length becomes increasingly linear. While for $H_0 \to 0$, which corresponds to a rougher surface, the the relation between the rugged and projected length becomes increasingly non-linear.

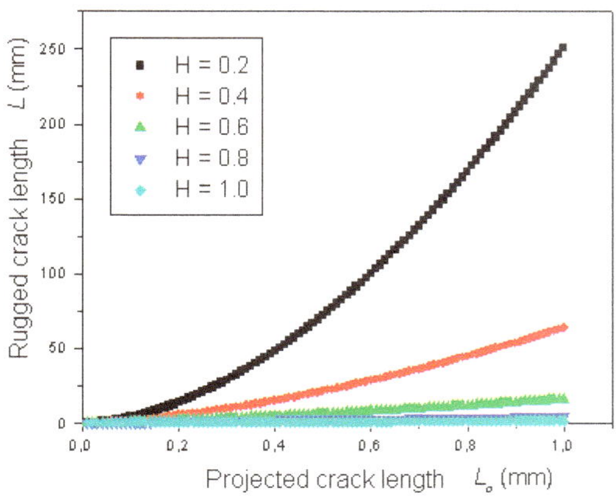

Figure 22. Graph of the rugged length, L, in function of the projected length, L_0, showing the influence of the Hurst exponent H, in the fractal model of the fracture surface.

Note that for $L_o = H_o$, one has, from the equation (62) and (68) the following relationship:

$$L_o = h_o \left(\frac{l_o}{L_o}\right)^{-H} , \tag{75}$$

which is a self-similar relation between the projected crack length, L_0, and height of the little box, h_0. This relationship shows that all self-affine fractal, in the approximation of a small scale, has a local self-similarity forming a fractal substructure, when is considered square portions, $L_0 \times L_0$, instead of rectangular portions, $L_0 \times H_0$.

It important to observe that L_0 denotes the distance between two points of the crack (the projected crack length). The self-affine measure, L of L_0, in the fractal dimension, D, is given by (72). l_0 is the possible minimum length of a micro-crack, which defines the scale l_0 / L_0 under which the crack profile is scrutinized, as discussed in previous section and will be discussed after in the section #.5.4.5. The Hurst exponent, H, is related to D by (60).

In the study of a self-affine fractal there are two extremes limits to be verified. One is the limit at which the boxes height is high in relation to its projected length, L_0, i.e. ($H_0 \rightarrow L_0$), which is also called local limit. The other limit is one in which the boxes height is low in relation to its projected length, L_0, i.e., ($H_0 \rightarrow h_0$) which is called global limit. It will be seen now each one of this limits case contained in the expression (72).

Case 1 : The self-similar or local limit of the fractality

Taking the local limit of the self-affine fractal measure as given by (72), i. e. for the case where, $H_0 = L_0 >> l_0$, one has:

$$L \cong L_o \left(\frac{l_o}{L_o} \right)^{H-1} \tag{76}$$

where

$$\frac{L}{L_o^{\,2-H}} \cong l_o^{\,H-1} = constant \tag{77}$$

This equation is analogous to self-similar mathematical relationship only that the exponent is $(1-H)$ instead of $(D-1)$, which satisfies the relation $H = 2 - D$ [3 , 40, 51 , 70].

According to these results it is observed that the relation (77) has a commitment to the Hurst exponent of the profiles on the considered observation scale $\varepsilon = l_0 / L_0$. It is observed that the consideration of a minimum fracture size l_{01} over a region, one must consider the local dimension of the fracture roughness on this scale. Similarly, if the considerations of a minimum fracture size are made in a scale that involves several regions, l_{02} this should take into account the value of the roughness global dimension on this scale, so that:

$$(2-H_1) l_{01}^{\,H_1-1} = (2-H_2) l_{02}^{\,H_2-1} = constant, \tag{78}$$

although $l_{01} \neq l_{02}$ e $H_1 \neq H_2$.

Case 2: The self-affine or global limit of fractality

Taking the global limit of the self-affine fractal measure given by (72), i.e. for the case in which: $H_0 = l_0 << L_0$. Therefore the length L is independently of H and $D = 1$, so

$$L \cong L_o \tag{79}$$

It must be noted that the ductile materials by having a high fractality have a crack profile which can be better fitted by the equation (76), while brittle materials by having a low fractality will be better fitted by the equation (79) corresponding the classical model, i.e., a flat geometry for the fracture surface. Furthermore, the cleavage which occurs on the microstructure of ductile materials tend to produce a surface, where $L \cong L_0$, which could be called smooth. However, this cleavage effect is just only local in these materials and therefore the resulting fracture surface is actually rugged.

4.3.2. Local Ruggedness of a fracture surface

Defining the local roughness of a fracture surface, as:

$$\xi \equiv \frac{dA}{dA_o} \Rightarrow A = \int \xi(A_0)dA_0. \tag{80}$$

where A is the rugged surface and A_0 is the projected surface. In the case of a rugged crack profile,one has:

$$\xi \equiv \frac{dL}{dL_o} \Rightarrow L = \int \xi(L_0)dL_0 \tag{81}$$

using (74) in (81), one has that:

$$\xi \equiv \frac{1+(2-H)\left(\frac{l_o}{L_o}\right)^{2H-2}}{\sqrt{1+\left(\frac{l_o}{L_o}\right)^{2H-2}}} \tag{82}$$

From (81) note that when there is no roughness on surfaces (flat fracture) one has that: $L = L_0$, thus

$$\frac{dL}{dL_o} = 1. \tag{83}$$

The quantity $\xi \equiv \frac{dL}{dL_o}$ seems be a good definition of ruggedness unlike the definition where the ruggedness is given by $\xi = L / L_{//}$ [56, 57] (where $L_{//} = L_{0M}\cos\theta$, see Figure 23) does not satisfy the requirement intuitive of the ruggednes when L_{0M} is only inclined with respect to L_0, while maintaining, $L_0 = L_{0M}$, as shown Figure 23.

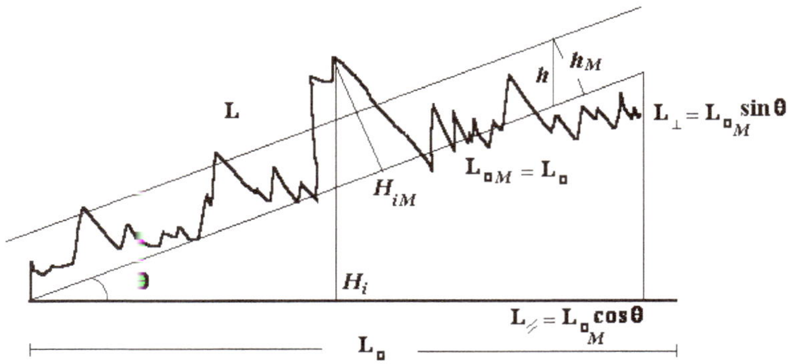

Figure 23. Schematization of a rugged surface which is inclined with respect to its projection.

The ruggedness must depend on infinitesimally of the projected length and its relative orientation to it. In this case, the surface roughness by the usual definition adds an error equal to the angle secant θ, or

$$\xi = L/L_{//} = \left(\frac{L}{L_{0M}}\right)\left(\frac{L_{0M}}{L_{//}}\right) = \left(\frac{L}{L_{0M}}\right)\frac{1}{\cos\theta}. \tag{84}$$

However, by the definition proposed herein, when only one inclines a smooth surface against to the horizontal, one has: $L = L_{0M}$ and $L_{0M} = L_0$ and again, $\xi = \frac{dL}{dL_o} = \frac{dL}{dL_{oM}}\frac{dL_{oM}}{dL_o} = 1$.

Within this philosophy will be considered as rugged any surface that presents in an infinitesimal portion a variation of their contour such that $\frac{dL}{dL_o} > 1$, therefore has to be:

$$\frac{dL}{dL_o} \geq 1. \tag{85}$$

4.3.3. *Preliminary considerations on the proposed model*

Considering a fractal model for the fracture surface given by the equation:

$$\Delta L = \Delta L_o\sqrt{1+\left(\frac{\Delta H_o}{l_o}\right)^2\left(\frac{l_o}{\Delta L_o}\right)^{2H-2}}, \tag{86}$$

it is possible to describe its ruggedness in order to include it in the mathematical formalism CFM to obtain a Fractal Fracture Mechanics (FFM). This derivative of equation (86) defines a fractal surface ruggedness, which for the case of a self-affine crack which grows with, $\Delta H_0 \rightarrow l_0$, is given by:

$$\xi \equiv \frac{1+(2-H)\left(\dfrac{l_o}{\Delta L_o}\right)^{2H-2}}{\sqrt{1+\left(\dfrac{l_o}{\Delta L_o}\right)^{2H-2}}} \geq 1. \tag{87}$$

such modifications were added to equations of the Irregular Fracture Mechanics to obtain a Fractal Fracture Mechanics as described below.

4.3.4. Comparison of fractal model with experimental results

In Figure 24 and Figure 25, a good agreement is observed in the curve fitting of equation (72) and equation (73) to the fractal analyses of the mortar specimen $A2$ side1 and the red ceramic specimen $A8$, respectively.

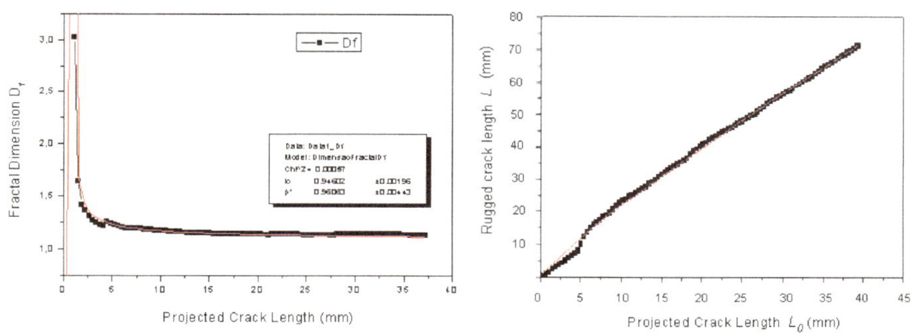

Figure 24. a) Fractal analysis of mortar specimen A2 side 1 – Fractal dimension x Projected length, L_0; b)Fractal analysis of mortar specimen A2 side 1 - rugged length L x projected length, L_0

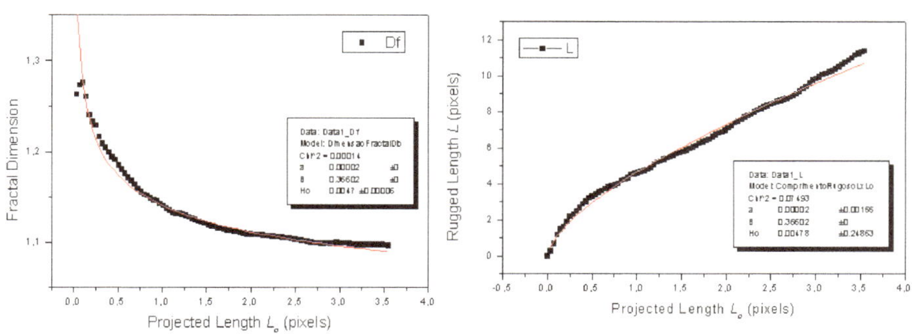

Figure 25. a) Fractal analysis of red clay A8 side 1 – Fractal dimension x Projected length, L_0; b) Fractal analysis of red ceramic specimen A8 - rugged length L x projected length, L_0

5. Conclusions

i. It is possible, in principle, mathematically distinguish a crack in different materials using geometric characteristics which can be portrayed by different values of roughness exponents in the relations (72) and (74).

ii. The fractal model of the rugged crack length, L in function of the projected crack length, L_0, suggested by Alves [72 73 ,74 , 75] seems have a good agreement with experimental results. This results allowed us to consolidate the model previously published in the literature on fracture [72 , 73 , 75].

iii. The rugged crack length is a response to its interaction with the microstructure. of the material. Therefore, mathematically is possible to portray the rugged peculiar behavior of a crack using fractal geometry

iv. The mathematical model presents a wealth (mathematical richness) that can still be explored in terms of determining the minimum crack length, l_0 for each material and the fractal dimension as a function of test parameters and material properties.

v. The mathematical model is sensitive to variations in the behavior of the crack length it is a linear or logarithmic with the projected crack length.

Comparing the experimental results with the model proposed in this chapter, it is concluded that one of the more important results obtained here are the equations (72), (74) and (82) leading to finding that the fracture surfaces of the materials analyzed are indeed (actually) self-affine fractals. Starting from this verification it becomes feasible to consider the fractal model of rugged fracture surface and its ruggedess inside the equations of the classical fracture mechanics, according to equation (74) and (82). As there is a close relationship between phenomenology and structure formed by virtue of its fractal geometry, the understanding of the formation processes of these dissipative structures, as the cracks, should be derived from their mathematical analysis, as the close relationship between the phenomenology of the formation process of dissipative structures and their fractal geometry. Therefore, the mathematical description of fractal structures must exceed a simple geometrical characterization, in order to correlate the pattern formed in the process of energy dissipation with the amount of energy dissipated in the process that generated it. Thus, it is possible to use the fractal geometry in order to understand other more and more complex processes inside the fracture mechanics. Therefore, the various mechanisms responsible by the crack deviation and by the formation of the rugged fracture surface can then, from the fractal model, be quantified in the fractal analysis of this surface.

The idea of obtaining a relationship between L and L_0 comes the need to maintain the present formalism used by the CFM, showing that fractal geometry can greatly contribute to the continued advancement of this science.

On the other hand, we are interested in developing a Fractal Thermodynamic for a rugged crack that will be related to the CFM and the Classical Fracture Thermodynamics when the crack ruggedness is neglected or the crack is considered smooth.

Author details

Lucas Máximo Alves
GTEME – Grupo de Termodinâmica, Mecânica e Eletrônica dos Materiais, Departamento de Engenharia de Materiais, Setor de Ciências Agrárias e de Tecnologia, Universidade Estadual de Ponta Grossa, Brazil

6. References

[1] Mandelbrot, Benoit B. (1982) The Fractal Geometry of Nature. San Francisco, Cal-Usa, New York: W. H. Freeman and Company. 472 p.
[2] Rodrigues, J. A. E V. C. Pandolfelli (1996) Dimensão Fractal E Energia Total de Fratura. Cerâmica, Maio/Junho. 42(275).
[3] Borodich, F. M. (1997) Some Fractals Models of Fracture. J. Mech. Phys. Solids. 45(2):239-259.
[4] Heping, Xie (1989) The Fractal Effect of Irregularity of Crack Branching on the Fracture Toughness of Brittle Materials. International Journal of Fracture. 41: 267-274.
[5] Mu, Z. Q. and C. W. Lung (1988) Studies on the Fractal Dimension and Fracture Toughness of Steel. J. Phys. D: Appl. Phys. 21: 848-850.
[6] Mecholsky, J. J.; D. E. Passoja and K. S. Feinberg-Ringel (1989) Quantitative Analysis of Brittle Fracture Surfaces Using Fractal Geometry. J. Am. Ceram. Soc. 72(1): 60-65.
[7] Lin, G. M.; J. K. L. Lai (1993) Fractal Characterization of Fracture Surfaces In a Resin-Based Composite. Journal of Materials Science Letters. 12: 470-472.
[8] Nagahama, Hiroyuki (1994) A Fractal Criterion for Ductile and Brittle Fracture. J. Appl. Phys. 15 March, 75(6): 3220-3222.
[9] Lei, Weisheng and Bingsen Chen (1994) Discussion on "The Fractal Effect of Irregularity of Crack Branching on the Fracture Toughness of Brittle Materials" By Xie Heping. International Journal of Fracture. 65: R65-R70.
[10] Lei, Weisheng and Bingsen Chen (1994) Discussion: "Correlation Between Crack Tortuosity and Fracture Toughness In Cementitious Material" By M. A. Issa, A. M. Hammad and A. Chudnovsky, A. International Journal of Fracture. 65: R29-R35.
[11] Lei, Weisheng and Bingsen Chen (1995) Fractal Characterization of Some Fracture Phenomena, Engineering Fracture Mechanics. 50(2): 149-155.
[12] Tanaka, M. (1996) Fracture Toughness and Crack Morphology In Indentation Fracture of Brittle Materials. Journal of Materials Science. 31: 749-755.
[13] Chelidze, T.; Y. Gueguen (1990) Evidence of Fractal Fracture, (Technical Note) Int. J. Rock. Mech Min. Sci & Geomech Abstr. 27(3): 223-225.
[14] Herrmann, H. J.; Kertész, J.; de Arcangelis, L. (1989) Fractal Shapes of Deterministic Cracks. Europhys. Lett. 10(2): 147-152.
[15] De Arcangelis, L.; Hansen A; Herrmann, H. J. (1989) Scaling Laws In Fracture. Phys. Review B. 1 July 40(1).
[16] Herrmann, Hans J. (1986) Growth: An Introduction. In: H. Eugene Stanley and Nicole Ostrowskym editors. On the Growth and Form, Fractal and Non-Fractal Patterns In Physics, Nato Asi Series, Series E: Applied Sciences N. 100 (1986), Proc. of the Nato Advanced Study Institute Ön Growth and Form", Cargese, Corsiva, France June 27-July 6 1985. Copyright By Martinus Nighoff Publishers, Dordrecht.
[17] Tsallis, C.; Plastino, A. R.; and Zheng, W.-M. (1997) Chaos, Solitons & Fractals 8: 885.

[18] Mccauley, Joseph L. (1993) Chaos, Dynamics and Fractals: An Algorithmic Approach To Deterministic Chaos, Cambridge Nonlinear Science Series, Vol. 2, Cambridge England: Cambridge University Press.

[19] Stanley, H. Eugene (1973) Introduction To Phase Transitions and Critical Phenomena. (Editors: Cooperative Phenomena Near Phase Transitions, a Bilbiography With Selected Readings, Mit,). Cambridge, Massachusetts : Claredon Oxford

[20] Uzunov, D. I. (1993) Theory of Critical Phenomena, Mean Field, Flutuactions and Renormalization. Singaore: World Scientific Publishing Co. Pte. Ltd.

[21] Beck, C. and F. Schlögl (1993) Thermodynamics of Chaotic Systems: An Introduction, Cambridge Nonlinear Science Series, Vol. 4, Cambridge England: Cambridge University Press,.

[22] Meakin, Paul (1995) Fractal Growth: , Cambridge Nonlinear Science Series, Vol. 5, England: Cambridge University Press.

[23] Vicsék, Tamás (1992) Fractal Growth Phenonmena. Singapore: World Scientific.

[24] Sander, L. M. (1984) Theory of Fractal Growth Process In: F. Family, D. P. Landau editors. Kinetics of Aggregation and Gelation. Amsterdam: Elsevier Science Publishers B. V. pp. 13-17.

[25] Meakin, Paul (1993) The Growth of Rough Surfaces and Interfaces. Physics Reports. December 235(485): 189-289.

[26] Pietronero, L.; Erzan, A.; Everstsz, C. (1988) Theory of Fractal Growth. Phys. Revol. Lett. 15 August 61(7): 861-864.

[27] Herrmann, Hans J.; Roux, Stéphane (1990) Statistical Models For the Fracture of Disordered Media. Random Materials and Processes Series. H. Eugene Stanley and Etienne Guyon editors., Amsterdam: North-Holland.

[28] Charmet, J. C. ; Roux, S and Guyon, E. (1990) Disorder and Fracture. New York: Plenum Press.

[29] Meakin, Paul; Li, G.; Sander, L. M.; Louis, E.; Guinea, F. (1989) A Simple Two-Dimensional Model For Crack Propagation. J. Phys. A: Math. Gen. 22: 1393-1403.

[30] Gross, Steven. P.; Jay. Fineberg, M. P. Marder, W. D. Mccormick and Harry. L. Swinney (1993) Acoustic Emissions From Rapidly Moving Cracks. Physical Review Letters. , 8 November 71(19): 3162-3165.

[31] Sharon, Eran; Steven Paul Gross and Jay Fineberg (1996) Energy Dissipation In Dynamic Fracture. Physical Review Letters. 18 March, 76(12): 2117-2120.

[32] Hausdorff F. (1919) Dimension Und Äußeres Maß. Mathematische Annalen March 79(1–2): 157–179. Doi:10.1007/Bf01457179.

[33] Besicovitch, A. S. (1929) On Linear Sets of Points of Fractional Dimensions. Mathematische Annalen 101.

[34] Besicovitch, A. S. H. D. Ursell. (1937) Sets of Fractional Dimensions. Journal of the London Mathematical Society 12, 1937. Several Selections From This Volume Are Reprinted In Edgar, Gerald A. (1993). Classics on Fractals. Boston: Addison-Wesley. Isbn 0-201-58701-7 See Chapters 9,10,11

[35] Mandelbrot, Benoit B (1975) Fractal..

[36] Mandelbrot, Benoit B (1977) Fractals: Form Chance and Dimension. San Francisco, Cal-USA: W. H. Freeman and Company.

[37] Alves, Lucas Máximo (2012) Application of a Generalized Fractal Model For Rugged Fracture Surface To Profiles of Brittle Materials. Paper in preparation.

[38] Bunde, Armin Shlomo Havlin (1994) Fractals In Science. Springer-Verlag.
[39] Family, Fereydoon; Vicsek, Tamás (1991) Dynamics of Fractal Surfaces. Singapore: World Scientific Publishing Co. Pte. Ltd. P.7-8, P. 73-77.
[40] Feder, Jens (1989) Fractals, New York: Plenum Press.
[41] Barnsley, Michael (1988) Fractals Everywhere, Academic Press, Inc, Harcourt Brace Jovanovich Publishers.
[42] Milman V. Yu., Blumenfeld R., Stelmashenko N. A. and Ball R. C. (1993) Phys. Rev. Lett. 71 , 204.
[43] Milman, Victor Y.; Nadia A. Stelmashenko and Raphael Blumenfeld (1994) Fracture Surfaces: a Critical Review of Fractal Studies and a Novel Morphological Analysis of Scanning Tunneling Microscopy Measurements, Progreess In Materials Science. 38: 425-474.
[44] Yamaguti, Marcos (1992). Doctoral Thesis - Universidade de São Paulo
[45] Allen, Martin; Gareth J. Brown; Nick J. Miles (1995) Measurements of Boundary Fractal Dimensions: Review of Current Techniques. Powder Technology. 84:1-14.
[46] Richardson, L. F. (19161) The Problem of Contiguity: An Appendix To Statistics of Deadly Quarrels. General Systems Yearbook. (6): 139-187.
[47] Alves, Lucas Máximo (1998) Escalonamento Dinâmico da Fractais Laplacianos Baseado No Método Sand-Box, In: Anais Do 42o Cong. Bras. de Cerâmica, Poços de Caldas de 3 a 6 de Junho,. Artigo Publicado neste Congresso Ref.007/1.
[48] Voss, Richard F. (1991) In: Family, Fereydoon. and Vicsék, Tamás, editors. Dynamics of Fractal Surfaces. Singapore: World Scientific. pp. 40-45.
[49] Morel, Sthéphane, Jean Schmittbuhl, Elisabeth Bouchaud and Gérard Valentin (2000) Scaling of Crack Surfaces and Implications on Fracture Mechanics Arxiv:Cond-Mat/0007100 6 Jul 2000, vol. 1, Or Phys. Rev. Lett. 21 August 85(8).
[50] Underwood, Erwin E. and Kingshuk Banerji (1996) Quantitative Fractography. Engineering Aspectes of Failure and Failure Analysis - Asm - Handbook - Vol. 12, Fractography - the Materials Information Society (1992). Astm 1996 , pp. 192-209
[51] Dauskardt, R. H.; F. Haubensak and R. O. Ritchie (1990) On the Interpretation of the Fractal Character of Fracture Surfaces; Acta Metall. Matter. 38(2): 143-159.
[52] Underwood, Erwin E. and Kingshuk Banerji (1986) Fractals In Fractography, Materials Science and Engineering, Ed. Elsevier. 80: pp. 1-14.
[53] Underwood, Erwin E. and Kingshuk Banerji (1996) Fractal Analysis of Fracture Surfaces. Engineering Aspectes of Failure and Failure Analysis - Asm - Handbook - Vol. 12, Fractography - the Materials Information Society (1992), Astm 1996. pp. 210-215
[54] Alves, L. M. ; Chinelatto, Adilson Luiz ; Chinelatto, Adriana Scoton Antonio ; Prestes, Eduardo (2004) Verificação de um modelo fractal de fratura de argamassa de cimento. In: Anais do 48º Congresso Brasileiro de Cerâmica, 28 de Junho a 1º de Julho de 2004, Em Curitiba – Paraná.
[55] Alves, L. M. ; Chinelatto, Adilson Luiz ; Chinelatto, Adriana Scoton Antonio ; Grzebielucka, Edson Cezar (2004) Estudo do perfil fractal de fratura de cerâmica vermelha. In: Anais do 48º Congresso Brasileiro de Cerâmica, 28 de Junho a 1º de Julho de 2004, Em Curitiba – Paraná.
[56] Dos Santos, Sergio Francisco (1999) Aplicação Do Conceito de Fractais Para Análise Do Processo de Fratura de Materiais Cerâmicos, Master Dissertation, Universidade Federal

de São Carlos. Centro de Ciências Exatas e de Tecnologia, Programa de Pós-Graduação em Ciência e Engenharia de Materiais, São Carlos.
[57] Anderson, T. L. (1995) Fracture Mechanics, Fundamentals and Applications. Crc Press, 2th Edition.
[58] Mandelbrot, Benoit B.; Dann E. Passoja & Alvin J. Paullay (1984) Fractal Character of Fracture Surfaces of Metals, Nature (London). 19 April, 308 [5961]: 721-722.
[59] Lung, C. W. and Z. Q. Mu, (1988)n Fractal Dimension Measured With Perimeter Area Relation and Toughness of Materials, Physical Review B. 1 December , 38(16): 11781-11784.
[60] Bouchaud, Elisabeth (1977) Scaling Properties of Crack. J. Phys: Condens. Matter. 9: 4319-4344.
[61] Bouchaud, E.; G. Lapasset and J. Planés (1990) Fractal Dimension of Fractured Surfaces: a Universal Value? Europhysics Letters. 13(1): 73-79.
[62] Bouchaud, E.; J. P. Bouchaud (1994-I) Fracture Surfaces: Apparent Roughness, Relevant Length Scales, and Fracture Toughness. Physical Review B. 15 December . 50(23): 17752 – 17755.
[63] Bouchaud, Elisabeth (1997) Scaling Properties of Cracks. J. Phy. Condens. Matter. 9: 4319-4344.
[64] Mosolov, A. B. (1993) Mechanics of Fractal Cracks In Brittle Solids, Europhysics Letters. 10 December, 24(8): 673-678.
[65] Family & Vicsek (1985), Scaling in steady-state cluster-cluster aggregate. *J. Phys. A* 18: L75.
[66] Barabási, Albert – László; H. Eugene Stanley (1995) Fractal Concepts In Surface Growth, Cambridge: Cambridge University Press.
[67] Lopez, Juan M. Miguel A. Rodriguez, and Rodolfo Cuerno (1997) Superroughening versus intrinsic anomalous scaling of surfaces, *Phys. Rev. E* 56(4): 3993-3998.
[68] Lopez, Juan M. and Schmittbuhl, Jean (1998) Anomalous scaling of fracture surfaces, *Phys. Rev. E*. 57(6): 6405-6408.
[69] Guy, A. G. (1986) Ciências Dos Materias, Editora Guanabara 435p.
[70] Mishnaevsky Jr., L. L. (1994) A New Approach To the Determination of the Crack Velocity Versus Crack Length Relation. Fatigue Fract. Engng. Mater. Struct. 17(10): 1205-1212.
[71] ASTM - E1737. (1996) Standard Test Method For J-Integral Characterization of Fracture Toughness. Designation Astm E1737/96,pp.1-24.
[72] Alves, Lucas Máximo; Rosana Vilarim da Silva and Bernhard Joachim Mokross (2000) In: New Trends In Fractal Aspects of Complex Systems – FACS 2000 – IUPAP International Conference October, 16, 2000m At Universidade Federal de Alagoas – Maceió, Brasil.
[73] Alves, Lucas Máximo; Rosana Vilarim da Silva, Bernhard Joachim Mokross (2001) The Influence of the Crack Fractal Geometry on the Elastic Plastic Fracture Mechanics. Physica A: Statistical Mechanics and Its Applications. 12 June 2001, 295,(1/2): 144-148.
[74] Alves, Lucas Máximo (2002) Modelamento Fractal da Fratura E Do Crescimento de Trincas Em Materiais. Relatório de Tese de Doutorado Em Ciência E Engenharia de Materiais, Apresentada À Interunidades Em Ciência E Engenharia de Materiais, da Universidade de São Paulo-Campus, São Carlos, Orientador: Bernhard Joachim Mokross, Co-Orientador: José de Anchieta Rodrigues, São Carlos – SP.
[75] Alves, Lucas Máximo (2005) Fractal Geometry Concerned with Stable and Dynamic Fracture Mechanics. Journal of Theorethical and Applied Fracture Mechanics. 44/1:pp: 44-57.

Chapter 2

Higher Order Weight Functions in Fracture Mechanics of Multimaterials

A. Yu. Belov

Additional information is available at the end of the chapter

1. Introduction

The quantities characterizing near-tip fields of cracks are generally recognized to play a crucial role in both linear and nonlinear fracture mechanics. Among various methods developed to analyze the structure of the near-tip fields, the weight function technique of Bueckner [4, 6] based on Betti's reciprocity theorem turned out to be especially promising. The concept of higher-order weight functions in mechanics of elastic cracks was introduced by Sham [20, 21] as an extension of the weight function approach. A historical introduction into the existing alternative formulations of the weight function theory and a review of its earlier development can be found in the papers by Belov and Kirchner [28, 31]. The theory of weight functions treats the stress intensity factor K, which is a coefficient normalizing the stress singularity $\sigma = K/(2\pi r)^{1/2}$ at the crack tip, as a linear functional of loadings applied to an elastic body. The kernel of the functional is however independent of loadings and, in this sense, universal for the given body geometry and crack configuration. To emphasize this fact, Bueckner [4] suggested that the kernel to be called 'universal weight function'. The weight functions play the role of influence functions for stress intensity factors, since the weight function value at a point situated inside the body or at its surface (including crack faces) is equal to the stress intensity factor, which is due to the unit concentrated force applied at this point. The weight function based functionals can be constructed not only for external forces but also for the dislocation distributions described by the dislocation density tensor, as it was shown by Kirchner [14]. The objective of the weight function theory is not to compute complete stress distributions in cracked bodies for an arbitrary loading, but to express only one parameter K characterizing the strength of the near-tip stress field as a functional (weighted average) of the loading. In particular, in the simplest case of a cracked body subjected to only surface loadings the functional has the form of a contour integral. However, in order to apply the weight function theory to practical situations, the kernel of the functional has to be evaluated and this can be done by solving a special elasticity problem, for instance, numerically by a finite element method. The stress singularities are inherent not only to cracks. Sharp re-entrant corners or notches that are encountered in a number of engineering structures can become likely sites of stress concentrations and therefore the potential sources of the crack initiation. At the tips (or vertices) of these notches the stress can be also singular $\sigma = K/r^{1-s}$, where $s > 0$ and K is a generalized stress intensity factor normalizing the stress singularity. An attractive feature of the approach based on Betti's reciprocity theorem is that it enables for the weight functions to be constructed not only for sharp cracks but also for notches of finite opening angle [19]. It is the purpose of this paper to review the main ideas underlying the higher-order weight function methodology and to consider its applications to elastically anisotropic multimaterials with notches or cracks. The present analysis is confined to the two-dimensional structures in the state of the generalized plane deformation, where considerable analytical advancement was demonstrated in the last two decades.

As is known, the stress field in a finite two-dimensional elastic body containing an edge crack can be represented in series form over homogeneous eigenfunctions of an infinite plane with a semi-infinite crack. Such a series representation was first utilized by Williams [1, 2] to describe stress distributions around the crack tip, and is commonly referred to as Williams' eigenfunction expansion, although Williams confined himself only to the case of isotropy and spacial homogeneity of the elastic constants tensor. The eigenfunction expansion of this type however exists whatever the body is elastically isotropic or anisotropic, homogeneous or angularly inhomogeneous (with elasticity constants dependent on the azimuth around the crack tip). The weight functions introduced by Bueckner enable to evaluate only the stress intensity K, that is the magnitude of the singular term, which close to the crack tip dominates other terms in the Williams' expansion. It is the purpose of higher order weight function theory to evaluate coefficients of non-singular terms in this expansion.

2. Symmetry in anisotropic theory of elasticity

If one exploits the linear elasticity theory, the tensor of the second order elastic constants $C_{ijkl}(\mathbf{r})$ of an anisotropic medium (both homogeneous and inhomogeneous) possesses the following types of symmetry:

a) due to the symmetry of stresses and strains

$$C_{ijkl}(\mathbf{r}) = C_{jikl}(\mathbf{r}) = C_{ijlk}(\mathbf{r}) \tag{1}$$

b) due to the existence of the elastic potential $W(\varepsilon_{kl})$

$$C_{ijkl}(\mathbf{r}) = C_{klij}(\mathbf{r}) \tag{2}$$

Owing to both properties given in Eqs. (1)-(2), one has

$$(ab) = (ba)^{\mathrm{t}}, \tag{3}$$

where the real 3×3 matrices are constructed according to the rule

$$(ab)_{jk} = a_i C_{ijkl}(\mathbf{r}) b_l \tag{4}$$

for two arbitrary vectors a_i and b_l. Although Eq. (3) looks rather simple, it underlies many fundamental results of the anisotropic elasticity theory. In particular, the proof of the orthogonality relation for the six-dimensional Stroh eigenvectors [3] is based only on Eq. (3), see [22] for further details. Here it is worth to mention that Betti's reciprocity theorem is based on the symmetry properties (1)-(2) as well. This fact was utilized in [5] to derive the aforementioned orthogonality relation for the Stroh eigenvectors from Betti's theorem. In fact, practically all significant analytical achievements in the anisotropic theory of elasticity employ the directly following from Eq. (3) symmetry relation

$$(\hat{\mathbf{T}}\hat{\mathbf{N}})^{\mathrm{t}} = \hat{\mathbf{T}}\hat{\mathbf{N}}, \tag{5}$$

where the 6×6 matrix $\hat{\mathbf{T}}$ is defined as

$$\hat{\mathbf{T}} = \begin{pmatrix} 0 & \mathbf{I} \\ \mathbf{I} & 0 \end{pmatrix}, \qquad \hat{\mathbf{T}}^2 = \begin{pmatrix} \mathbf{I} & 0 \\ 0 & \mathbf{I} \end{pmatrix} = \hat{\mathbf{I}}, \tag{6}$$

$\hat{\mathbf{I}}$ is the 6×6 unit matrix, and the 6×6 matrix $\hat{\mathbf{N}} = \hat{\mathbf{N}}(\mathbf{r})$ is the well-known matrix of Stroh

$$\hat{\mathbf{N}}(\mathbf{r}) = \begin{pmatrix} \hat{\mathbf{N}}_1 & \hat{\mathbf{N}}_2 \\ \hat{\mathbf{N}}_3 & \hat{\mathbf{N}}_4 \end{pmatrix}, \tag{7}$$

consisting of the 3×3 blocks

$$\begin{aligned} \hat{\mathbf{N}}_1 &= -(nn)^{-1}(nm), & \hat{\mathbf{N}}_2 &= -(nn)^{-1}, \\ \hat{\mathbf{N}}_3 &= (mm) - (mn)(nn)^{-1}(nm), & \hat{\mathbf{N}}_4 &= -(mn)(nn)^{-1}. \end{aligned} \tag{8}$$

The blocks of the Stroh matrix are formed by a convolution of the elastic constants tensor $C_{ijkl}(\mathbf{r})$ with two unit vectors $\mathbf{m}$ and $\mathbf{n}$ forming together with the unit vector $\mathbf{t}$ the right-handed basis $(\mathbf{m}, \mathbf{n}, \mathbf{t})$. Eq. (5) is easily proved by direct inspection.

3. The consistency equation

Here, we review the fundamentals of the weight function theory in inhomogeneous elastic media, following the method of Belov and Kirchner [28]. Let us consider a two-dimensional (that is infinite along the axis x_3) notched body A of finite size in the x_1x_2-plane, as shown in Fig. 1. The body is supposed to be loaded such that a state of generalized plane strains occurs, that is the displacement vector $\mathbf{u}$ remains invariant along x_3 and has both plane (u_1 and u_2) and anti-plane (u_3) components. We deal with a special class of multimaterials, which are composed from the elastically anisotropic homogeneous wedge-like regions with a common apex, as shown Fig. 2. The wedges differ in their elastic constants. In fact, the multimaterial structures discussed in this chapter are a particular case of the elastic media with angular inhomogeneity of the elastic properties. Therefore they can be treated within the framework of the general formalism developed by Kirchner [17] for elastically anisotropic angularly inhomogeneous media, where the elasticity constants $C_{jikl}(\omega)$ depend on the azimuth ω counted around the axis x_3 from which the radius r is counted, as illustrated in Fig. 3. The essence of this approach is to employ a six-dimensional consistency equation for the field variable $(\mathbf{u},\boldsymbol{\phi})$ formed by the displacement vector $\mathbf{u}$ and the Airy stress function vector $\boldsymbol{\phi}$. The

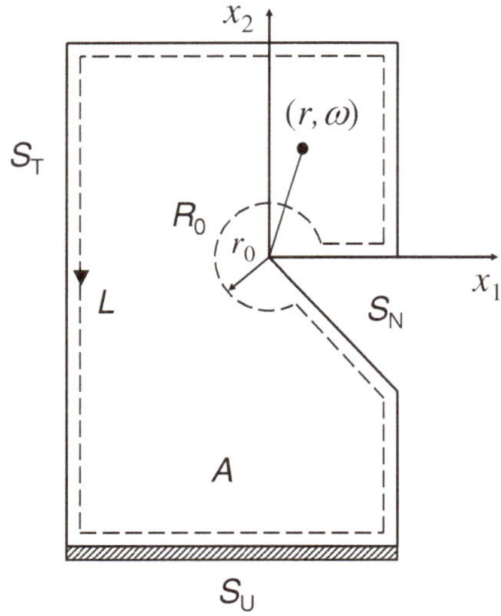

Figure 1. Finite specimen A with a notch. The tractions are prescribed on S_T and the displacements on S_U; the notch faces S_N are traction free. The reciprocity theorem is applied to the dashed contour L.

consistency equation results from the fact that some linear forms consisting of the first-order spacial derivatives of the displacements and stress functions must represent components of the same stress tensor. Consequently the stresses σ_{ij} can be equally derived from $\mathbf{u}$ via Hook's law as

$$\sigma_{ij} = C_{ijkl}(\omega)\partial_k u_l \tag{9}$$

or from $\boldsymbol{\phi}$ according to

$$\sigma_{i1} = \partial_2\phi_i, \qquad \sigma_{i2} = -\partial_1\phi_i. \tag{10}$$

Direct comparison of Eq. (9) and Eq. (10) yields a first-order differential equation

$$\left\{\hat{\mathbf{N}}(\omega)\frac{\partial}{\partial r} - \hat{\mathbf{I}}\,\frac{1}{r}\frac{\partial}{\partial\omega}\right\}\begin{pmatrix}\mathbf{u}(r,\omega)\\ \boldsymbol{\phi}(r,\omega)\end{pmatrix} = \mathbf{0}, \tag{11}$$

where the matrix $\hat{\mathbf{N}}(\omega)$ is defined by Eq. (7) and Eq. (8) and the unit vectors $\mathbf{m}$ and $\mathbf{n}$ are rotated counterclockwise by an angle ω against a fixed basis $\{\mathbf{m}_0, \mathbf{n}_0\}$, as shown in Fig. 3. The consistency condition given in Eq. (11) ensures that any its solution corresponds to equilibrated stresses (because they are derived from the stress functions) and compatible strains (because they are derived from the displacements). Therefore, the solutions of Eq. (11) describe states free of body forces and dislocation distributions. As it was emphasized in [28], the consistency equation (11) remains valid for arbitrary inhomogeneity, where the matrix $\hat{\mathbf{N}}(r,\omega)$ depends also on the radius r via $C_{ijkl}(r,\omega)$, and provides an extension of the well-known result [9] obtained under assumption of elastic homogeneity. The examples of successful application of the consistency equation to the analysis of the stress state due to linear defects such as dislocations, line forces, and disclinations in angularly inhomogeneous

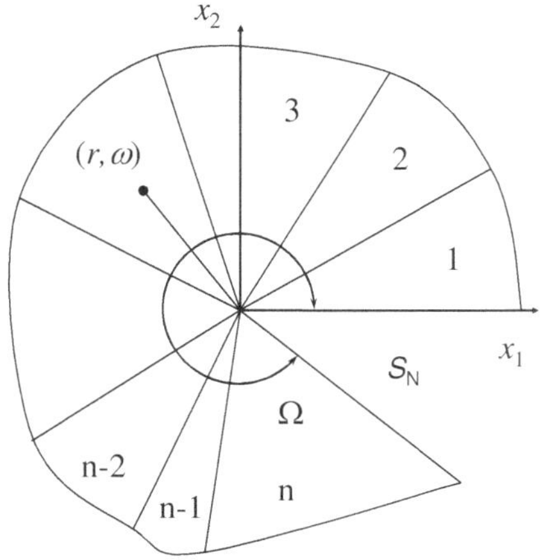

Figure 2. Multimaterial consisting of n bonded together elastic wedges with different elastic constants $C_{ijkl}^{(m)}$, $(m = 1, \ldots, n)$.

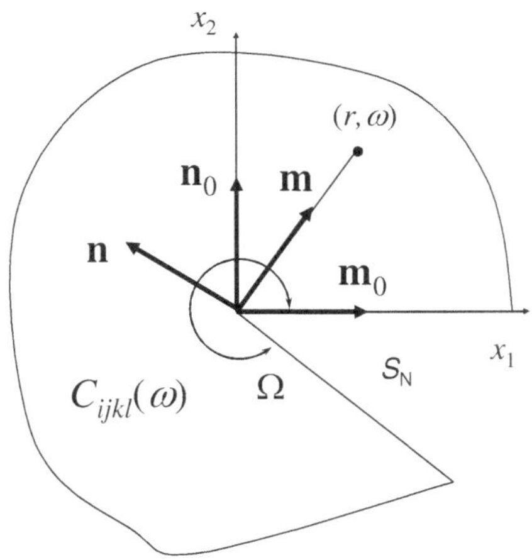

Figure 3. Elastic plane with a notch and the elastic constants $C_{ijkl}(\omega)$ continuously dependent on the azimuth ω. Basis (**m**,**n**,**t**) is rotated counterclockwise by an angle ω against a fixed basis ($\mathbf{m}_0$,$\mathbf{n}_0$,**t**).

anisotropic media can be found in [16] and [23, 24], respectively. The consistency equation was further applied in [29] to study the stress behavior in the angularly inhomogeneous elastic wedges near and at the critical wedge angle.

4. Eigenfunction expansions

According to [28], an extension of the Williams' eigenfunction expansion [1] to the notched body shown in Fig. 3 can be constructed from homogeneous solutions of Eq. (11). A suitable separable solution varying as a power of distance r has the form

$$\begin{pmatrix} \mathbf{u}(r,\omega) \\ \boldsymbol{\phi}(r,\omega) \end{pmatrix} = r^s \hat{\mathbf{V}}^{(s)}(\omega) \begin{pmatrix} \mathbf{h} \\ \mathbf{g} \end{pmatrix}, \tag{12}$$

where $(\mathbf{h}, \mathbf{g})$ is a constant six-dimensional vector and

$$\hat{\mathbf{V}}^{(s)}(\omega) = \begin{pmatrix} \hat{\mathbf{V}}_1^{(s)} & \hat{\mathbf{V}}_2^{(s)} \\ \hat{\mathbf{V}}_3^{(s)} & \hat{\mathbf{V}}_4^{(s)} \end{pmatrix}, \tag{13}$$

is a 6×6 matrix function of the azimuth ω, which is sometimes also referred to as transfer matrix. It is to be found by inserting the separable solution (12) into the consistency equation (11). As was shown in [17], this procedure results in the first-order ordinary differential equation

$$\frac{d\hat{\mathbf{V}}^{(s)}(\omega)}{d\omega} = s\hat{\mathbf{N}}(\omega)\hat{\mathbf{V}}^{(s)}(\omega) \tag{14}$$

with the initial condition

$$\hat{\mathbf{V}}^{(s)}(0) = \hat{\mathbf{I}}. \tag{15}$$

The solution of Eq. (14) is the ordered exponential defined as

$$\begin{aligned} \hat{\mathbf{V}}^{(s)}(\omega) &= \text{Ordexp}\left(s\int_0^{\omega} \hat{\mathbf{N}}(\theta)\mathrm{d}\theta\right) \\ &= \prod_{i=1}^{k} \exp\left(s\hat{\mathbf{N}}(\theta_{i-1})\delta\theta\right), \end{aligned} \tag{16}$$

where $\theta_0 = 0$, $\delta\theta = \omega/k$, and $k \to \infty$. A representation of the ordered exponential as a series expansion useful for approximate calculations within the framework of the perturbation theory can be found in [17]. The six-dimensional field (12) satisfies the consistency equation (11) and provides the solution, which is both compatible and equilibrated in the bulk for any value of the parameter s, which may be real or complex. Hence, the bulk operator $\hat{\mathbf{N}}(\omega)$ in Eq. (11) itself doesn't impose any restrictions on the admissible values of s. However, in order for (12) to become an eigenfunction of the angularly inhomogeneous notched plane (see Fig. 3), the appropriate boundary condition at the notch faces must be satisfied, and it is the boundary condition that results in the discrete spectrum of the eigenvalues $\{s_n\}$ and the corresponding eigenvectors $(\mathbf{h}_n, \mathbf{g}_n)$. If both notch faces, $\omega = 0$ and $\omega = \Omega$, are traction free, the boundary condition reads as

$$\boldsymbol{\phi}(r,0) = \boldsymbol{\phi}(r,\Omega) = \mathbf{0}. \tag{17}$$

In view of Eq. (15), the condition at the notch face $\omega = 0$ implies that $\mathbf{g} = \mathbf{0}$. The condition at the other face, $\omega = \Omega$, gives a linear homogeneous algebraic system of equations for the three components of $\mathbf{h}$,

$$\hat{\mathbf{V}}_3^{(s)}(\Omega) \cdot \mathbf{h} = \mathbf{0}. \tag{18}$$

The system (18) has a non-trivial solution for $\mathbf{h}$ only provided that the parameter s satisfies the eigenvalue equation

$$\|\hat{\mathbf{V}}_3^{(s)}(\Omega)\| = 0, \tag{19}$$

where the symbol $\| \dots \|$ stands for determinant. Equation (19) yields an infinite set of roots $\{s_n\}$, each of which generates an eigenfunction. With a positive real part Re s_n, the eigenfunction (12) has bounded elastic energy in any neighborhood of the notch tip, although this requirement doesn't exclude the existence of the stress singularity at $r = 0$.

Finally, an inner Williams' expansion for the notch is given by

$$\begin{pmatrix} \mathbf{u}(r,\omega) \\ \boldsymbol{\phi}(r,\omega) \end{pmatrix} = \sum_{\text{Re } s_n \geq 0} K^{(n)} r^{s_n} \hat{\mathbf{V}}^{(s_n)}(\omega) \begin{pmatrix} \mathbf{h}_n \\ \mathbf{0} \end{pmatrix}, \tag{20}$$

with the coefficients $K^{(n)}$ being the eigenfunction amplitudes, which characterize the fine structure of the near-tip fields. Since the eigenvalue problem resulting from the traction free boundary conditions (17) is invariant with respect to the rigid body translations and rotations, the eigenvalues $s = 0$ and $s = 1$ are roots of equation (19), whatever the angular dependence the elastic constants $C_{ijkl}(\omega)$ have. Hence, two terms in Eq. (20) require special consideration. These two terms describe a rigid body translation and a rotation and they are ordered in the expansion (20) by $n = 0$ and $n = 1$ respectively. At this point, it is worth noting that the expansion coefficients $K^{(0)}$ and $K^{(1)}$ of both rigid body motion terms can be uniquely defined only if $S_{\text{U}} \neq 0$, where S_{U} is a part of the body surface S (see, Fig. 1 for details), at which the displacements are prescribed. Otherwise the two coefficients remain arbitrary and the corresponding terms in the expansion (20) can be omitted.

In the case of $S_{\text{U}} \neq 0$ the coefficients $K^{(0)}$ and $K^{(1)}$ become important, especially in the numerical analysis. In order to reveal their geometrical interpretation, let us consider the corresponding eigenfunctions explicitly. Rigid body translations are generated by the eigenvalue $s_0 = 0$. Since $\hat{\mathbf{V}}^{(0)}(\omega)$ reduces to the unit matrix, the eigenfunction associated with this eigenvalue takes the form

$$\begin{pmatrix} \mathbf{u}^{(0)}(r,\omega) \\ \boldsymbol{\phi}^{(0)}(r,\omega) \end{pmatrix} = K^{(0)} \begin{pmatrix} \mathbf{h}_0 \\ \mathbf{0} \end{pmatrix}. \tag{21}$$

Thereby the coefficient $K^{(0)}$ describes the notch tip (and the body as a whole) displacement magnitude in the direction of the vector $\mathbf{h}_0$, which length is assumed to be normalized to unity,

$$\mathbf{u}(\mathbf{0}) = K^{(0)} \mathbf{h}_0. \tag{22}$$

In turn the rigid body rotation term is generated by the eigenvalue $s_1 = 1$ and the eigenvector $\mathbf{h}_1 = \mathbf{n}(0)$. Using the properties of the ordered exponential $\hat{\mathbf{V}}^{(1)}(\omega)$ (consult with [24, 28] for further details), the corresponding eigenfunction can be found in explicit form for an arbitrary rotational inhomogeneity

$$\begin{pmatrix} \mathbf{u}^{(1)}(r,\omega) \\ \boldsymbol{\phi}^{(1)}(r,\omega) \end{pmatrix} = K^{(1)} r \hat{\mathbf{V}}^{(1)}(\omega) \begin{pmatrix} \mathbf{n}(0) \\ \mathbf{0} \end{pmatrix} = \theta r \begin{pmatrix} \mathbf{n}(\omega) \\ \mathbf{0} \end{pmatrix}. \tag{23}$$

The coefficient $K^{(1)} = \theta$ represents a rigid body rotation by an angle θ.

In general, for some particular angular dependencies of the elastic constants $C_{ijkl}(\omega)$ or for some values of the notch angle the eigenvalue equation (19) can have multiple roots. This case needs special treatment since the expansion (20) over the power-law eigenfunctions is no longer complete and must be completed by the power-logarithmic solutions. The necessary modifications can be done by taking into consideration some general properties of solutions of elliptic problems in domains with piecewise smooth boundaries [26]. In fact, such degeneracies are of minor practical importance for fracture mechanics and are not discussed in this paper. The only exception is the root $s = 1$ associated with rigid body rotation as well as the complementary root $s = -1$ generating a solution for a concentrated couple applied at the noth tip. This case is analyzed in detail in [29], where also analytical expressions for power-logarithmic solutions in elastically anisotropic angularly inhomogeneous media were presented (see also [24]).

5. Complementary eigenfunctions

The eigenfunction expansion (20) of the near-tip field contains only the terms of bounded elastic energy. However, the eigenvalue problem (18) admits solutions of unbounded energy as well. The latter correspond to self-equilibrated loadings applied at the notch tip. The eigenfunctions of bounded and unbounded elastic energy are not independent and there exists an intrinsic symmetry between them, which follows from the invariance of Eq. (19) with respect to the index inversion $s \to -s$. As has been proved by Belov and Kirchner [31], for any angular inhomogeneity $C_{ijkl}(\omega)$, whenever Eq. (19) is satisfied,

$$||\hat{\mathbf{V}}_3^{(-s)}(\Omega)|| = 0 \tag{24}$$

is also valid. Hence, for any eigenfunction (12) generated by a positive real part root s there exists a complementary eigenfunction generated by an eigenvalue $-s$ with negative real part and unbounded elastic energy. This symmetry between the positive and negative real part solutions of Eq. (19) is the cornerstone of the weight function theory.

Since eigenvalues s and $-s$ appear always in pairs, the complementary eigenfunction

$$\begin{pmatrix} \mathbf{u}(r,\omega) \\ \boldsymbol{\phi}(r,\omega) \end{pmatrix} = r^{-s}\hat{\mathbf{V}}^{(-s)}(\omega)\begin{pmatrix} \mathbf{h}^* \\ \mathbf{g}^* \end{pmatrix}, \tag{25}$$

where $(\mathbf{h}^*, \mathbf{g}^*)$ is a constant vector, is also a solution without body forces and dislocations that obeys the traction free boundary condition (17) at the notch faces. The vector $(\mathbf{h}^*, \mathbf{g}^*)$ excites this field just as $(\mathbf{h}, \mathbf{g})$ excited (12).

6. Pseudo-orthogonality relations

The second property of the eigenfunctions (12), which underlies the weight function theory, is their six-dimensional orthogonality. The paper by Chen [13] appears to be the first work where the orthogonality property of the eigenfunctions along with Betti's reciprocity theorem were applied to compute the coefficients in the Williams' eigenfunction expansion for an edge crack. The case an elastically isotropic medium considered in [13] is rather simple, since the existing analytical expressions for the eigenfunctions enable for the orthogonality property to be easily proved by direct calculation. Using the same method, Chen and Hasebe [25, 27] derived the orthogonality property for an interface crack in an isotropic bimaterial and also for an orthotropic material with pure imaginary roots of the Stroh matrix. The cumbersome direct calculations [13, 25, 27] are possible only for very simple cases and reveal neither the nature of the orthogonality relations nor their connection with the symmetry of the elasticity equations. Belov and Kirchner [28] suggested a proof of the orthogonality property for both cracks and notches of finite opening angle in an elastically anisotropic media possessing arbitrary inhomogeneity of the elastic constants $C_{ijkl}(\omega)$. In contrast to [13, 25], the proof given in [28] shows that the orthogonality property of the eigenfunctions (12) directly follows from the symmetry (5) of the operator $\hat{\mathbf{N}}(\mathbf{r})$.

The idea of the proof [28] consists in the following. Integrating by parts an average of the weighted product of two ordered exponentials of arbitrary indices s and q, one finds

$$\begin{aligned} q\int_0^{\Omega}[\hat{\mathbf{V}}^{(s)}(\omega)]^{\mathrm{t}}\hat{\mathbf{T}}\hat{\mathbf{N}}(\omega)\hat{\mathbf{V}}^{(q)}(\omega)\mathrm{d}\omega &= \int_0^{\Omega}[\hat{\mathbf{V}}^{(s)}(\omega)]^{\mathrm{t}}\hat{\mathbf{T}}\,\frac{\mathrm{d}}{\mathrm{d}\omega}\hat{\mathbf{V}}^{(q)}(\omega)\mathrm{d}\omega \\ &= [\hat{\mathbf{V}}^{(s)}(\Omega)]^{\mathrm{t}}\hat{\mathbf{T}}\,\hat{\mathbf{V}}^{(q)}(\Omega) - \hat{\mathbf{T}} \\ &\quad - s\int_0^{\Omega}[\hat{\mathbf{V}}^{(s)}(\omega)]^{\mathrm{t}}\hat{\mathbf{N}}^{\mathrm{t}}(\omega)\hat{\mathbf{T}}\hat{\mathbf{V}}^{(q)}(\omega)\mathrm{d}\omega \end{aligned} \tag{26}$$

So far only the fact that the ordered exponential $\hat{\mathbf{V}}^{(q)}(\omega)$ satisfies equation (14) has been used. Taking into account that the 'bulk' operator $\hat{\mathbf{T}}\hat{\mathbf{N}}(\mathbf{r})$ is symmetric (according to Eq.(5)), we obtain an important property of the ordered exponentials

$$(s+q)\int_0^{\Omega}[\hat{\mathbf{V}}^{(s)}(\omega)]^{\mathrm{t}}\hat{\mathbf{T}}\hat{\mathbf{N}}(\omega)\hat{\mathbf{V}}^{(q)}(\omega)\mathrm{d}\omega = [\hat{\mathbf{V}}^{(s)}(\Omega)]^{\mathrm{t}}\hat{\mathbf{T}}\,\hat{\mathbf{V}}^{(q)}(\Omega) - \hat{\mathbf{T}}. \tag{27}$$

This result is independent of the boundary conditions (17) specified at the notch faces. It takes place for any indices s and q, which are not necessary roots of Eq. (19). Let us now consider two roots s_n and s_p satisfying the condition $s_n + s_p \neq 0$. Then, according to Eq. (27), the weighted average can be represented as

$$\int_0^{\Omega}[\hat{\mathbf{V}}^{(s_p)}(\omega)]^{\mathrm{t}}\hat{\mathbf{T}}\hat{\mathbf{N}}(\omega)\hat{\mathbf{V}}^{(s_n)}(\omega)\mathrm{d}\omega = \frac{1}{s_p+s_n}\left([\hat{\mathbf{V}}^{(s_p)}(\Omega)]^{\mathrm{t}}\hat{\mathbf{T}}\,\hat{\mathbf{V}}^{(s_n)}(\Omega) - \hat{\mathbf{T}}\right). \tag{28}$$

Now let $(\mathbf{h}_p, \mathbf{0})$ and $(\mathbf{h}_n, \mathbf{0})$ be corresponding eigenvectors. Multiplying equation (28) from the right and from the left by these eigenvectors, one obtains the orthogonality relation in the form

$$(\mathbf{h}_p,\mathbf{0})^{\mathrm{t}}\int_0^{\Omega}[\hat{\mathbf{V}}^{(s_p)}(\omega)]^{\mathrm{t}}\hat{\mathbf{T}}\hat{\mathbf{N}}(\omega)\hat{\mathbf{V}}^{(s_n)}(\omega)\mathrm{d}\omega\begin{pmatrix}\mathbf{h}_n\\ \mathbf{0}\end{pmatrix} = 0, \qquad (s_n+s_p\neq 0). \tag{29}$$

Details of the proof are clear from Eq. (28) and the identity

$$\left(\left[\hat{\mathbf{V}}^{(s_p)}(\Omega)\begin{pmatrix}\mathbf{h}_p\\ \mathbf{0}\end{pmatrix}\right]^{\mathrm{t}}\hat{\mathbf{T}}\hat{\mathbf{V}}^{(s_n)}(\Omega) - (\mathbf{h}_p,\mathbf{0})\hat{\mathbf{T}}\right)\begin{pmatrix}\mathbf{h}_n\\ \mathbf{0}\end{pmatrix} = (\mathbf{h}_p\hat{\mathbf{V}}_1^{(s_p)}(\Omega),\mathbf{0})\begin{pmatrix}\mathbf{0}\\ \hat{\mathbf{V}}_1^{(s_n)}(\Omega)\mathbf{h}_n\end{pmatrix} = 0. \tag{30}$$

Finally, the orthogonality relation can be rewritten explicitly in terms of the eigenfunctions (12) as

$$\int_0^{\Omega} (\mathbf{u}_p, \boldsymbol{\phi}_p)^{\mathrm{t}} \hat{\mathbf{T}}\hat{\mathbf{N}}(\omega) \begin{pmatrix} \mathbf{u}_n \\ \boldsymbol{\phi}_n \end{pmatrix} \mathrm{d}\omega = 0, \qquad (s_n + s_p \neq 0). \tag{31}$$

The six-dimensional orthogonality (or pseudo-orthogonality) is not an orthogonality in the generally accepted sense, because it takes place not only for different eigenvalues, but also when $s_n = s_p$. This means that all eigenfunctions are 'self-orthogonal'. The pseudo-orthogonality property fails only for the pairs $s_p \neq -s_n$, which have special status in the higher-order weight function theory.

7. Fundamental field and weight function of higher-order

Following [28], we consider a notched body A shown in Fig. 1 and subject it to an external surface loading system which includes prescribed surface tractions $\mathbf{F}$ on the boundary S_{T} and imposed displacements $\mathbf{U}$ at the remainder S_{U} of the body surface $S = S_{\mathrm{T}} + S_{\mathrm{U}}$. We further suppose that A is free from body forces and dislocations. The notch faces S_{N} are assumed to be traction free. This system of loadings leads to the boundary conditions

$$\begin{cases} T_k = \mathrm{F}_k \ \text{ on } \ S_{\mathrm{T}}, \ \text{ and } \ T_k = 0 \ \text{ on } \ S_{\mathrm{N}}, \\ u_k = \mathrm{U}_k \ \text{ on } \ S_{\mathrm{U}}, \end{cases} \tag{32}$$

where $T_k = \sigma_{ij}\nu_j$ and ν_j is an outer unit normal to the body surface. Inside A the elastic field produced by the loading system (32) is represented by the eigenfunction expansion (20). This field is called regular, while it can result in a stress singularity at the notch tip. As already noted, the elastic energy associated with the regular field remains bounded in any neighborhood of the notch tip.

In order to derive weight functions for the coefficients $K^{(n)}$ in the series expansion (20), it is convenient to consider the cases $S_{\mathrm{U}} = 0$ and $S_{\mathrm{U}} \neq 0$ separately.

(I) If $S_{\mathrm{U}} = 0$, except for $n = 0$ and 1, all coefficients $K^{(n)}$ are defined uniquely. In order to find a coefficient $K^{(m)}$, one needs to apply Betti's reciprocity theorem to the regular field and to a specially chosen auxiliary field called the fundamental field of order m. It consists of a complementary solution (25) for the term of order m and a regular part,

$$\begin{pmatrix} \mathbf{u}^{*(m)}(r,\omega) \\ \boldsymbol{\phi}^{*(m)}(r,\omega) \end{pmatrix} = r^{-s_m} \hat{\mathbf{V}}^{(-s_m)}(\omega) \begin{pmatrix} \mathbf{h}^*_m \\ \mathbf{0} \end{pmatrix} + \sum_{\mathrm{Re}\, s_p > 0} k_p r^{s_p} \hat{\mathbf{V}}^{(s_p)}(\omega) \begin{pmatrix} \mathbf{h}_p \\ \mathbf{0} \end{pmatrix}, \tag{33}$$

where the sum is extended over all eigenfunctions of bounded elastic energy. The fundamental field of the mth order corresponds to a certain source placed at $r = 0$ and thereby provides zero body forces and dislocation density in the bulk. This justifies introduction of both the displacement (no dislocations) and the Airy stress function (no body forces) anywhere inside the body. The coefficients k_p must be chosen so as to subject the solution (33) to the traction free boundary conditions

$$T_k^{*(m)} = 0 \quad \text{on} \quad S_{\mathrm{T}} + S_{\mathrm{N}}. \tag{34}$$

Because $S_U = 0$, the representation (33) of the mth order fundamental field is possible for $m \neq 0$ and 1. The terms corresponding to rigid body motions can be chosen arbitrary.

For a subdomain $A' \subset A$, bounded as shown in Fig. 1 by a closed contour L which consists of a circular arc R_0 of radius r_0 around the notch tip, the body surface $S = S_T$, and the remaining part $S_{N'}$ of the notch faces S_N, application of Betti's reciprocity theorem yields

$$\Gamma^{(m)} = -\int_{S_T+S_{N'}} (T_k^{*(m)} u_k - T_k u_k^{*(m)}) \mathrm{d}s = -\int_{S_T} u_k^{*(m)} \mathrm{F}_k \mathrm{d}s, \tag{35}$$

where u_k and T_k are the displacement and traction due to the regular field, and

$$\Gamma^{(m)} = \int_{R_0} (T_k u_k^{*(m)} - T_k^{*(m)} u_k) \mathrm{d}s. \tag{36}$$

Integrating Eq.(36) by part, one obtains

$$\Gamma^{(m)} = \boldsymbol{\phi}^{*(m)}(r,0)\mathbf{u}^{*(m)}(r,0) - \boldsymbol{\phi}^{*(m)}(r,\Omega)\mathbf{u}^{*(m)}(r,\Omega) + \int_0^{\Omega} (\boldsymbol{\phi}^{*(m)}, \mathbf{u}^{*(m)}) \frac{d}{d\omega} \begin{pmatrix} \mathbf{u} \\ \boldsymbol{\phi} \end{pmatrix} \mathrm{d}\omega, \tag{37}$$

with the two first terms vanishing owing to the traction free boundary conditions (17) imposed on all eigenfunctions at the notch faces. Hence, substituting the explicit expressions for the regular and fundamental field into (37), one finds

$$\begin{aligned} \Gamma^{(m)} &= \sum_n K^{(n)} s_n r_0^{s_n - s_m} (\mathbf{h}_m^*, \mathbf{0})^{\mathrm{t}} \int_0^{\Omega} [\hat{\mathbf{V}}^{(-s_m)}(\omega)]^{\mathrm{t}} \hat{\mathbf{T}} \hat{\mathbf{N}}(\omega) \hat{\mathbf{V}}^{(s_n)}(\omega) \mathrm{d}\omega \begin{pmatrix} \mathbf{h}_n \\ \mathbf{0} \end{pmatrix} \\ &+ \sum_{n,p} K^{(n)} k_p s_n r_0^{s_n + s_p} (\mathbf{h}_p, \mathbf{0})^{\mathrm{t}} \int_0^{\Omega} [\hat{\mathbf{V}}^{(s_p)}(\omega)]^{\mathrm{t}} \hat{\mathbf{T}} \hat{\mathbf{N}}(\omega) \hat{\mathbf{V}}^{(s_n)}(\omega) \mathrm{d}\omega \begin{pmatrix} \mathbf{h}_n \\ \mathbf{0} \end{pmatrix}. \end{aligned} \tag{38}$$

As r_0 shrinks to zero, the second sum in Eq. (38) vanishes due to the fact that the real parts of all eigenvalues s_n and s_p are positive. However, there is also another reason for this term in Eq. (38) to vanish. In fact, it must vanish due to the pseudo-orthogonality property (29). As concerns the first sum in Eq. (38), it contains the terms formally divergent as the radius $r_0 \to 0$. However, owing to the pseudo-orthogonality property (29), these terms drop out of Eq. (38) and finally only one term for which $s_n = s_m$ remains non-vanishing. This term is independent of r_0 and remains constant as it shrinks. Note also that the second sum in Eq. (38) is not sensitive to the rigid body motion terms in the fundamental field.

According to Eq. (38), the reciprocity theorem relates the expansion coefficient of order m directly with external loading as

$$K^{(m)} \Upsilon^{(m)} = -\int_{S_T} u_k^{*(m)} \mathrm{F}_k \mathrm{d}s, \tag{39}$$

with a normalizing geometry factor

$$\Upsilon^{(m)} = -s_m (\mathbf{h}_m^*, \mathbf{0})^{\mathrm{t}} \int_0^{\Omega} [\hat{\mathbf{V}}^{(-s_m)}(\omega)]^{\mathrm{t}} \hat{\mathbf{T}} \hat{\mathbf{N}}(\omega) \hat{\mathbf{V}}^{(s_m)}(\omega) \mathrm{d}\omega \begin{pmatrix} \mathbf{h}_m \\ \mathbf{0} \end{pmatrix} \tag{40}$$

Correspondingly, an expansion coefficient $K^{(m)}$ is available via the mth order weight function,

$$h_k^{(m)}(x_1, x_2) = u_k^{*(m)}(x_1, x_2)/Y^{(m)}, \tag{41}$$

as a functional

$$K^{(m)} = \int_{S_T} h_k^{*(m)}(x_1, x_2)\mathrm{F}_k \mathrm{d}s \tag{42}$$

of the surface loading. Thus the mth order weight function differs from the corresponding fundamental field (33) only in a constant geometry factor (40).

(II) If $S_U \neq 0$, all expansion coefficients in the series (20) are defined unambiguously, including $K^{(0)}$ and $K^{(1)}$. For $m \neq 0$ the fundamental field of the mth order is still given by the solution (33) provided that its bounded energy part is completed by the rigid body motion terms. The coefficients k_p in (33) are now chosen to subject it to the boundary conditions

$$\begin{cases} T_k^{*(m)} = 0 & \text{on} \quad S_T + S_N, \\ u_k^{*(m)} = 0 & \text{on} \quad S_U. \end{cases} \tag{43}$$

Modifying the reciprocal relation (35) to include S_U, one obtains

$$\Gamma^{(m)} = -\int_{S_T} u_k^{*(m)}\mathrm{F}_k \mathrm{d}s + \int_{S_T} T_k^{*(m)}\mathrm{U}_k \mathrm{d}s. \tag{44}$$

The remaining calculations are similar to those performed in the case of vanishing S_U. Weight functions of the mth order are introduced according to

$$h_k^{(m)}(x_1, x_2) = u_k^{*(m)}(x_1, x_2)/Y^{(m)}, \tag{45}$$

$$H_k^{(m)}(x_1, x_2) = T_k^{*(m)}(x_1, x_2)/Y^{(m)}. \tag{46}$$

The coefficients $K^{(m)}$ in the eigenfunction expansion are now expressed via the mth order weight functions as

$$K^{(m)} = \int_{S_T} h_k^{*(m)}(x_1, x_2)\mathrm{F}_k \mathrm{d}s - \int_{S_U} H_k^{*(m)}(x_1, x_2)\mathrm{U}_k \mathrm{d}s. \tag{47}$$

As it was noted above in the case of non-vanishing S_U the coefficient $K^{(0)}$ needs a special treatment. The complementary field for the rigid body translation (21) has a logarithmic rather than power-law functional form. Indeed, the auxiliary source generating this complementary solution is a concentrated force applied at the notch tip. Unlike other eigenfunctions it is not self-equilibrated. The complementary logarithmic solution can be constructed by means of the analytical expression for the elastic field of a force at the tip of a notch in an angularly inhomogeneous plane [16, 18]. Details of further development and the corresponding 0th order fundamental field for the calculation of the rigid body translation term are available from [28].

8. Multimaterials

A continuously inhomogeneous elastic material is actually only a useful tool, which considerably simplifies the establishing of important properties of elastic fields involved in weight function theory. Nowadays functionally graded materials with continuous angular inhomogeneity of elastic properties are still exotic and in engineering structures we deal mostly with piecewise homogeneous media (junctions of a finite number of dissimilar materials) called multimaterials. In the case of multimaterials further analytical advancement in the weight function theory becomes possible. The ordered exponentials are known to appear in Eq. (16) instead of the conventional exponentials since the angular inhomogeneity causes non-commutability of the matrices $\hat{\mathbf{N}}(\omega)$ for different values of the argument ω. However, when the medium is piecewise homogeneous, the matrices $\hat{\mathbf{N}}(\omega)$ commute within each homogeneous wedge-like region [15] and the integration in the ordered exponentials can be performed analytically. For example, in the case of a multimaterial composed from three wedges (triple junction)

$$C_{ijkl}(\omega) = \begin{cases} C_{ijkl}^{(1)} & 0 < \omega < \alpha, \\ C_{ijkl}^{(2)} \quad \text{for} & \alpha < \omega < \beta, \\ C_{ijkl}^{(3)} & \beta < \omega < \Omega, \end{cases} \tag{48}$$

the ordered exponential admits factorization and reduces to (for details, see [17, 23, 24])

$$\operatorname{Ordexp}\left(s\int_0^{\omega}\hat{\mathbf{N}}(\theta)d\theta\right) = \begin{cases} \hat{\mathbf{V}}_1^s(\omega) & 0 < \omega < \alpha, \\ \hat{\mathbf{V}}_2^s(\omega)\hat{\mathbf{V}}_2^{-s}(\alpha)\hat{\mathbf{V}}_1^s(\alpha) \quad \text{for} & \alpha < \omega < \beta, \\ \hat{\mathbf{V}}_3^s(\omega)\hat{\mathbf{V}}_3^{-s}(\beta)\hat{\mathbf{V}}_2^s(\beta)\hat{\mathbf{V}}_2^{-s}(\alpha)\hat{\mathbf{V}}_1^s(\alpha) & \beta < \omega < \Omega, \end{cases} \tag{49}$$

where $\hat{\mathbf{V}}_i^s(\omega)$ and $\hat{\mathbf{V}}_i^{-s}(\omega)$ denote for each homogeneous region powers of

$$\hat{\mathbf{V}}_i(\omega) = \operatorname{Ordexp}\left(\int_0^{\omega}\hat{\mathbf{N}}(\theta)d\theta\right) = \hat{\mathbf{I}}\cos\omega + \hat{\mathbf{N}}_i(0)\sin\omega \tag{50}$$

and

$$\hat{\mathbf{V}}_i^{-1}(\omega) = \hat{\mathbf{I}}\cos\omega - \hat{\mathbf{N}}_i(\omega)\sin\omega, \tag{51}$$

The matrix $\hat{\mathbf{N}}_i(\omega)$ for each homogeneous wedge-like region of a multimaterial is constructed by replacing $C_{ijkl}(\omega)$ in the definition (4) by $C_{ijkl}^{(i)}$. If s is not an integer, the powers of the matrices (50) and (51) should be defined in terms of their spectral decompositions over the eigenvectors of the matrices $\hat{\mathbf{N}}_i(\omega)$ (for details, see [22]).

9. Conclusions

Here, it was shown that the established in [28] pseudo-orthogonality property of the power eigenfunctions follows directly from the symmetry of the operator $\hat{\mathbf{N}}(\mathbf{r})$, which is commonly referred to as Stroh matrix [3, 22] of anisotropic elasticity theory. In the last decade the proof of the pseudo-orthogonality property was republished in a large number of papers [32–38], where however only trivial particular cases of anisotropy and inhomogeneity were analyzed. The general proof by Belov and Kirchner [28] is not cited in these papers, which are to be considered as plagiarism, although some of them contain further development, in particular, by taking into account piezoelectricity. Here, it is worth to mention that the proof of the pseudo-orthogonality property remains valid for the general case of piezoelectric piezomagnetic magnetoelectric anisotropic media, provided that the dimension of both the matrix $\hat{\mathbf{N}}(\mathbf{r})$ and the field variables is increased to include these effects (for details, see [30]). In conclusion, it may be also said that the pseudo-orthogonality property allows for a set of path-independent integrals similar to H-integral [7, 8, 10–12] to be introduced for multimaterials with notches or cracks. This is achieved by applying Betti's reciprocity theorem to the complementary field (25) rather than to the fundamental field (33). The contour L must be properly shifted from the surface S to interior domain of A.

Author details

Alexander Yu. Belov
Institute of Crystallography RAS, Moscow, Russian Federation

10. References

[1] Williams, M.L. (1952). Stress singularities resulting from various boundary conditions in angular corners of plates in extension, ASME Journal of Applied Mechanics, 19:526-528.
[2] Williams, M.L. (1957). On the stress distribution at the base of a stationary crack, ASME Journal of Applied Mechanics, 24:109-114.
[3] Stroh, A.N. (1962). Steady State Problems in Anisotropic Elasticity, Journal of Mathematical Physics, 41:77-103.
[4] Bueckner, H.F. (1970). A Novel Principle for the Computation of Stress Intensity Factors. Zeitschrift für Angewandte Mathematik und Mechanik, 50:529-546.
[5] Malén, K. & Lothe, J. (1970). Explicit Expressions for Dislocation Derivatives. Physica Status Solidi, 39:287-296.
[6] Bueckner, H.F. (1971). Weight Functions for the Notched Bar. Zeitschrift für Angewandte Mathematik und Mechanik, 51:97-109.
[7] Stern, M.; Becker, E.H. & Dunham, R.S. (1976). A Contour Integral Computation of Mixed-Mode Stress Intensity Factors. International Journal of Fracture, 12:359-368.
[8] Stern, M. & Soni, M.L. (1976). On the Computation of Stress Intensities at Fixed-Free Corners. International Journal of Solids and Structures, 12:331-337.
[9] Chadwick, P. & Smith, G.D. (1977). Foundations of the Theory of Surface Waves in Anisotropic Elastic Materials. Advances in Applied Mechanics, 17:303-376.
[10] Hong, C.-C. & Stern, M. (1978). The Computation of Stress Intensity Factors in Dissimilar Materials. Journal of Elasticity, 8:21-34.
[11] Stern, M. (1979). The Numerical Calculation of Thermally Induced Stress Intensity Factors. Journal of Elasticity, 9:91-95.
[12] Sinclair, G.B.; Okajima, M. & Griffin, J.H. (1984). Path Independent Integrals for Computing Stress Intensity Factors at Sharp Notches in Elastic Plates. International Journal for Numerical Methods in Engineering, 20:999-1008.

[13] Chen, Y.Z. (1986). New Path Independent Integrals in Linear Elastic Fracture Mechanics. Engineering Fracture Mechanics, 22:673-686.
[14] Kirchner, H.O.K. (1986). Description of loadings and screenings of cracks with the aid of universal weight functions. International Journal of Fracture, 31:173-181.
[15] Kirchner, H.O.K. & Lothe, J. (1986). On the redundancy of the $\bar{N}$ matrix of anisotropic elasticity. Philosophical Magazine Part A, 53:L7-L10.
[16] Kirchner, H.O.K. (1987). Line defects along the axis of rotationally inhomogeneous media. Philosophical Magazine Part A, 55:537-542.
[17] Kirchner, H.O.K. (1989). Elastically anisotropic angularly inhomogeneous media I. A new formalism. Philosophical Magazine Part B, 60:423-432.
[18] Ting, T.C.T. (1989). Line forces and dislocations in angularly inhomogeneous anisotropic elastic wedges and spaces. Quaterly Applied Mathematics, 47:123-128.
[19] Bueckner, H.F. (1989). Observations on weight functions. Engineering Analysis with Boundary Elements, 6:3-18.
[20] Sham, T.L. (1989). The theory of higher order weight functions for linear elastic plane problems. International Journal of Solids and Structures, 25:357-380.
[21] Sham, T.L. (1991). The determination of the elastic T-term using higher order weight functions. International Journal of Fracture, 48:81-102.
[22] Indenbom, V.L. & Lothe, J. (Eds.) (1992). Elastic Strain Fields and Dislocation Mobility, Amsterdam: North-Holland.
[23] Belov, A.Yu. (1992). A wedge disclination along the vertex of the wedge-like inhomogeneity in an elastically anisotropic solid. Philosophical Magazine Part A, 65:1429-1444.
[24] Belov, A.Yu. (1993). Scaling regimes and anomalies of wedge disclination stresses in anisotropic rotationally inhomogeneous media. Philosophical Magazine Part A, 68:1215-1231.
[25] Chen, Y.Z. & Hasebe, N. (1994). Eigenfunction Expansion and Higher Order Weight Functions of Interface Cracks. ASME Journal of Applied Mechanics 61:843-849.
[26] Nazarov, S.A. & Plamenevsky, B.A. (1994). Elliptic Problems in Domains with Piecewise Smooth Boundaries, volume 13 of de Gruyter Expositions in Mathematics. Berlin, New York: Walter de Gruyter and Co.
[27] Chen, Y.H. & Hasebe, N. (1995). Investigation of EEF Properties for a Crack in a Plane Orthotropic Elastic Solid with Purely Imaginary Characteristic Roots. Engineering Fracture Mechanics, 50:249-259.
[28] Belov, A.Yu. & Kirchner, H.O.K. (1995). Higher order weight functions in fracture mechanics of inhomogeneous anisotropic solids. Philosophical Magazine Part A, 72:1471-1483.
[29] Belov, A.Yu. & Kirchner, H.O.K. (1995). Critical angles in bending of rotationally inhomogeneous elastic wedges. ASME Journal of Applied Mechanics, 62: 429-440.
[30] Alshits, V.I.; Kirchner, H.O.K. & Ting, T.C.T. (1995). Angularly inhomogeneous piezoelectric piezomagnetic magnetoelectric anisotropic media. Philosophical Magazine Letters, 71:285-288.
[31] Belov, A.Yu. & Kirchner, H.O.K. (1996). Universal weight functions for elastically anisotropic, angularly inhomogeneous media with notches or cracks. Philosophical Magazine Part A, 73:1621-1646.

[32] Qian, J. & Hasebe, N. (1997). Property of Eigenvalues and Eigenfunctions for an Interface V-Notch in Antiplane Elasticity. Engineering Fracture Mechanics, 56:729-734.

[33] Chen, Y.H. & Ma, L.F. (2000). Bueckner's Work Conjugate Integrals and Weight Functions For a Crack in Anisotropic Solids. Acta Mechanica Sinica (English Series), 16:240-253.

[34] Ma, L.F. & Chen, Y.H. (2001). Weight Functions for Interface cracks in Dissimilar Anisotropic Piezoelectric Materials. International Journal of Fracture 110:263-279.

[35] Chen, Y.H. & Ma, L.F. (2004). Weight Functions for Interface Cracks in Dissimilar Anisotropic Materials. Acta Mechanica Sinica (English Series), 16:82-88.

[36] Ou, Z.C. & Chen, Y.H. (2004). A New Method for Establishing Pseudo Orthogonal Properties of Eigenfunction Expansion Form in Fracture Mechanics. Acta Mechanica Solida Sinica, 17:283-289.

[37] Ou, Z.C. & Chen, Y.H. (2006). A New approach to the Pseudo-Orthogonal Properties of Eigenfunction Expansion Form of the Crack-Tip Complex Potential Function in Anisotropic and Piezoelectric Fracture Mechanics. European Journal of Mechanics A/Solids, 25:189-197.

[38] Klusák, J.; Profant, T. & Kotoul, M. (2009). Various Methods of Numerical Estimation of Generalized Stress Intensity Factors of Bi-Material Notches. Applied and Computational Mechanics, 3:297-304.

Chapter 3

Fractal Fracture Mechanics Applied to Materials Engineering

Lucas Máximo Alves and Luiz Alkimin de Lacerda

Additional information is available at the end of the chapter

1. Introduction

The Classical Fracture Mechanics (CFM) quantifies velocity and energy dissipation of a crack growth in terms of the projected lengths and areas along the growth direction. However, in the fracture phenomenon, as in nature, geometrical forms are normally irregular and not easily characterized with regular forms of Euclidean geometry. As an example of this limitation, there is the problem of stable crack growth, characterized by the *J-R* curve [1, 2]. The rising of this curve has been analyzed by qualitative arguments [1, 2, 3, 4] but no definite explanation in the realm of EPFM has been provided.

Alternatively, fractal geometry is a powerful mathematical tool to describe irregular and complex geometric structures, such as fracture surfaces [5, 6]. It is well known from experimental observations that cracks and fracture surfaces are statistical fractal objects [7, 8, 9]. In this sense, knowing how to calculate their true lengths and areas allows a more realistic mathematical description of the fracture phenomenon [10]. Also, the different geometric details contained in the fracture surface tell the history of the crack growth and the difficulties encountered during the fracture process [11]. For this reason, it is reasonable to consider in an explicit manner the fractal properties of fracture surfaces, and many scientists have worked on the characterization of the topography of the fracture surface using the fractal dimension [12, 13]. At certain point, it became necessary to include the topology of the fracture surface into the equations of the Classical Fracture Mechanics theory [6, 8, 14]. This new "Fractal Fracture Mechanics" (FFM) follows the fundamental basis of the Classical Fracture Mechanics, with subtle modifications of its equations and considering the fractal aspects of the fracture surface with analytical expressions [15, 16].

The objective of this chapter is to include the fractal theory into the elastic and plastic energy released rates G_0 and J_0, in a different way compared to other authors [8, 13, 14, 17, 18, 19]. The non-differentiability of the fractal functions is avoided by developing a differentiable analytic function for the rugged crack length [20]. The proposed procedure changes the classical G_0, which is linear with the fracture length, into a non-linear equation. Also, the same approach is extended and applied to the Eshelby-Rice non-linear *J*-integral. The new equations reproduce accurately the growth process of cracks in brittle and ductile materials. Through algebraic manipulations, the energetics of the geometric part of the fracture process in the *J*-integral are separated to explain the registered history of strains left on the fracture surfaces. Also, the micro and macroscopic parts of the *J*-integral are distinguished. A generalization for the fracture resistance *J-R* curve for different materials is presented, dependent only on the material properties and the geometry of the fractured surface.

Finally, it is shown how the proposed model can contribute to a better understanding of certain aspects of the standard ASTM test [15].

2. Literature review of fractal fracture mechanics

2.1. Background of the fractal theory in fracture mechanics

Mandelbrot [21] was the first to point out that cracks and fracture surfaces could be described by fractal models. Mecholsky *et al.* [12] and Passoja and Amborski [22] performed one of the first experimental works reported in the literature, using fractal geometry to describe the fracture surfaces. They sought a correlation of the roughness of these surfaces with the basic quantity D called fractal dimension.

Since the pioneering work of Mandelbrot *et al.* [23], there have been many investigations concerning the fractality of crack surfaces and the fracture mechanics theory. They analyzed fracture surfaces in steel obtained by Charpy impact tests and used the "slit island analysis" method to estimate their fractal dimensions. They have also shown that D was related to the toughness in ductile materials.

Mecholsky *et al.* [12, 24] worked with brittle materials such as ceramics and glass-ceramics, breaking them with a standard three point bending test. They calculated the fractal dimension of the fractured surfaces using Fourier spectral analysis and the "slit island" method, and concluded that the brittle fracture process is a self-similar fractal.

It is known that the roughness of the fracture surface is related to the difficulty in crack growth [25] and several authors attempted to relate the fractal dimension with the surface energy and fracture toughness. Mecholsky *et al.* [24] followed this idea and suggested the dependence between fracture toughness and fractal dimension through

$$K_{IC} = E\left(D^* a_0\right)^{1/2} \tag{1}$$

where E is the elastic modulus of the material, a_0 is its lattice parameter, $D^* = D - d$ is the fractional part of the fractal dimension and d is the Euclidean projection dimension of the fracture.

Mu and Lung [26] suggested an alternative equation, a power law mathematical relation between the surface energy and the fractal dimension. It will be seen later in this chapter that both suggestions are complementary and are covered by the model proposed in this work.

2.2. The elasto-plastic fracture mechanics

There have been several proposals for including the fractal theory into de fracture mechanics in the last three decades. Williford [17] proposed a relationship between fractal geometric parameters and parameters measured in fatigue tests. Using Williford's proposal Gong and Lai [27] developed one of the first mathematical relationships between the *J-R* curve and the fractal geometric parameters of the fracture surface. Mosolov and Borodich [32] established mathematical relations between the elastic stress field around the crack and the rugged exponent of the fracture surface. Later, Borodich [8, 29] introduced the concept of specific energy for a fractal measurement unit. Carpinteri and Chiaia [30] described the behavior of the fracture resistance as a consequence of its self-similar fractal topology. They used Griffith's theory and found a relationship between the *G*-curve and the advancing crack length and the fractal exponent. Despite the non-differentiability of the fractal functions, they were able to obtain this relationship through a renormalizing method. Bouchaud and Bouchaud [31] also proposed a formulation to correlate fractal parameters of the fracture surface.

Yavari [28] studied the *J*-integral for a fractal crack and showed that it is path-dependent. He conjectured that a *J*-integral fractal should be the rate of release of potential energy per unit of measurement of the fractal crack growth.

Recently, Alves [16] and Alves *et al.* [20] presented a self-affine fractal model, capable of describing fundamental geometric properties of fracture surfaces, including the local and global ruggedness in Griffith's criterion. In their formulations the fractal theory was introduced in an analytical context in order to establish a mathematical expression for the fracture resistance curve, putting in evidence the influence of the crack ruggedness.

3. Postulates of a fracture mechanics with irregularities

To adapt the CFM, starting from the smooth crack path equations to the rugged surface equations, and using the fractal geometry, it is necessary to establish in the form of postulates the assumptions that underlie the FFM and its correspondence with the CFM.

I. Admissible fracture surfaces

Consider a crack growing along the x-axis direction (Figure 1), deviating from the x-axis path by floating in y-direction. The trajectory of the crack is an admissible fractal if and only if it represents a single-valued function of the independent variable x.

II. Scale limits for a fractal equivalence of a crack

The irregularities of crack surfaces in contrast to mathematical fractals are finite. Therefore, the crack profiles can be assumed as fractals only in a limited scale $l_0 \leq L_0 \leq L_{0\max}$ [36]. The lower limit l_0 is related to the micro-mechanics of the cracked material and the upper limit $L_{0\max}$ is a function of the geometric size of the body, crack length and other factors.

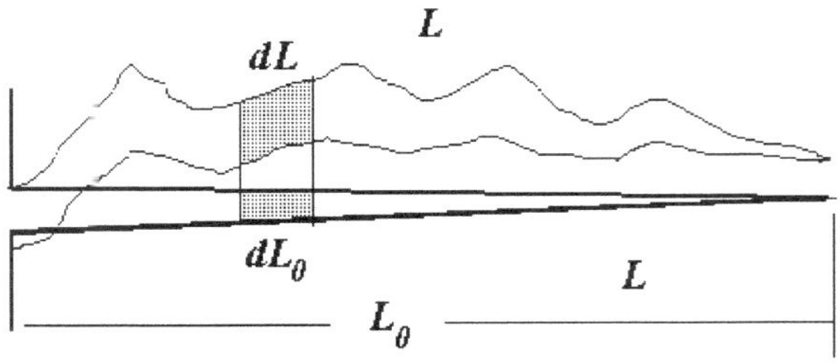

Figure 1. Rugged crack and its projection in the plan of energetic equivalence.

III. Energy equivalence between the rugged crack surface and its projection

Irwin *apud* Cherepanov *et al.* [36] realized the mathematical complexity of describing the fracture phenomena in terms of the complex geometry of the fracture surface roughness in different materials. For this reason, he proposed an energy equivalence between the rough surface path and its projection on the Euclidean plane.

In the energetic equivalence between rugged and projected crack surfaces it is considered that changes in the elastic strain energy introduced by a crack are the same for both rugged and projected paths,

$$U_{L0} = U_L \tag{2}$$

where the subscript "0" denotes quantities in the projected plane. Consequently, the surface energy expended to form rugged fracture surfaces or projected surfaces are also equivalent,

$$U_{\gamma 0} = U_\gamma \tag{3}$$

IV. Invariance of the equations

Consider a crack of length L and the quantities that describe it. Assuming the existence of a geometric operation that transforms the real crack size L to an apparent projected size L_0, the length L may be described in terms of L_0 by a fractal scaling equation, as presented in a previous chapter.

It is claimed that the classical equations of the fracture mechanics can be applied to both rugged and projected crack paths, i.e., they are invariant under a geometric transformation between the rugged and the projected paths. In the crack wrinkling operation (smooth to rough) it is desired to know what will be the form of the fracture mechanics equations for the rough path as a function of the projected length L_0, and their behavior for different roughness degrees and observation scales.

V. Continuity of functions

It is considered that the scalar and vector functions that define the irregular surfaces $\vec{A} = \vec{A}(x,y)$ are described by a model (as the fractal model) capable of providing analytical and differentiable functions in the vicinity of the generic coordinate points $P = P(x,y,z)$, so that it is possible to calculate the surface *roughness*. Thus, it is always possible to define a normal vector in corners.

VI. Transformations from the projected to rugged path equations

As a consequence of the previous two postulates, it can be shown using the chain rule that the relationship between the rates for projected and rugged paths are given by

$$\frac{df(L_0)}{dL_0} = \frac{df(L)}{dL}\frac{dL}{dL_0} \tag{4}$$

This result is used to transform the equations from the rugged to the projected path.

4. Energies in linear elastic fracture mechanics for irregular media

The study of smooth, rough, fractal and non-fractal cracks in Fracture Mechanics requires the development of their respective equations of strain and surface energies.

4.1. The elastic strain energy U_L for smooth, rugged and fractal cracks

Consider three identical plates of thickness t, with Young's modulus E', subjected to a stress σ, each of them cracked at its center with a smooth, a rugged and a fractal crack as shown in Figure 2. The area of the unloaded elastic energy due to the introduction of the crack with length L_l is

$$A_l = m_l L_l^2 \tag{5}$$

where $m_l = \pi$ is the shape factor for the smooth crack. The accumulated elastic energy is

$$\Delta U_e = \int \frac{\sigma^2}{2E'} dV \tag{6}$$

Thus, the elastic energy released by the introduction of a smooth crack with length L_l is

$$\Delta U_l = -m_l t \frac{\sigma_l^2 L_l^2}{2E'_l} \tag{7}$$

For an elliptical crack the unloaded region can be considered almost elliptical and the shape factor is $m_l = \pi$, thus

$$\Delta U_l = -t\frac{\pi\sigma_l^{\,2}L_l^2}{2E'_l} \tag{8}$$

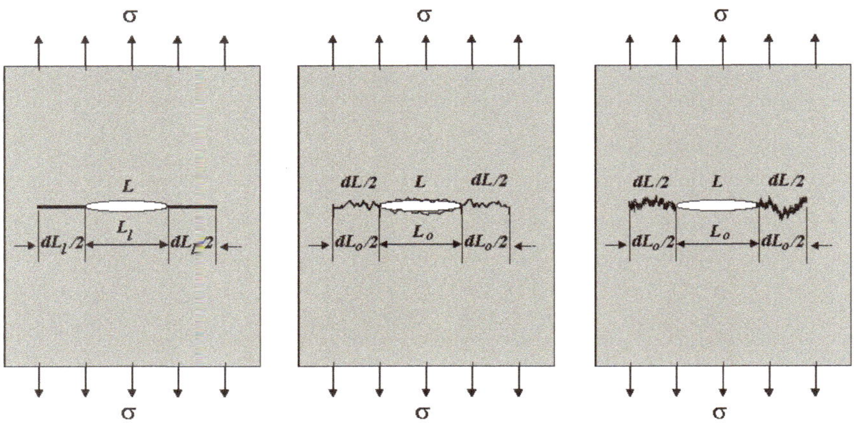

Figure 2. Griffith model for the crack growth introduced in a plate under σ stress: a) flat crack and initial length L_l with increase dL_l in size; b) rugged crack and initial length L with increase dL in size; c) fractal crack, showing increase dL in size.

Analogously, the area of the unloaded elastic energy due to the introduction of a rugged crack of length L is given by

$$A = m^* L^2 \tag{9}$$

where m^* is a shape factor for the rugged crack. Thus, the elastic energy released by the introduction of a rugged crack with length L is

$$\Delta U_L = -m^* t\frac{\sigma_r^{\,2}L^2}{2E'} \tag{10}$$

Considering that the rugged crack is slightly larger than its projection, then

$$L = \varsigma L_0 \tag{11}$$

Consequently, the change of elastic strain energy from the point of view of the projected length L_0 can be expressed as:

$$\Delta U_{L0} = -m^* t\frac{\sigma_0^{\,2}L_0^{\,2}}{2E'} \tag{12}$$

where $\sigma_0 = \sigma_r\varsigma$.

4.2. A self-affine fractal model for a crack - LEFM

To take the roughness into account, it will be inserted in the CFM equations a self-affine fractal model developed in a previous chapter of this book.

4.2.1. The relationship between strain energies for rugged U_L and projected U_{L0} cracks in terms of fractal geometry

The crack length of the self-affine fractal can be expressed as

$$L = L_0\sqrt{1+\left(\frac{H_0}{l_0}\right)^2\left(\frac{L_0}{l_0}\right)^{2(H-1)}} \tag{13}$$

where H_0 is the vertically projected crack length and the unloading fractal area of the elastic energy can be expressed as a function of the apparent length,

$$A_0 = m_0 L_0^{\,2} \tag{14}$$

And results that

$$\Delta U_{L0} = 2m^*t\int\frac{\sigma_r^{\,2}}{2E'_0}\left(1+(2-H)\left(\frac{H_0}{l_0}\right)^2\left(\frac{l_0}{L_0}\right)^{2H-2}\right)L_0 dL_0 \tag{15}$$

Therefore, the elastic energy released by the introduction of a crack length L_0 is

$$\Delta U_{L0} = -m^*t\frac{\sigma_0^{\,2}L_0^{\,2}}{2E'_0} \tag{16}$$

where $\sigma_0 = \sigma_r\sqrt{1+\left(\frac{H_0}{l_0}\right)^2\left(\frac{l_0}{L_0}\right)^{2H-2}}$

Observe that equation (12) is recovered from equation (16) applying the limits $H_0 \to l_0 \ll L_0$ and $H \to 1.0$ with $\sigma_r = \sigma_0$ and $E' = E'_0$.

To understand the effect of crack roughness on the change of elastic strain energy, one may consider postulates III and IV, thus

$$\Delta U_{Lo} = \Delta U_L = -\frac{m^*\sigma_r^{\,2}L_0^{\,2}}{2E'}\left[1+\left(\frac{H_0}{l_0}\right)^2\left(\frac{l_0}{L_0}\right)^{2H-2}\right] \tag{17}$$

It can be noticed that for $H \to 1$, which corresponds to a smoother surface, the relationship between the strain energy and the projected length L_0 is more linear. While for $H \to 0$, which corresponds to a rougher surface, this relationship is increasingly non-linear. This is reasonable since the more ruggedness, more elastic strain per unit of crack length.

4.2.2. Relationship between the applied stress on the rough and projected crack lengths

Comparing (8), (10) and (12), one has

$$\Delta U_{L0} = \Delta U_{Ll}\left(\frac{m^*}{\pi}\right) = \Delta U_L \tag{18}$$

Then, from postulate II, i.e., the following relationship is valid only for the situation of free loading without crack growth.

$$\frac{\sigma_0{}^2}{E'_0} = \frac{\sigma_r{}^2}{E'}\left(\frac{L}{L_0}\right)^2 \tag{19}$$

Using equation (13) in (19), one has the resilience as a function of the projected length L_0

$$\frac{\sigma_0{}^2}{E_0} = \frac{1}{\sqrt{2}}\frac{\sigma^2}{E}\left[1+\left(\frac{H_0}{l_0}\right)^2\left(\frac{l_0}{L_0}\right)^{2H-2}\right] \tag{20}$$

Or, the rugged length L can be written in terms of the projected length L_0, thus

$$L = \sqrt{\frac{E'}{E'_0}}\left(\frac{\sigma_0}{\sigma_r}\right)L_0 \tag{21}$$

Since the elasticity modulus is independent of the crack path, one has

$$\sigma_0 L_0 = \sigma_r L \tag{22}$$

Substituting equation (13) in equation (22), one has the relationship between stresses on the rough and projected surfaces,

$$\sigma_0 = \frac{\sigma_r}{\sqrt{2}}\left[1+\left(\frac{H_0}{l_0}\right)^2\left(\frac{l_0}{L_0}\right)^{2H-2}\right]^{1/2} \tag{23}$$

This last result is still incomplete since it is not valid for crack propagation. For its correction it will be considered that the elastic energy released rate G can be expressed as a function of G_0 according to equation (4).

4.2.3. The surface energy $U_{\gamma 0}$ for smooth, rugged and projected cracks in accordance with fractal geometry

The surface energy of a smooth and a rugged crack are, respectively, given by

$$\begin{aligned} \Delta U_{\gamma l} &= 2(L_l t)\gamma_l \\ \Delta U_{\gamma l} &= 2\gamma_l t L_l \end{aligned} \tag{24}$$

and

$$\begin{aligned} \Delta U_{\gamma L} &= 2(Lt)\gamma_r \\ \Delta U_{\gamma L} &= 2\gamma_r t L \end{aligned} \tag{25}$$

Using equation (11), the surface energy of the projected length L_0 is given by

$$\begin{aligned} \Delta U_{\gamma 0} &= 2(tL_0)\gamma_0 \\ \Delta U_{\gamma 0} &= 2\gamma_0 t L_0 \end{aligned} \tag{26}$$

where $\gamma_0 = \gamma_r \varsigma$. The surface energy equation (25) can be rewritten in terms of the projected length L_0 of a self-affine fractal crack

$$\Delta U_{\gamma 0} = 2\gamma_r t L_0 \sqrt{1 + \left(\frac{H_0}{l_0}\right)^2 \left(\frac{l_0}{L_0}\right)^{2H-2}} \tag{27}$$

To see the influence of crack roughness on the surface energy, one may consider postulates III and IV, thus

$$U_{\gamma 0} = U_\gamma = \frac{2\gamma_r L_0}{\sqrt{2}} \sqrt{1 + \left(\frac{H_0}{l_0}\right)^2 \left(\frac{l_0}{L_0}\right)^{2H-2}} \tag{28}$$

5. Stable or quasi-static fracture mechanics to the rough path

In this section, a review of the conceptual changes introduced by Irwin (1957) in Griffith's theory (1920) is presented considering an irregular fracture surface, taking into account the postulates previously proposed. The purpose of this section is to use the mathematical formalism of Linear Elastic Fracture Mechanics for stable growth of smooth cracks, generalizing it to the case of an irregular rough crack.

5.1. The Griffith energy balance in terms of fractal geometry

According to Griffith´s energy balance, one has

$$dU_T = d\left(U_i + U_L - F + U_\gamma\right) \le 0 \tag{29}$$

whilst

$$F - U_L \ge U_\gamma \tag{30}$$

Where U_T is the total energy, U_i is the initial potential elastic energy, F is the work done by external forces, U_L is the change of elastic energy stored in the body caused by the introduction of the crack length L_0 and U_γ is the energy released to form the fracture surfaces.

One can now add the contributions of ΔU_{L0} and $\Delta U_{\gamma 0}$ to reproduce Griffith´s energy balance in a fractal vision. In other words,

$$\Delta U_T = U_i + \Delta U_L + \Delta U_\gamma - F \tag{31}$$

and

$$\frac{d}{dL_l}(U_i + -\frac{\pi\sigma^2 L_0^2}{2E}\left[1+\left(\frac{H_0}{l_0}\right)^2\left(\frac{l_0}{L_0}\right)^{2H-2}\right] + \frac{2\gamma L_o}{\sqrt{2}}\sqrt{1+\left(\frac{H_0}{l_0}\right)^2\left(\frac{l_0}{L_0}\right)^{2H-2}} - F) \leq 0 \tag{32}$$

This new result is shown in Figure 3, which is analogous to the traditional Griffith energy balance graphs, but distorted due to the roughness of the fracture surface. Observe that for a reference total energy value the roughness of the crack surface tends to increase the critical size of the fracture L_{0C} compared to a material with a smooth fracture $\left(L_{IC} \leq L_C\right)$. This is due to the roughness being a result of the interaction of the crack with the microstructure of the material.

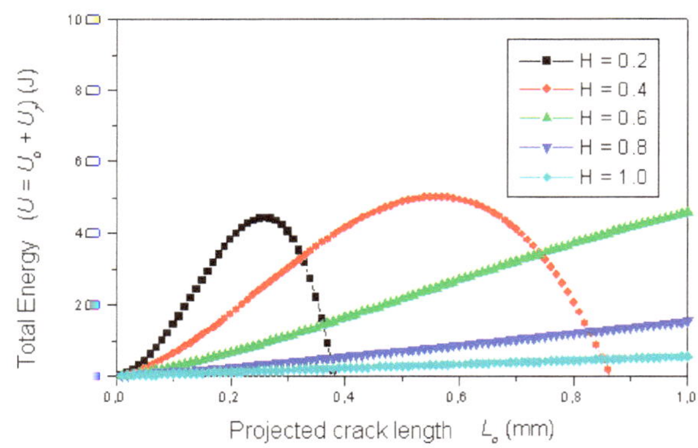

Figure 3. Griffith´s energy balance in the view of the fractal geometry of fracture surface roughness.

5.2. The modification of Irwin in Griffith´s energy balance theory for smooth, rugged and projected cracks

Irwin found from Griffith´s instability equation, given by (29), that this instability should take place by varying the crack length, so

$$\frac{d}{dL}\left(U_i + U_L + U_\gamma - F\right) \le 0 \tag{33}$$

which can be rewritten as

$$\frac{d}{dL}(F - U_L) \ge \frac{dU_\gamma}{dL} \tag{34}$$

since U_i is constant. On the left hand side of equation (34), $dF/dL - dU_L/dL$ is the amount of energy that remains available to increase crack extension by an amount dL. On the right hand side of equation (34), dU_γ/dL is the surface energy that must be released to form the rugged crack surfaces. This energy is the crack growth resistance.

Deriving equation (30) with respect to the projected crack length L_0, one has

$$\frac{d}{dL_0}(F - U_L) \ge \frac{dU_\gamma}{dL_0} \tag{35}$$

Considering postulate II, one can apply the derivation chain rule and obtain

$$\frac{d}{dL}(F - U_L)\frac{dL}{dL_0} \ge \frac{dU_\gamma}{dL}\frac{dL}{dL_0} \tag{36}$$

Considering the following cases:

i. Fixed grips condition with $F = constant$: since $U_L = U_{L0} = -m^* \sigma^2 L_0^2 / 2E'$ decreases with the crack length, and using equations (10) and (25) in (36), one can derive

$$\frac{m^* \sigma_r{}^2 L}{E'} \ge 2\gamma_r. \tag{37}$$

Or, by using equations (17) and (26) in (35), one finds

$$\frac{m^* \sigma_r{}^2 L_0}{2E'}\left(1 + (2-H)\left(\frac{H_0}{l_0}\right)^2\left(\frac{l_0}{L_0}\right)^{2H-2}\right) \ge 2\gamma_0. \tag{38}$$

ii. Condition of constant loading or stress, where necessarily $|F| = 2|U_L|$, since $U_L = U_{L0} = m^* \sigma^2 L_0^2 / 2E'$ increases with the work of external forces, and using equations (10) and (25) in (36), one can find

$$\frac{m^* \sigma_r{}^2 L}{E'} \ge 2\gamma_r. \tag{39}$$

Or, by using equations (17) and (26) in (35), one has

$$\frac{m^* \sigma_r^2 L_0}{2E'}\left(1+(2-H)\left(\frac{H_0}{l_0}\right)^2\left(\frac{l_0}{L_0}\right)^{2H-2}\right) \geq 2\gamma_0. \tag{40}$$

Irwin defined the elastic energy released rate G and the fracture resistance R in equation (34), like

$$G \equiv \frac{d(F-U_L)}{dL} \tag{41}$$

and

$$R \equiv \frac{dU_\gamma}{dL}. \tag{42}$$

These definitions can be extended to the terms in equation (35), so

$$G_0 \geq R_0 \qquad\qquad G_0 \equiv \frac{d(F-U_{L0})}{dL_0} \tag{43}$$

and

$$R_0 \equiv \frac{dU_{\gamma 0}}{dL_0}. \tag{44}$$

Notice that the proposal made by Irwin extended the concept of specific energy γ_{eff} to the concept of R-curve given by equation (42), allowing to consider situations where the microstructure of the material interacts with the crack tip. In this way, it is assumed that the surface energy is dependent on the direction of crack growth.

Finally, using equations (41) and (42) in (36), the Griffith-Irwin criterion is obtained,

$$G\frac{dL}{dL_0} \geq R\frac{dL}{dL_0}. \tag{45}$$

5.3. Comparative analysis between smooth, projected and rugged fracture quantities

Based on the results of the previous section, further analyses of the magnitudes of the Fracture Mechanics are performed in order to obtain a mathematical reformulation for an irregular or rugged Fracture Mechanics.

5.3.1. Relationship between the elastic energy released rate rates for smooth, projected and rugged cracks

Using the chain rule, it is possible to write G_0 in terms of G,

$$G_0 = G\frac{dL}{dL_0} \tag{46}$$

The energetics equivalence between the rugged surface and its projection establishes that the energy per unit length along the rugged path is equal to the energy per unit length along the projected path. Notice that

$$\frac{dU_{L0}}{dL_0} \geq \frac{dU_L}{dL} \tag{47}$$

since $dL/dL_0 \geq 1$, therefore,

$$G_0 \geq G. \tag{48}$$

The elastic energy released rates for the projected and rugged paths are, respectively

$$G_0 = \frac{dU_{L0}}{dL_0} = \frac{m^* \sigma_0{}^2 L_0}{E'_0} \tag{49}$$

and

$$G = \frac{dU_L}{dL} = \frac{m^* \sigma_r{}^2 L}{E'}. \tag{50}$$

Combining these expressions and including, for comparison, the elastic energy released rate for a smooth path, one has for infinitesimal crack lengths,

$$G_0 = G_l \frac{dL_l}{dL_0}\left(\frac{m^*}{\pi}\right) = G\frac{dL}{dL_0} \tag{51}$$

Considering that the smooth crack length is equal to the projected crack length, one has

$$G_0 = G_l\left(\frac{m^*}{\pi}\right) = G\frac{dL}{dL_0} \tag{52}$$

Observe that the difference between the elastic energy released rate for the smooth, rugged and projected cracks is the ruggedness added on crack during its growth. Using a thermodynamic model for the crack propagation, it can be concluded that a rugged crack dissipates more energy than a smooth crack propagating at the same speed.

The elastic energy released rate G_0 can be written in terms of a fractal geometry,

$$G_0 = \frac{m^* \sigma_r^{\ 2}}{E'} L_0 \left[1 + (2 - H) \left(\frac{H_0}{l_0} \right)^2 \left(\frac{l_0}{L_0} \right)^{2H-2} \right] \tag{53}$$

5.4. The crack growth resistance R for smooth, projected and rough paths

Considering a plane strain condition, crack growth resistance for a smooth crack is given by

$$R_l = \frac{dU_{\gamma l}}{dL_l} \tag{54}$$

Substituting equation (24) in equation (54), one finds

$$R_l = 2\gamma_l \tag{55}$$

Observe that if the fracture path is smooth, the specific surface energy γ_l is a cleavage surface energy and does not necessarily depend on the crack length. This model is only valid for brittle crystalline materials where the plastic strain at the crack tip does not absorb sufficient energy to cause dependence between fracture toughness and crack length.

Similarly, for a rugged crack, the fracture resistance to propagation is given by

$$R = 2\gamma_r \tag{56}$$

The concept of fracture growth resistance for the projected surface is given by

$$R_0 = \frac{dU_{\gamma}}{dL_0} \tag{57}$$

and substituting equation (26) in equation (57), one has

$$R_0 = 2\gamma_0 \tag{58}$$

Again, this model is valid for ideally brittle materials where there is almost no plastic strain at the crack tip. It basically corresponds to the model presented by Griffith, with a modified interpretation introduced by Irwin with the $G - R$ curve concept.

5.5. Relationship between rugged R and projected R_0 fracture resistances

Using the chain rule, and admitting Irwin's energetic equivalence represented by equation (3), the projected fracture resistance can be written on the basis of the resistance of the real surface,

$$R_0 = R\frac{dL}{dL_0} \tag{59}$$

where dL/dL_0 is derived from equation (13),

$$\frac{dL}{dL_0} = \frac{1+(2-H)\left(\frac{H_0}{l_0}\right)^2\left(\frac{l_0}{L_0}\right)^{2H-2}}{\sqrt{2\left(1+\left(\frac{H_0}{l_0}\right)^2\left(\frac{l_0}{L_0}\right)^{2H-2}\right)}} \geq 1 \tag{60}$$

Therefore, the crack growth resistance (R-curve), which is defined for a flat projected surface, is given substituting equation (56) and equation (60) in equation (59),

$$R_0 = 2\gamma_r \frac{1+(2-H)\left(\frac{H_0}{l_0}\right)^2\left(\frac{l_0}{L_0}\right)^{2H-2}}{\sqrt{2\left(1+\left(\frac{H_0}{l_0}\right)^2\left(\frac{l_0}{L_0}\right)^{2H-2}\right)}} \tag{61}$$

5.6. Final remarks about equivalent quantities of smooth, rugged and projected fracture surfaces

It is important to emphasize that the energetic equivalence between the rugged surface crack path and its projection was considered such that the developed equations of the Fracture Mechanics for the flat plane path are still valid in the absence of any roughness.

However, if a flat and smooth fracture L_l is considered with the same length of a projected fracture L_0, the energetic quantities and their derivatives have the following relationship,

$$U_{Ll} \leq U_{L0} \rightarrow \frac{dU_L}{dL_l} \leq \frac{dU_{L0}}{dL_0} \rightarrow G_l \leq G_0 \tag{62}$$

and

$$U_{\gamma l} \leq U_{\gamma 0} \rightarrow \frac{dU_{\gamma l}}{dL_l} \leq \frac{dU_{\gamma 0}}{dL_0} \rightarrow R_l \leq R_{\gamma 0}, \tag{63}$$

which have produced conflicting conclusions in the literature [37, 38, 46]. Since the energy for the smooth length L_0^l is smaller than the energy for the projected L_0 or rough L lengths, one has

$$U_{Ll} \leq U_L \rightarrow G_l \leq G\frac{dL}{dL_0} \tag{64}$$

and

$$U_{\gamma l} \leq U_{\gamma} \rightarrow R_{\gamma l} \leq R\frac{dL}{dL_0} \tag{65}$$

In postulate III it was assumed that the rugged crack path satisfies the same energetic conditions of the plan path, but in the LEFM this roughness is not taken into account, causing discrepancies between theory and experiments. For example, it has not been possible to explain by an analytical function in a definitive way the growth of the $G-R$ curve. The proposed introduction of the term dL/dL_0 allows correcting this problem.

6. The elastic-plastic fractal fracture mechanics

The non-linear elastic plastic energy released rate J_0 for a crack of plane projected path can be extended from the Irwin-Orowan approach. They introduced the specific energy of plastic strain γ_p on the elastic energy released rate G_0 to describe the fracture phenomenon with considerable plastic strain at the crack tip. Thus, it is possible to define the elastic plastic energy released rate in an analogous way to the definition of the elastic energy released rate,

$$J_o \equiv \frac{d(F-U_{Vo})}{dL_o} \tag{66}$$

where U_{Vo} is the volumetric strain energy given by the sum of the elastic and plastic (U_{pl}) contributions to the strain energy in the material.

6.1. Influence of ruggedness in elastic plastic solids with low ductility

Considering elastic plastic materials with low ductility where the effect of the plastic term is small compared to the elastic term, one can define a crack growth resistance as

$$J_{Ro} = \frac{K_{Ro}{}^2 f(v)}{E}, \tag{67}$$

where $f(v)$ is a function that defines the testing condition. For plane stress $f(v)=1$, and for plane strain $f(v)=1-v^2$ and K_{Ro} is the fracture toughness resistance curve.

Due to the ruggedness the crack grows an amount $dL > dL_0$ and correcting equation (59), one has

$$R_o = \frac{dU_{\gamma}}{dL}\frac{dL}{dL_o} = \left(2\gamma_e + \gamma_p\right)\frac{dL}{dL_o}. \tag{68}$$

Similarly,

$$J_o = \frac{d(F - U_V)}{dL}\frac{dL}{dL_o}. \tag{69}$$

The energy balance proposed by Griffith-Irwin-Orowan, for stable fracture, is

$$J_o = R_o. \tag{70}$$

Therefore, for plane stress or plane strain conditions, one can write from equation (61) that,

$$J_{Ro} = \left(2\gamma_e + \gamma_p\right)\frac{dL}{dL_o} = \frac{{K_{Ro}}^2 f(v)}{E} \tag{71}$$

Thus,

$$K_{Ro} = \sqrt{\frac{\left(2\gamma_e + \gamma_p\right)E}{f(v)}\frac{dL}{dL_o}}. \tag{72}$$

Knowing that fracture toughness is given by

$$K_{Co} = \sqrt{\frac{\left(2\gamma_e + \gamma_p\right)E}{f(v)}}, \tag{73}$$

one has,

$$K_{Ro} = K_{Co}\sqrt{\frac{dL}{dL_o}}. \tag{74}$$

From the Classical Fracture Mechanics, the fracture resistance for the loading mode I, is given by

$$K_{IRo} = Y_o\left(\frac{L_o}{w}\right)\sigma_f\sqrt{L_o}, \tag{75}$$

where $Y_o\left(\frac{L_o}{w}\right)$ is a function that defines the shape of the specimen (CT, SEBN, etc) and the type of test (traction, flexion, etc), and σ_f is the fracture stress. Considering the case when $L_0 = L_{0C}$, then $K_{IR0} = K_{IC0}$ and the fracture toughness for the loading mode I is given by

$$K_{ICo} = Y_o\left(\frac{L_{oc}}{w}\right)\sigma_f\sqrt{L_{oc}}. \tag{76}$$

Therefore, from equation (72) the fracture toughness curve for the loading mode I is given by

$$K_{IRo} = K_{ICo}\sqrt{\frac{dL}{dL_o}}. \tag{77}$$

Substituting equation (75) and equation (76) in equation (77), one has

$$\frac{dL}{dL_o} = Y_o^{\,2}\left(\frac{L_o}{w}\right)\frac{\sigma_f^{\,2} L_o f(v)}{\left(2\gamma_e + \gamma_p\right)E}, \tag{78}$$

Observe that according to the right hand side of equation (78), the ruggedness dL/dL_0 is determined by the condition of the test (plane strain or stress), the shape of the sample (CT, SEBN, etc), the type of test (traction, flexion, etc) and kind of material.

Considering the fracture surface as a fractal topology, one observes that the characteristics of the fracture surface listed above in equation (78) are all included in the ruggedness fractal exponent *H*. Substituting equation (60) in equation (71), one obtains

$$J_{Ro} = \left(2\gamma_e + \gamma_p\right)\frac{1+(2-H)\left(\frac{H_0}{l_0}\right)^2\left(\frac{l_0}{\Delta L_0}\right)^{2H-2}}{\sqrt{1+\left(\frac{H_0}{l_0}\right)^2\left(\frac{l_0}{\Delta L_0}\right)^{2H-2}}}. \tag{79}$$

which is non-linear in the crack extension ΔL_0. It corresponds to the classical equation (70) corrected for a rugged surface with Hurst's exponent *H*. Experimental results [1, 2] show that J_0 and the crack resistance R_0 rise non-linearly and it is well known that this rising of the *J-R* curve is correlated to the ruggedness of the cracked surface [3, 4].

6.2. The J_0 Eshelby-Rice integral for rugged and plane projected crack paths

The *J*-integral concept of Eshelby-Rice is a non-linear extension of the definition given by Irwin-Orowan, for the linear elastic plastic energy released rate. In this context the potential energy Π_0 is defined as

$$\Pi_0 = \int_{V_0} W dV_0 - \int_C \vec{T}.\vec{u} ds, \tag{80}$$

where *W* the energy density integral in the in the volume V_0 encapsulated by the boundary C with tractions $\vec{T}$ and displacements $\vec{u}$, and s is the distance along the boundary C, as shown in Figure 4.

Accordingly,

$$J_0 = -\frac{d\Pi_0}{dL_0} = -\frac{d}{dL_0}\left(\int_V W dV_0 - \int_C \vec{T}.\vec{u} ds\right) \tag{81}$$

where dL_0 is the incremental growth of the crack length. In the two-dimensional case, where the fracture surface is characterized by a crack with length ΔL_0 and a unit thickness body, one has $dV = dxdy$ and

$$J_0 \equiv -\frac{d\Pi_0}{dL_0} = -\left(\int_V W \frac{dx}{dL_0} dy - \int_C \vec{T}.\frac{\partial \vec{u}}{\partial L_0} ds \right). \tag{82}$$

For a fixed boundary C, $d/dL_0 = -d/dx$, and the J_0 -integral for the plane projected crack path can be written only in terms of the boundary,

$$J_0 = \int_V W dy - \int_C \vec{T}.\frac{\partial \vec{u}}{\partial x} ds. \tag{83}$$

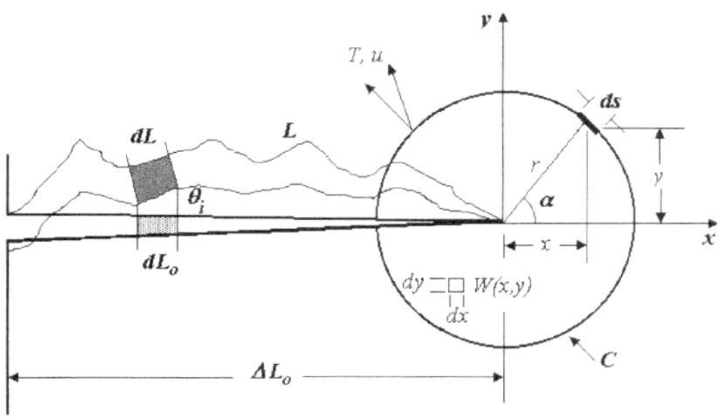

Figure 4. Boundary around to the rugged crack tip where is defined the *J*-Integral [43].

Now, the *J-R* Eshelby-Rice integral theory is modified to include the fracture surface ruggedness. Initially, equation (82) is rewritten,

$$J_0 = -\left(\int_V W \frac{dx}{dL} \frac{dL}{dL_0} dy - \int_C \vec{T}.\frac{\partial \vec{u}}{\partial L} \frac{dL}{dL_0} ds \right). \tag{84}$$

From postulate IV, the new *J*-integral on the rugged crack path is given by

$$J \equiv -\frac{d\Pi}{dL} = -\left(\int_V W \frac{dx^*}{dL} dy^* - \int_C \vec{T}.\frac{\partial \vec{u}}{\partial L} ds \right) \tag{85}$$

where the * symbol represents coordinates with respect to the rugged path. So, in an analogous way to the *J*-integral for the projected crack path given by equation (85), since $d/dL = -d/dx^*$, one has

$$J = \int_V W dy^* - \int_C \vec{T}.\frac{\partial \vec{u}}{\partial x^*} ds. \tag{86}$$

Returning to equation (82) and considering postulate III along with the derivative chain rule and substituting equation (85), one has

$$J_0 \equiv -\frac{d\Pi}{dL}\frac{dL}{dL_0} = -\left(\int_V W\frac{dx^*}{dL}dy^* - \int_C \vec{T}.\frac{\partial \vec{u}}{\partial L}ds\right)\frac{dL}{dL_0}. \tag{87}$$

Comparing (84) with equation (87) and considering that the rugged crack is a result of a transformation in the volume of the crack, analogous to the "*bakers´ transformation*" of the projected crack over the Euclidian plane, it can be concluded that

$$\begin{aligned} dx^*dy^* &= \frac{\partial(x^*,y^*)}{\partial(x,y)}dxdy \\ \frac{dx^*}{dL}dy^*\frac{dL}{dL_0} &= \frac{dx}{dL}\frac{dL}{dL_0}dy \end{aligned} \tag{88}$$

which show the equivalence between the volume elements,

$$dV = dx^*dy^* = dxdy. \tag{89}$$

Therefore, the ruggedness dL/dL_0 of the rugged crack path does not depend on the volume V, nor on the boundary C and nor on the infinitesimal element length ds or dy. Thus, it must depend only on the characteristics of the rugged path described by the crack on the material. Finally, the integral in equation (84) can be written as

$$J_0 = -\left(\int_V W\frac{dx}{dL}dy - \int_C \vec{T}.\frac{\partial \vec{u}}{\partial L}ds\right)\frac{dL}{dL_0} \tag{90}$$

where the infinitesimal increment $dx/dL = -\cos\theta_i$ accompanies the direction of the rugged path L, as show in Figure 4. Thus,

$$J = \int_V Wdy\cos\theta_i - \int_C \vec{T}.\frac{\partial \vec{u}}{\partial x}\cos\theta_i ds. \tag{91}$$

Observe that the J-integral for the rugged crack path given by equation (91) differs from the J-integral for the plane projected crack path given by equation (83) by a fluctuating term, $\cos\theta_i$ inside the integral. It can be observed that the energetic and geometric parts of the fracture process are separated and put in evidence the influence of the ruggedness of the material in the elastic plastic energy released rate,

$$J_0 \equiv J\frac{dL}{dL_0}. \tag{92}$$

It must be pointed out that this relationship is general and the introduction of the fractal approach to describe the ruggedness is just a particular way of modeling.

6.3. Fractal theory applied to J-R curve model for ductile materials

This section includes the formalism of fractal geometry in the EPFM to describe the roughness effects on the fracture mechanical properties of materials. For this purpose the classical expression of the elastic-plastic energy released rate was modified by introducing the fractality (roughness) of the cracked surface. With this procedure the classical expression (49) of LEFM, linear with the crack length, is changed into a non-linear equation (53), which reproduces with precision the quasi-static crack propagation process in ductile materials.

Observe that the quasi-static crack growth condition is obtained with Griffith fracture criterion, doing $J_0 = R_0$ and $dJ_0 / dL_0 = dR_0 / dL_0$. In this case, it is concluded that the *J-R* curve is given by Griffith criterion $J = 2\gamma_{eff}$ in equations (92) and (59). Therefore, for a self-affine crack with $H_0 \to l_0$, one has

$$J_0 = 2\gamma_{eff} \frac{1 + (2-H)\left(\dfrac{l_0}{L_0}\right)^{2H-2}}{\sqrt{2\left(1 + \left(\dfrac{H_0}{l_0}\right)^2 \left(\dfrac{l_0}{L_0}\right)^{2H-2}\right)}} \tag{93}$$

This model shows in unambiguous way how different morphologies (roughness) are correlated with the *J-R* curve growth. Given the energy equivalence between rough and projected surfaces for the crack path, the *J-R* curve increases due to the influence of the roughness, which has not been computed previously with the classical equations of EPFM.

The *J*-integral on the rugged crack path is a specific characteristic of the material and can be considered as being proportional to J_C [15], on the onset of crack extension, since in this case it has the rugged crack length greater than the projected crack length $(L \gg L_0)$. Thus,

$$J_o \sim J_C \frac{dL}{dL_o}. \tag{94}$$

Substituting the fractal crack model proposed in equation (60), one has

$$J_o \sim J_C \frac{1 + (2-H)\left(\dfrac{H_0}{l_0}\right)^2 \left(\dfrac{l_o}{\Delta L_o}\right)^{2H-2}}{\sqrt{1 + \left(\dfrac{H_0}{l_0}\right)^2 \left(\dfrac{l_o}{\Delta L_o}\right)^{2H-2}}}, \tag{95}$$

corroborating that the surface specific energy is related to the critical fracture resistance.

$$J_C \sim \left(2\gamma_e + \gamma_p\right). \tag{96}$$

6.3.1. Case – 1. Ductile self-similar limit

The local self-similar limit can be calculated applying the condition $H_0 \rightarrow L_0 \gg l_0$ in equation (79), obtaining

$$J_{Ro} = \left(2\gamma_e + \gamma_p\right)\left(2 - H\right)\left(\frac{l_o}{\Delta L_o}\right)^{H-1} \tag{97}$$

or, with $D = 2 - H$, one has

$$J_0 = 2\gamma_{eff} D\left(\frac{l_0}{L_0}\right)^{1-D}. \tag{98}$$

This result corresponds to the one found by Mu and Lung [26, 37] for ductile materials. Equation (98) is shown in Figure 5, where *J-R* curves are calculated for different values of the fractal dimension D. $2\gamma_{eff} = 10.0KJ/m^2$ is adopted and L_0/l_0 is the crack length in l_0 units. This figure shows very clearly how the surface morphology (characterized by D) determines the shape of the *J-R* curve at the beginning of the crack growth.

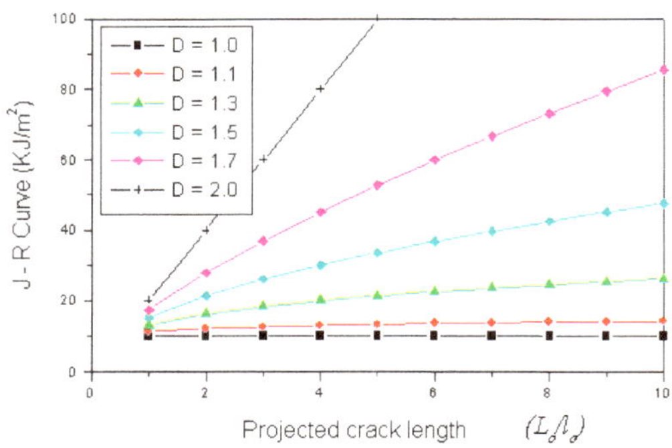

Figure 5. *J-R* curves calculated according to the projected crack length L_0, for a fracture of unit thickness, and fractal dimensions $D = 1.0, 1.1, 1.3, 1.5, 1.7$ and 2.0 with $2\gamma_e = 10KJ/m^2$.

In Figure 6, *J-R* curves with fractal dimension $D = 1.3$ are calculated according to the projected length L_0 for different measuring rulers l_0, showing how the morphology of rugged surface cracks is best described for small values of l_0, causing the pronounced rising of *J-R* curve. Figure 6 and equation (98) show that the initial crack resistance is correlated to the surface morphology characterized by dimension D, in accordance with the literature.

The self-similar limit of *J-R* curve, given by equation (98), is valid only for regions near the onset of the crack growth in brittle materials ($H_0 \rightarrow L_0$). This is due to the hardening of the material, which gives rise to ruggedness of the fracture surface.

In the case of ductile materials, the length of the work hardening zone H_0 affects an increasingly greater area of the material as the crack propagates, but the self-similar limit $(H_0 = L_0 \gg l_0)$ is still valid.

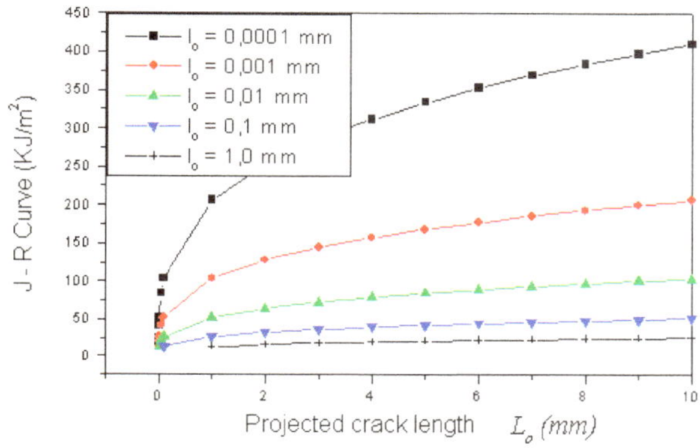

Figure 6. *J-R* curves calculated in function of the projected crack length L_0 with different ruler lengths $l_0 = 0.0001, 0.001, 0.01, 0.1$ and $1.0mm$, for a fracture of unit thickness, fractal dimension $D = 1.3$ and $2\gamma_e = 10KJ/m^2$.

However, in the case of brittle materials (ceramics), after the initial stage of hardening, the crack maintains this state in a region of length H_0, very short if compared to the crack length L_0, generating a self-similar fractal structure only when the crack length L_0 is small, in the order of l_0, i.e., $H_0 = L_0 = l_0$. When the crack length L_0 becomes much larger than the initial size of the hardening region H_0 present at the onset of crack growth, the self-similar limit is not valid, and the self-affine (or global) limit of fracture becomes valid.

6.3.2. Case – 2. Brittle self-affine limit

It is easy verify that in stable crack growth, where $J_0 = R_0$, using equations (59) and (79), one has $dL/dL_0 = 1$ when $L \to \infty$. The global self-affine limit of J_0 can be calculated applying the condition when the observation scale corresponds to a rather small amplitude of the crack, similar in size to the crack increment, i.e., when $H_0 \to l_0 \ll L_0$ in equation (79), resulting the linear elastic expression

$$J_0 = 2\gamma_{eff} \tag{99}$$

where $J_0 = G_0$ and

$$G_{Ro} = 2\gamma_e + \gamma_p. \tag{100}$$

This result corresponds to a classic one in Fracture Mechanics, which is the general case valid for brittle materials as glass and ceramics.

7. Experimental analyses

7.1. Ceramic, metallic and polyurethane samples

The analyzed ceramic samples were produced by Santos [19] and Mazzei [41]. The raw material used for its production was an alumina powder A-1000SG by ALCOA with 99% purity. Specimens of dimensions $52mm \times 8mm \times 4mm$ were sintered at 1650 °C for 2 hours, showing average 7 mm grain sizes. Their average mechanical properties are shown in Table 1 with elastic modulus $E = 300\ GPa$ and rupture stress $\sigma_f = 340MPa$.

The analyzed metallic samples were multipass High Strength Low Alloy (HSLA) steel weld metals and standard DCT specimens. HSLA are divided in two groups based on the welding process utilized and the microstructural composition. The first group (*A1* and *A2* welds) is composed of *C-Mn Ti-Killed* weld metals and were joined by a manual metal arc process. The second group (*B1* and *B2* welds), joined by a submerged arc welding process, is also a *C-Mn Ti-Killed* weld metal, but with different alloying elements added to increase the hardenability. Mechanical properties of both welds and DCT metals are listed in Table 1.

Material	*Sample*	σ_f *(MPa)*	*E (GPa)*	J_{IC}*(exp)(KJ/m²)*	L_{OC}*(exp)(mm)*	K_{IC} *(MPa.m^1/2)*	*H (exp)*
Ceramic	Alumina	340	300	0,030	0.4956	424,2477056	0,7975± 0,0096
Metals	A1CT2	516,00	1,34	291,60	0,48256	635,3313677	0,71 ± 0,01
	A2SEB2	537,00	3,63	174,67	0,36264	573,1747828	0,77 ± 0,01
	B1CT6	771,00	16,64	40,61	0,22634	650,1446157	0,77 ± 0,02
	B2CT2	757,00	1,96	99,22	0,26553	691,3971955	0,58 ± 0,05
	DCT1	554,00	1,7197	227,00	0,40487	624,8021278	-
	DCT2	530,00	1,6671	211,47	0,3995	593,7576222	-
	DCT3	198,75	0,3902	318,00	1,00000	352,2752029	-
Polymers	PU0,5	40,70	0.8 ± 0.0	8,10	0,29951	39,47980593	0,47 ± 0,07
	PU1,0	40,70	0.8 ± 0.0	3,00	0,23685	35,10799599	0,50 ± 0,05

Table 1. Data extracted from experimental testing of *J-R* curves obtained by compliance method.

The analyzed polymeric samples are a two-component Polyurethane, consisting of 1:1 mixture of polyol and prepolymer. The polyol was synthesized from oil and the prepolymer from diphenyl methane diisocyanate (MDI). Their mechanical properties are shown in Table 1.

7.2. Fracture tests

A standard three-point bending test was performed on alumina specimens, SE(B), notched plane. Low speed and constant prescribed displacement 1 *mm/min* was employed to obtain stable propagation. The *R*-curve was obtained using LEFM equations and fracture results are shown in Table 1.

The fracture toughness evaluation of metallic samples was executed using the *J*-integral concept and the elastic compliance technique with partial unloadings of 15% of the maximum load. For weld metals the *J-R* curve tests were performed by the compliance and multi-test techniques. Tests were executed in a MTS810 (Material Test System) system at ambient temperature, according to standard ASTM E1737-96 [15]. A single edge notch bending SENB and compact tension CT were used. One *J-R* curve for each tested specimen was retrieved and fracture results are shown in Table 1.

To obtain the fractured surfaces of polymeric materials, fracture toughness tests were performed by multiple specimen technique using the concept of *J-R* curve according to ASTM D6068-2002 [42]. However, these tests were different from the ones used for weld metals, due to the viscoelasticity of the polymers. The used nomenclatures PU0,5 and PU1,0 mean the loading rate used during the test, 0,5 *mm/min* and 1,0 *mm/min*, respectively. Fracture results are shown in Table 1.

7.3. Fractal analyses of fractured specimens

The fractured surfaces of ceramic samples were obtained with a Rank Taylor Hobson profilometer (Talysurf model 120) and an HP 6300 scanner. The fractal analyses to obtain the Hurst dimensions were made by methods, such as Counting Box, Sand Box and Fourier transform. The fracture surface analysis of metallic and polymeric samples were executed using scanning electronic microscopy SEM and the analyses to obtain the Hurst exponents were made with the Contrast Islands Fractal Analysis. Fractal dimension results are shown in the last column of Table 1.

7.4. G-R and J-R curve tests and fitting with self-similar and self-affine fractal models

A characteristic load-displacement result in the Alumina ceramic sample is shown in Figure 7. Observe that the stiffness of the material at the first deflection region is constant, corresponding to the elastic modulus of the material. However, as the crack propagates, the stiffness varies significantly.

The corresponding *G-R* curve test is shown in Figure 8. It can be seen that at the onset of crack growth ($L_0 \approx L_{0C}$), the behavior of this material is self-similar, as previously discussed. However, the results in the wider range of crack lengths ($L_{0C} < L_0 < L_{0\max}$) show that this material behave according to the self-affine model. Finally, at the end of *G-R* curve

($L_0 \to \infty$) the behavior is explained by the influence of the shape function $Y(L_0 / w)$ used in the testing methodology [41].

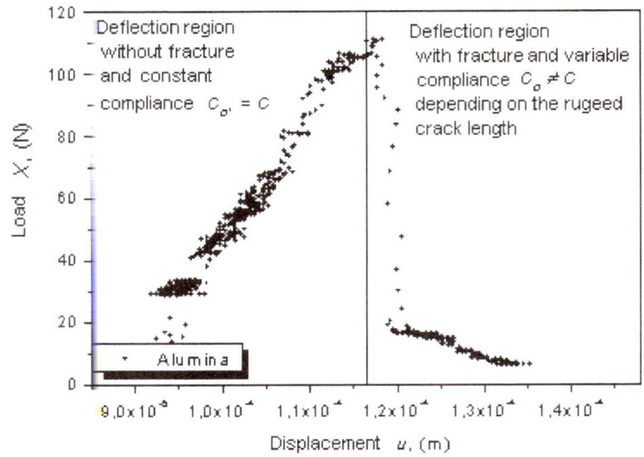

Figure 7. Load (X) versus displacement (u) for a G-R curve test in a ceramic sample [41].

J-R curves obtained from standard metallic specimens provided by ASTM standard testing are shown in Figure 9 along with the fitting with the proposed fractal models. Fitting results with these samples, named DCT1, DCT2 and DCT3, are a consistent validation of the applied fractal models The fitting results of the self-similar and self-affine models coincide and are not distinguishable in Figure 9.

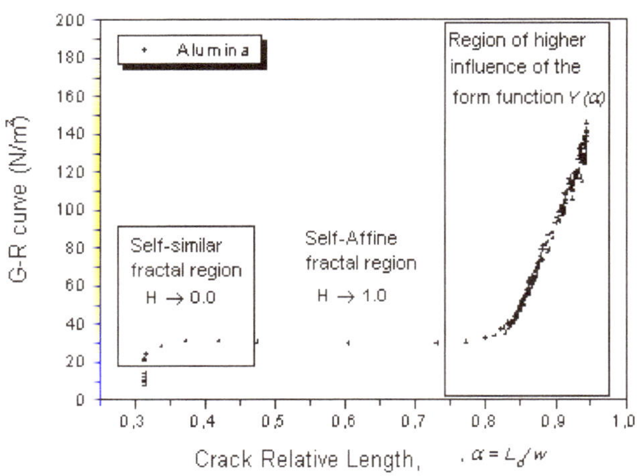

Figure 8. G-R curve fitted with the self-similar model (equation (97)) and the self-affine model (equation (100)) for the Alumina sample [41].

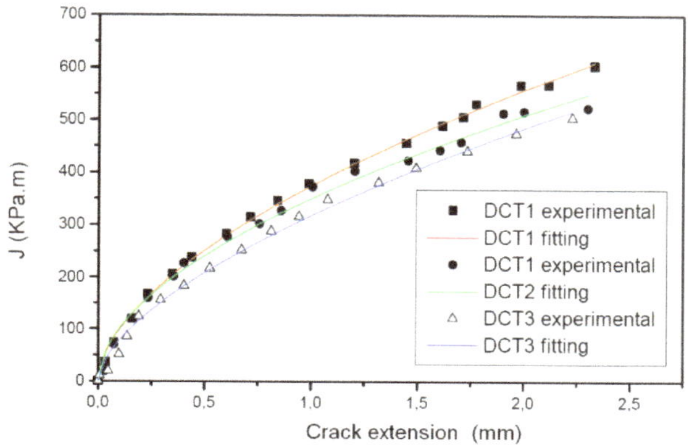

Figure 9. *J-R* curve fitted with the self-similar model shown in equation (97) and the self-affine model shown in equation (93) for steel samples DCT1, DCT2 and DCT3 [43].

Typical testing results performed to obtain *J-R* curves of metallic weld materials are shown in Figure 10 and Figure 11. In all results, *J-R* curves measured experimentally were fitted using models given by equations (93) and (97), where the factor $2\gamma_e + \gamma_p$ was obtained by adjusting the l_0 and H values for each different sample, by the self-similar and the self-affine models.

The *J-R* curves for the tested polymeric specimens are shown in Figure 12 and Figure 13. Reasonably good results were obtained despite the greater dispersion of data.

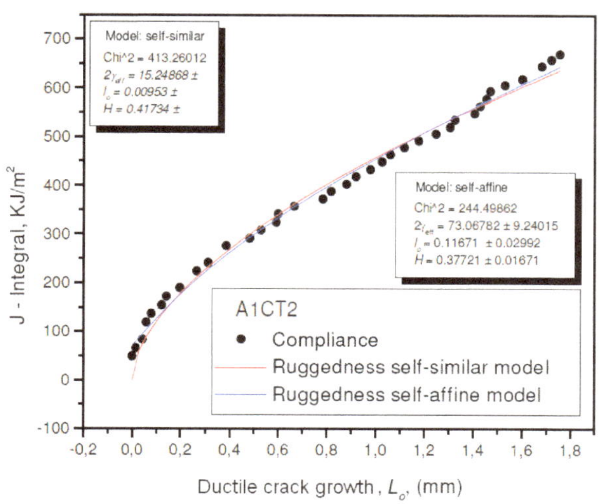

Figure 10. *J-R* curve fitted with the self-similar model shown in equation (97) and the self-affine model shown in equation (93) for HSLA-Mn/Ti steel (sample A1CT2).

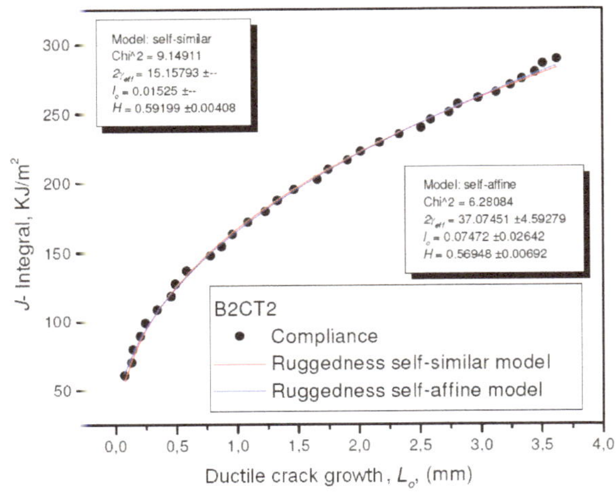

Figure 11. *J-R* curve fitted with the self-similar model shown in equation (97) and the self-affine model shown in equation (93) for HSLA-Mn/Ti steel (sample B2CT2) killed with titanium and other alloy elements to increase hardenability [43].

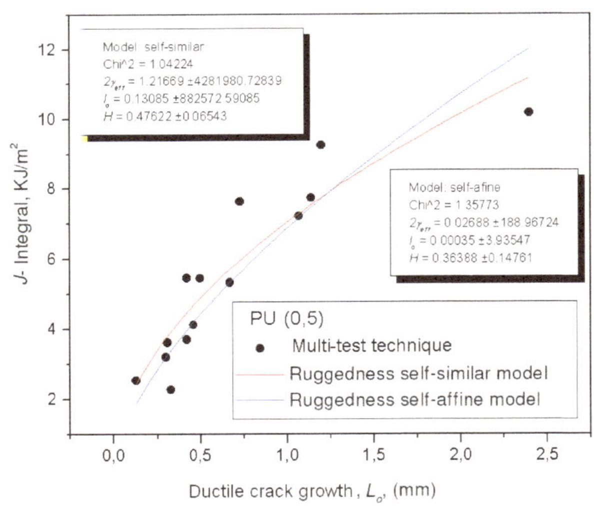

Figure 12. *J-R* curve fitted with the self-similar model shown in equation (97) and the self-affine model shown in equation (93) for the poliurethane polymer PU0,5.

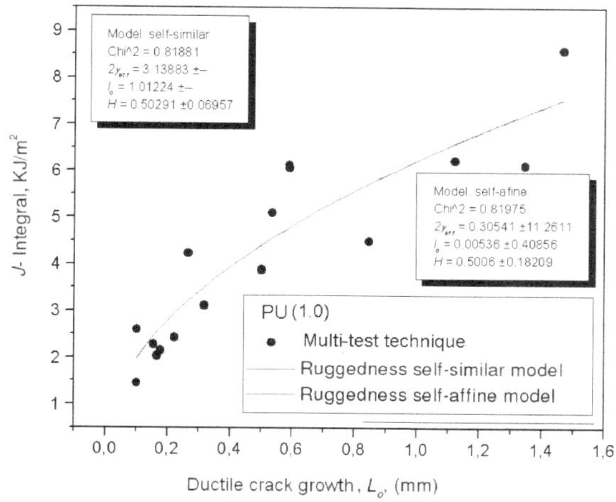

Figure 13. *J-R* curve fitted with the self-similar model shown in equation (97) and the self-affine model shown in equation (93) for the poliurethane polymer PU1,0.

After the experimental *J-R* curves were fitted using equation (79) and equation (97), values of $2\gamma_{eff}$, H and l_0 were determined and are shown in Table 2 and Table 3. With $J_{R0} = 2\gamma_{eff}$, the value of the crack size $L_{0\gamma_{eff}}$ was calculated and it corresponded to the specific surface energy. Using the experimental values of J_{IC}, L_{0C} and H given in Table 1, the values of the constants in the last column of Table 2 and Table 3 were calculated.

Mate-rial	Sample	$2\gamma_{eff}(KJ/m^2)$	$H(theo)$	$l_0(mm)$	$L_{0\gamma_{eff}} = l_0(2-H)^{1/(H-1)}$	$C_1/2\gamma_{eff} = (2-H)l_0^{H-1}$	$J_C L_C^{H-1} = constant$
Cera-mic	Alumi-na	0,0301871	1,000	0,2493645	0,2493645	1,00000	0,03018707
Metals	A1CT2	283,247	0,417 ± 0,018	1,00944	0,459079	1,57411	445,862579
	A2SEB2	187,639	0,208 ± 0,057	0,82912	0,396956	2,07868	390,042318
	B1CT6	40,514	0,573 ± 0,038	0,51758	0,225086	1,89071	76,600193
	B2CT2	101,204	0,592 ± 0,0041	0,64484	0,278764	1,68407	170,433782
	DCT1	230,843	0,426	0,91887	0,416893	1,65219	381,397057
	DCT2	209,127	0,461	0,87082	0,391328	1,65806	346,745868
	DCT3	317,819	0,393	2,18249	0,999062	1,00057	318,000000
Poly-mers	PU0,5	17,4129	0,476	2,88612	1,291434	0,87464	15,230001
	PU1,0	2,95252	0,503	0,51653	0,229374	2,079	6,138287

Table 2. Fitting data of *J-R* curves with the self- similar model [43].

A good level of agreement is seen between measured Hurst's exponents H at Table 1 and theoretical ones shown in Table 2 and Table 3. Larger differences in metals can be attributed to the quality of the fractographic images, which did not present well defined "Contrast Islands".

Material	Sample	$2\gamma_{eff}\left(KJ/m^2\right)$	$H(theo)$	$l_0(mm)$	$L_{0\gamma_{eff}} = l_0(2-H)^{1/(H-1)}$	$C_1/2\gamma_{eff} = (2-H)l_0^{H-1}$	$J_C L_C^{H-1}$ = *constant*
Ceramic	Alumina	0,0301871	1,000	0,2493645	0,2493645	1,00000	0,03018707
Metals	A1CT2	160,640	0,609	0,24422	0,105004	2,413408	387,700806
	A2SEB2	102,750	0,442	0,31002	0,140040	2,993092	307,535922
	B1CT6	2[illegible],980	0,700	0,08123	0,033873	2,757772	63,385976
	B2CT2	5[illegible],978	0,705	0,10304	0,042893	2,529433	146,651006
	DCT1	129,850	0,599	0,23309	0,100540	2,511844	326,184445
	DCT2	113,850	0,624	0,20167	0,086294	2,512302	298,592197
	DCT3	173,810	0,612	0,5282	0,226901	1,778386	318,000000
Polymers	PU0,5	7,500	0,664	0,56541	0,238775	1,618852	12,150370
	PU1,0	1,690	0,649	0,10898	0,046244	2,938220	4,971102

Table 3. Fitting data of *J-R* curves with the self- affine model [43].

7.5. Complementary discussion

The proposed fractal scaling law (self-affine or self-similar) model is well suited for the elastic-plastic experimental results. However, the self-similar model in brittle materials appears to underestimate the values of specific surface energy γ_{eff} and the minimum size of the microscopic fracture l_0, although not affecting the value of the Hurst exponent H.

For a self-affine natural fractal such as a crack, the self-similar limit approach is only valid at the beginning of the crack growth process [39], and the self-affine limit is valid for the rest of the process. It can be observed from the results that the ductile fracture is closer to self-similarity while the brittle fracture is closer to self-affinity.

Equation (79) represents a self-affine fractal model and demonstrates that apart from the coefficient H, there is a certain "universality" or, more accurately, a certain "generality" in the *J-R* curves. This equation can be rewritten using a factor of universal scale, $\varepsilon = l_0 / L_0$, as

$$\underbrace{f(2\gamma_e + \gamma_p, J_0) = \frac{J_o}{2(2\gamma_e + \gamma_p)}}_{energetic} = \underbrace{\frac{1+(2-H)\varepsilon^{2H-2}}{\sqrt{2\left(1+\varepsilon^{2H-2}\right)}} = g(\varepsilon, H)}_{geometric} \tag{101}$$

which is a valid function for all experimental results shown in Figure 14. It shows the existent relation between the energetic and geometric components of the fracture resistance of the material. The greater the material energy consumption in the fracture, straining it plastically, the longer will be its geometric path and more rugged will be the crack.

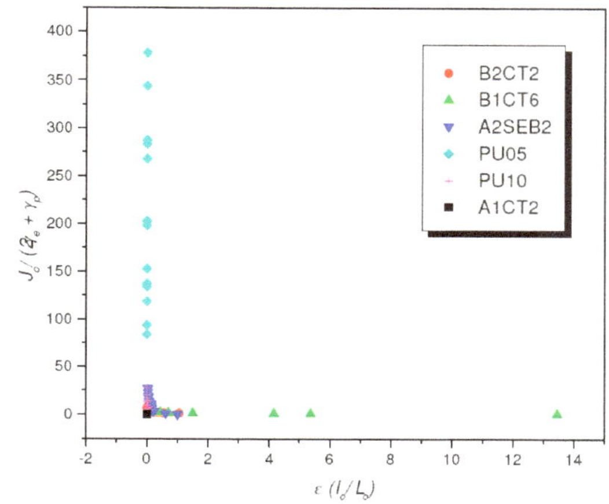

Figure 14. Generalized *J-R* curves for different materials, modelled using the self-affine fractal geometry, in function of the scale factor $\varepsilon = l_0/L_0$ of the crack length [43].

In the self-similar limit $(l_0 \ll L_0 = H_0)$, equation (97) is applicable and the energetic and geometric components are put in evidence in the equation below,

$$J_0 = \underbrace{(2\gamma_{eo} + \gamma_p)}_{energetic}\underbrace{(2-H)\left(\frac{l_0}{\Delta L_0}\right)^{H-1}}_{geometric} \tag{102}$$

From equation (102), an expression can be derived which results in a constant value associated to each material,

$$\underbrace{J_0 \Delta L_0}_{macroscopic}{}^{H-1} = \underbrace{(2\gamma_{e0} + \gamma_p)(2-H)l_0{}^{H-1}}_{microscopic} = (const)_{material} \tag{103}$$

It is possible to conclude that the macroscopic and microscopic terms on the left and right-hand sides of equation (103) are both equal to a constant, suggesting the existence of a fracture fractal property valid for the beginning of crack growth, and justified experimentally and theoretically. These constant values were calculated for each point in each *J-R* curve for the tested materials. The average value for each material is listed in the last column of Table 2 and Table 3. Observe that this new property is uniquely determined by the process of crack growth, depending on the exponent H, the specific surface energy $2\gamma_e + \gamma_p$ and the minimum crack length l_0.

This new constant can be understood as a "fractal energy density" and it is a physical quantity that takes into account the ruggedness of the fracture surface and other physical properties. Its existence can explain the reason for different problems encountered when defining the value of fracture toughness K_{IC}. This constant can be used to complement the information yielded by the fracture toughness, which depends on several factors, such as the thickness B of the specimen, the shape or size of the notch, etc. To solve this problem, ASTM E1737-96 [15] establishes a value for the crack length a (approximately $0.5 < a / W < 0.7$ and, $B = 0.5W$, where W is the width of the specimen) for obtaining the fracture toughness K_{IC}, in order to maintain the small-scale yielding zone.

As shown in equation (103), a relationship exists between the specific surface energy $2\gamma_{eff}$ and the minimum crack size l_0 in the considered observation scale $\varepsilon = l_0 / L_0$. In Figure 15, it can be observed that the consideration of a minimum size for the fracture l_{01} on a grain should mean the effective specific energy of the fracture $2\gamma_{eff1}$ in this scale. In a similar way, the consideration of a minimum size of fracture in a different scale, like one that involves several polycrystalline grains l_{02}, l_{03} etc.., should take into account the value of an effective specific energy in this other scale, $2\gamma_{eff2}, 2\gamma_{eff3}$, etc., in such a way that

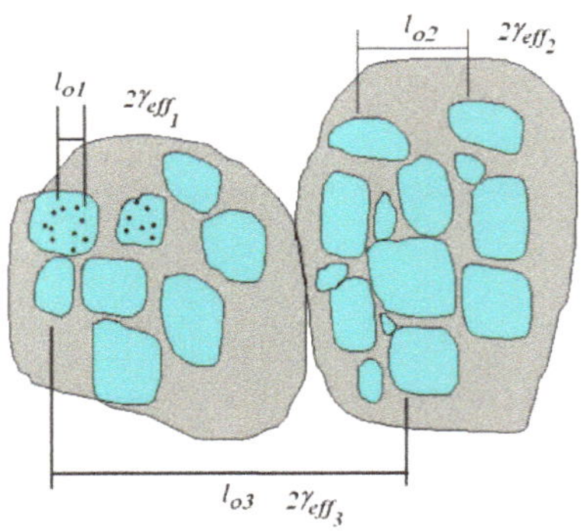

Figure 15. Microstructural aspects of the observation scale with different l_0 ruler sizes, for the fractal scaling of fracture [43].

$$2\gamma_{ef1}(2 - H_1) l_{o1}^{H_1 - 1} = 2\gamma_{eff2}(2 - H_2) l_{o2}^{H_2 - 1} = const, \tag{104}$$

although $l_{01} \neq l_{02} \neq l_{03}$ and $2\gamma_{eff1} \neq 2\gamma_{eff2} \neq 2\gamma_{eff3}$. So, the constant does not depend on the single rule of measurement l_0 used in the fractal model, but it depends on the kind of material used in the testing.

Another interpretation of equation (102) can be made by splitting the elastic and plastic terms,

$$J_0 = \underbrace{2\gamma_e(2-H)\left(\frac{l_0}{\Delta L_0}\right)^{H-1}}_{elastic} + \underbrace{\gamma_p(2-H)\left(\frac{l_0}{\Delta L_0}\right)^{H-1}}_{plastic}, \tag{105}$$

For the particular situation where $J_0 = J_{IC}$ and $\Delta L_0 = \Delta L_{0C}$, it can be derived from equation (97),

$$J_{IC} = \left(2\gamma_e + \gamma_p\right)\left(2-H\right)\left(\frac{l_0}{L_{0C}}\right)^{H-1} \tag{106}$$

and from equation (72),

$$K_{IC} = \sqrt{(2\gamma_e + \gamma_p)E(2-H)\left(\frac{l_o}{L_{oc}}\right)^{H-1}} \tag{107}$$

Therefore, using the fact that once the experimental value of J_{IC} is determined and the fitting of *J-R* curve has already yielded the values $2\gamma_e + \gamma_p, l_0$ and H for the material, the value L_{0C} can be calculated.

Fracture Mechanics science was originally developed for the study of isotropic situations and homogeneous bodies.

At the microscopic level, the elastic material is modeled considering Einstein's solid harmonic approximation where Hooke's law is employed for the force between the chemical bonds of the atoms or molecules [48]. Therefore, the elastic theory is used to make linear approximations and it does not involve micro structural effects of the material.

At the mesoscopic level the equation of energy used for the fracture does not take into account effects at the atomic scale involving non-homogeneous situations [47]. Based on the arguments of the last paragraphs, it becomes clear why Herrmman *et al.* [49] needed to include statistical weights, as a crack growth criterion, for the break of chemical bonds in fracture simulations, as a form of portraying micro structural aspects of the fracture (defects) when using finite difference and finite element methods in computational models.

At the macroscopic level, on the other hand, Griffith's theory uses a thermodynamic energy balance. It is important to remember that the linear elastic theory of fracture developed by Irwin and Westergaard and the Griffith's theory are differential theories for the macroscopic scale, which means they are punctual in their local limit. These two approaches involve the micro structural aspects of the fracture, since they take a larger infinitesimal local limit than the linear elastic theory at the atomic and mesoscopic scales. This infinitesimal macroscopic scale is big enough to include 10^{15} particles as the lower thermodynamic limit, where the physical quantity Fracture Resistance (*J-R* Curve) portrays aspects of the interaction of the crack with the microstructure of the material.

In this chapter, Classical Fracture Mechanics was modified directly using fractal theory, without taking into account more basic formulations, such as the interaction force among particles, or Lamé's energy equation in the mesoscopic scale as a form to include the ruggedness in the fracture processes.

The use of the fractality in the fracture surface to quantify the physical process of energy dissipation was approached with two different proposals. The first was given by Mu and Lung [26, 37], who proposed a phenomenological exponential relation between crack length and the elastic energy released rate in the following form

$$G_{IC} = G_{I0}\varepsilon^{1-D}, \tag{108}$$

where ε is the length of the measurement rule. The second proposal was given by Mecholsky *et al.* [24] and Mandelbrot *et al.* [23], who suggested an empirical relation between the fractional part of the fractal dimension D^* and fracture toughness K_{IC},

$$K_{IC} \sim A\left(D^*\right)^{1/2} \tag{109}$$

where $A = E_0\sqrt{l_0}$ is a constant and E_0 is the stiffness modulus and l_0 is a parameter that has a unit length (an atomic characteristic length). The elastic energy released rate is then given by,

$$G_0 = El_0D^* \tag{110}$$

where $G_{0C} = K_{IC}^2 / E$ is the critical energy released rate.

The authors cited above used the Slit Island Method in their measurements of the fractal dimension D and it is important to emphasize that both proposals have plausible arguments, in spite of their mathematical differences. Observe that in the proposal of Mu and Lung [26, 37] the fractal dimension appears in the exponent of the scale factor, while in the proposal of Mecholsky *et al.* [24] and Mandelbrot *et al.* [23] the fractal dimension appears as a multiplying term of the scale factor.

The mathematical expression proposed in this work, equation (93) and equation (97), for the case $J_0 \equiv G_0$, is compatible with the two proposals above and can be seen as a unification of these two different approaches in a single mathematical expression. In other words, the two previous proposals are complementary views of the problem according to the expression deduced in this chapter.

A careful experimental interpretation must be done from results obtained in a *J-R* curve test. The authors mentioned above worked with the concept of G, valid for brittle materials, and not with the concept of J valid for ductile materials. The experimental results show that for the case of metallic materials the fitting with their expressions are only valid in the initial development of the crack because of the self-similar limit, while self-affinity is a general characteristic of the whole fracture process [39].

The plane strain is a mathematical condition that allows defining a physical quantity called K_{IC}, which doesn't depend on the thickness of the material. The measure of an average crack size along the thickness of the material, according to ASTM E1737-96 [15], is taken as an average of the crack size at a certain number of profiles along the thickness. In this way, any self-affine profile, among all the possible profiles that can be obtained in a fracture surface, are statistically equivalent to each other, and give a representative average for the Hurst exponent.

The crack height (corresponding to the opening crack test CTOD) follows a power law with the scales, $\varepsilon_h = \varepsilon_v = \varepsilon = l_0 / \Delta L_0$ and can be written as,

$$\frac{\Delta H_0}{h_0} = \left(\frac{\Delta L_0}{l_0} \right)^{1-H} \tag{111}$$

This relation shows that, while the measurement of the number of units of the crack length $N_h = \Delta L_0 / l_0$ in the growth direction grows linearly, the number of units of the crack height units $N_v = \Delta H_0 / l_0$ grows with a power of $1 - H$. If it is considered that the inverse of the number of crack increments in the growth direction $N_h^{-1} = l_0 / \Delta L_0$ is also a measure of strain of the material, as the crack grows, and considering that the number of crack height increments can be a measure of the amount of the piling up dislocation, in agreement with equation (111), then the normal stress is of the type [44, 45]

$$\sigma \sim \varepsilon^{-H} \tag{112}$$

Observe that this relation shows a homogeneity in the scale of deformations, similar to the power law hardening equation [34]. This shows that the fractal scaling of a rugged fracture surface is related to the power law of the hardening. It is possible that the fractality of the rugged fracture surface is a result of the accumulation of the pilled up dislocations in the hardening of the material before the crack growth.

In all three situations (metallic, polymer and ceramic) the presence of microvoids, or other microstructural defects, cooperate with the formation of ruggedness on the fracture surface. This ruggedness on the way it was modeled records the "history" of crack growth being responsible for the difficulty encountered by the crack to propagate, thus defining the crack growth resistance. In EPFM literature, the rising of *J-R* curve for a long time has been associated with the interposition of plane stress and plane strain conditions generating the unique morphology of the fracture surface ruggedness [1, 2]. In metals this rising has been associated with the growth and coalescence of microvoids [2]. However, the Fractal EPFM has proposed that the morphology of the fracture surface, characterized by parameters of fractal geometry, explains in a simple and direct way the rising of the *J-R* curves.

The success of fracture fractal modeling between the *J-R* curve and the exponent H can be attributed to the following fact: a fracture occurs only after a process of hardening in the material, even minimal. Such a process follows a power law [35], self-similar [33], of the stress applied, σ with the strain ε, as shown in equation (166). It is therefore possible to associate the *elasto-pla tic energy released rate* J which is an energetic quantity with the applied stress σ, which is an *energy density*, and the *fracture length* L_0 with strain, and $\varepsilon = \Delta l / l$ and the ruggedness exponent H with the strain hardening exponent "n" [15]. As the strain hardening occurs before the onset of crack growth, it is evident that its physical result appears registered in the fracture surface in terms of ruggedness, created in the process of crack growth. This process of crack growth admits a fractal scaling in terms of the projected surface L_0, so it is possible that the effect of its prior work hardening is responsible for the further self-affinity of fracture valid at the beginning of crack growth. This is because in the limit of the beginning of crack growth, the fractal scaling relationship is a self-similar power law, analogous to the power law hardening relationship [8, 33].

The technical standards ASTM E813 [40] and ASTM E1737-96 [15] suggest an exponential fitting of the type

$$J_0 = C_1 \Delta L_0^{\ C_2} \tag{113}$$

for the *J-R* curves. They do not supply any explanation for the nature of the coefficients for this fitting. However, by comparing equation (113) with equation (97), it can be concluded that $C_1 = 2\gamma_{eff}(2-H) l_C^{F-1}$ and $C_2 = 1-H$, which explains the physical nature of this parameters;

8. Conclusions

The theory presented in this chapter introduces fractal geometry (to describe ruggedness) in the formalism of classical EPFM. The resulting model is consistent with the experimental results, showing that fractal geometry has much to contribute to the advance of this particular science.

It was shown that the rising of the *J-R* curve is due to the non-linearity in Griffith-Irwin-Orowan's energy balance when ruggedness is taken into account. The idea of connecting the morphology of a fracture with physical properties of the materials has been done by several authors and this connection is shown in this chapter with mathematical rigor.

It is important to emphasize that the model proposed in this chapter illuminates the nature of the coefficients for the fitting proposed by the fractal model, which is the true influence of ruggedness in the rising of the *J-R* curve. The application of this model in the practice of fracture testing can be used in future, since the techniques for obtaining the experimental parameters, l_0, H, and γ_{eff} can be accomplished with the necessary accuracy.

The method for obtaining the *J-R* curves proposed in this chapter does not intend to substitute the current experimental method used in Fracture Mechanics, as presented by the ASTM standards. However, it can give a greater margin of confidence in experimental results, and also when working with the microstructure of the materials. For instance, in search of new materials with higher fracture toughness, once the model explains micro and macroscopically the behavior of *J-R* curves.

It is well known that the fracture surfaces in general are multifractal objects [9] and the treatment presented here applies only to monofractals surfaces. However, for purposes of demonstrating the ruggedness influence on the phenomenology of Fracture Mechanics, through the models presented in this chapter, the obtained results were satisfactory. The generalization by multifractality is a matter to be discussed in future work.

Author details

Lucas Máximo Alves
GTEME – Grupo de Termodinâmica, Mecânica e Eletrônica dos Materiais,
Departamento de Engenharia de Materiais, Universidade Estadual de Ponta Grossa, Uvaranas,
Ponta Grossa – PR, Brazil

Luiz Alkimin de Lacerda
LACTEC – Instituto de Tecnologia para o Desenvolvimento, Departamento de Estruturas Civis,
Centro Politécnico da Universidade Federal do Paraná, Curitiba – PR, Brazil

9. References

[1] Kraff, J.M.; Sullivan, A.M.; Boyle, R.W. (1962) Effect of Dimensions on Fast Fracture Instability of Notched Sheets. In: Proceedings of the Cracks Propagation Symposium Cranfield. England: The College of Aeronautics, Cranfield. 1: pp.8-28.

[2] Ewalds, H.L.; Wanhill, R.J.H. (1986) Fracture Mechanics. Netherlands: Delftse Uitgevers Maatschappij, Third Edition, Co-Publication of Edward Arnold Publishers, London 1993.

[3] Hübner, H.; Jillek, W. (1977) Subcritical Crack Extension and Crack Resistance In Polycrystaline Alumina. J. Mater. Sci. 12(1): 117-125.

[4] Swanson, P.L.; Fairbanks, C.J.; Lawn, B.R.; Mai, Y-M.; Hockey, B.J. (1987) Crack-Interface Grain Bridging as a Fracture Resistance Mechanism In Ceramics: I, Experimental Study on Alumina, J. Am. Ceram. Soc. 70(4): 279-289.

[5] Mandelbrot, B.B. (1982) The Fractal Geometry of Nature, San Francisco, Cal-USA, New York: W. H. Freeman and Company.

[6] Underwood, E.E.; Banerji, K. (1992) Quantitative Fractography,. Engineering Aspectes of Failure and Failure Analysis. In: ASM - Handbook Fractography - The Materials Information Society. ASTM 1996. 12: pp. 192-209

[7] Dauskardt, R. H.; Haubensak, F.; Ritchie, R.O. (1990) On the Interpretation of the Fractal Character of Fracture Surfaces; Acta Metall. Matter. 38(2): 143-159.

[8] Borodich, F. M. (1997) Some Fractal Models of Fracture. J. Mech. Phys. Solids. 45(2): 239-259.

[9] Xie, H.; Wang, -A.; Stein, E. (1998) Direct Fractal Measurement and Multifractal Properties of Fracture Surfaces, Physics Letters A. 242: 41-50.

[10] Herrmann, H.J.; Stéphane, R. (1990) Statistical Models For the Fracture of Disordered Media, Random Materials and Processes. In: Series Editors: H. Eugene Stanley and Etienne Guyon editors. Amsterdam: North-Holland.

[11] Rodrigues, J.A.; Pandolfelli, V.C (1998) Insights on the Fractal-Fracture Behaviour Relationship. Materials Research. 1(1): 47-52.

[12] Mecholsky, J. J.; Passoja, D.E.; Feinberg-Ringel, K.S. (1989) Quantitative Analysis of Brittle Fracture Surfaces Using Fractal Geometry, J. Am. Ceram. Soc. 72(1): 60-65.

[13] Tanaka, M. (1996) Fracture Toughness and Crack Morphology in Indentation Fracture of Brittle Materials. Journal of Materials Science. 31: 749-755.

[14] Xie, H. (1989) The Fractal Effect of Irregularity of Crack Branching on the Fracture Toughness of Brittle Materials. International Journal of Fracture. 41: 267-274.

[15] ASTM E1737 (1996) Standard Test Method For J-Integral Characterization of Fracture Toughness. pp.1-24.

[16] Alves, L.M. (2005) Fractal Geometry Concerned with Stable and Dynamic Fracture Mechanics. Journal of Theoretical and Applied Fracture Mechanics. 44(1): 44-57.

[17] Williford, R. E. (1990) Fractal Fatigue. Scripta Metallurgica et Materialia. 24: 455-460.

[18] Chelidze, T.; Gueguen, Y. (1990) Evidence of Fractal Fracture, (Technical Note) Int. J. Rock. Mech Min. Sci & Geomech Abstr. 27(3): 223-225.

[19] Dos Santos, S.F. (1999) Aplicação do Conceito de Fractais para Análise do Processo de Fratura de Materiais Cerâmicos, Dissertação de Mestrado, Universidade Federal de São Carlos, São Carlos.

[20] Alves, L.M.; Silva, R.V.; Mokross, B.J. (2001) The Influence of the Crack Fractal Geometry on the Elastic Plastic Fracture Mechanics. Physica A: Statistical Mechanics and Its Applications. 295(1/2): 144-148.

[21] Mandelbrot, B.B. (1977) Fractals: Form Chance and Dimension, San Francisco, Cal-USA: W. H. Freeman and Company.

[22] Passoja, D.E.; Amborski, D.J. (1978) In Microsstruct. Sci. 6: 143-148.

[23] Mandelbrot, B.B. Passoja, D.E.; Paullay, A.J. (1984) Fractal Character of Fracture Surfaces of Metals, Nature (London), 308 [5961]: 721-722.

[24] Mecholsky, J.J.; Mackin, T.J.; Passoja, D.E. (1988) Self-Similar Crack Propagation In Brittle Materials. In: Advances In Ceramics, Fractography of Glasses and Ceramics, the American Ceramic Society, Inc. J. Varner and V. D. Frechette editors. Westerville, Oh: America Ceramic Society 22: pp. 127-134.

[25] Rodrigues, J.A.; Pandolfelli, V.C. (1996) Dimensão Fractal e Energia Total de Fratura. Cerâmica 42(275).

[26] Mu, Z.Q.; Lung, C.W. (1988) Studies on the Fractal Dimension and Fracture Toughness of Steel, J. Phys. D: Appl. Phys. 21: 848-850.

[27] Gong, B.; Lai, Z.H. (1993) Fractal Characteristics of *J-R* Resistance Curves of Ti-6Al-4V Alloys, Eng. Fract. Mech. 44(6): 991-995.

[28] Yavari, A. (2002) The Mechanics of Self-Similar and Self-Afine Fractal Cracks, Int. Journal of Fracture. 114: 1-27.

[29] Borodich, F. M. (1994) Fracture energy of brittle and quasi-brittle fractal cracks. Fractals in the Natural and Applied Sciences(A-41), Elsevier, North-Holland, 61–68.
[30] Carpinteri, A.; Chiaia, B. (1996) Crack-Resistance as a Consequence of Self-Similar Fracture Topologies, International Journal of Fracture, 76: 327-340.
[31] Bouchaud, E.; Bouchaud, J.P. (1994) Fracture Surfaces: Apparent Roughness, Relevant Length Scales, and Fracture Toughness. Physical Review B, 50(23): 17752–17755.
[32] Mosolov, A.B.; Borodich, F.M. (1992) Fractal Fracture of Brittle Bodies During Compression, Sovol. Phys. Dokl., May. 37(5): 263-265.
[33] Mosolov, A.B. (1993) Mechanics of Fractal Cracks In Brittle Solids, Europhysics Letters, 10 December. 24(8): 673-678.
[34] Anderson, T.L. (1995) Fracture Mechanics, Fundamentals and Applications. CRC Press, 2th Edition.
[35] Kanninen, M.F.; Popelar, C.H. (1985) Advanced Fracture Mechanics, the Oxford Engineering Science Series 15, Editors: A. Acrivos, et al. Oxford: Oxford University Press. Chapter 7, p. 437.
[36] Cherepanov, G.P.; Balankin, A.S.; Ivanova, V.S. (1995) Fractal fracture mechanics–A review. Engineering Fracture Mechanics, 51(6): 997-1033.
[37] Lung, C.W.; Mu, Z.Q. (1988) Fractal Dimension Measured with Perimeter Area Relation and Toughness of Materials, Physical Review B, 38(16): 11781-11784.
[38] Lei, W.; Chen, B. (1995) Fractal Characterization of Some Fracture Phenomena, Eng. Fract. Mechanics. 50(2): 149-155.
[39] Mandelbrot, B.B. (1991) Self-affine Fractals and Fractal Dimension. In: Family, Fereydoon. and Vicsék, Tamás editors. Dynamics of Fractal Surfaces. Singapore: World Scientific. pp.19-39.
[40] ASTM E813, (1989) Standard Test Method For J_{Ic}, A Measure of Fracture Toughness.
[41] Mazzei, A.C.A. (1999) Estudo sobre a determinação de curva-R de compósitos cerâmica-cerâmica. Tese de Doutorado, DEMA-UFScar.
[42] ASTM D6068 - 10 (2002) Standard Test Method for Determining *J-R* Curves of Plastic Materials, crack growth resistance, fracture toughness, JR curves, plastics, 96.
[43] Alves, L.M.; Da Silva, R.V.; De Lacerda, L.A. (2010) Fractal Modeling of the *J-R* Curve and the Influence of the Rugged Crack Growth on the Stable Elastic-Plastic Fracture Mechanics, Engineering Fracture Mechanics, 77, pp. 2451-2466.
[44] Zaiser, M.; Grasset, F.M.; Koutsos, V.; Aifantis, E.C. (2004) Self-Affine Surface Morphology of Plastically Deformed Metals, Phys. Rev. Lett. 93: 195507.
[45] Weiss, J. (2001) Self-Affinity of Fracture Surfaces and Implications on a Possible Size Effect on Fracture Energy. International Journal of Fracture. 109: 365–381.
[46] Mishnaevsky Jr, L. (2000) Optimization of the Microstructure of Ledeburitic Tool Steels: a Fractal Approach. Werkstoffkolloquium (MPA, University of Stuttgart).
[47] Fung, Y.C. (1969) A first course in continuum mechanics. N. J: Prentice-Hall, INC, Englewood Criffs.
[48] Holian, B.L.; Blumenfeld, R.; Gumbsch, P. (1997) An Einstein Model of Brittle Crack Propagation. Phys. Rev. Lett. 78: 78–81, DOI: 10.1103/PhysRevLett.78.78.

[49] Herrmann, H.J., Kertész, J.; De Arcangelis, L. (1989) Fractal Shapes of Deterministic Cracks, Europhys. Lett. 10(2): 147-152.

Section 2

Fracture of Biological Tissues

Chapter 4

Fracture of Dental Materials

Karl-Johan Söderholm

Additional information is available at the end of the chapter

1. Introduction

Finding a material capable of fulfilling all the requirements needed for replacing lost tooth structure is a true challenge for man. Many such restorative materials have been explored through the years, but the ideal substitute has not yet been identified. What we use today for different restorations are different metals, polymers and ceramics as well as combinations of these materials. Many of these materials work well even though they are not perfect. For example, by coating and glazing a metal crown shell with a ceramic, it is possible to make a strong and aesthetic appealing crown restoration. This type of crown restoration is called a porcelain-fused-to-metal restoration, and if such crowns are properly designed, they can also be soldered together into so called dental bridges. The potential problem with these crowns is that the ceramic coating may chip with time, which could require a complete remake of the entire restoration. Another popular restorative material consists of a mixture of ceramic particles and curable monomers forming a so called dental composite resin. These composites resins can be bonded to cavity walls and produce aesthetic appealing restorations. A potential problem with these restorations is that they shrink during curing and sometimes debond and fracture. In addition to porcelain-fused-to-metal crowns and composites, all-ceramic and metallic restorations as well as polymer based dentures are also commonly used. These constructions have their inherent limitations too.

The reason it is difficult to make an ideal dental material is because such a material has to be biocompatible, strong, aesthetic, corrosion resistant and reasonable easy to process, properties that are difficult to find in one single material. Besides, material as well as processing costs of such a material should be relatively low in order to make the use of the material wide among all social-economical groups. That demand makes the ideal material identification process even more challenging.

Today, dentistry to a great deal is driven by aesthetic demands, restricting the selectable materials mainly to tooth colored materials. Because of that demand, dentists are moving away from traditional metallic restorations with high fracture toughness values, toward resin based composites and all-ceramic restorations with rather low fracture toughness values. Modern all-ceramic restorations consist of core structures made by fracture tough ceramics such as alumina and partly stabilized zirconia. However, the rather opaque appearance of these two ceramics often requires that they are veneered with less fracture tough but more aesthetic appealing ceramics. The use of more aesthetic appealing materials has not increased the longevity of dental restorations, but in some cases when composite resins are being used, the move toward bonded composites might have increased the way tooth structures can be preserved. The benefit of such usage is that it decreases the amount of tooth structure needed to be removed during preparation and can therefore increase the longevity of the tooth.

The intention with this chapter is to give an overview of some fundamental fracture mechanics aspect of aesthetic restorative materials such as dental ceramics and dental composite resins, as well as some fracture mechanics considerations related to the way ceramics and composites are bonded to the tooth via a cement/adhesive. However, before addressing these man-made materials, the two most important dental materials, the biologically developed materials, enamel and dentin, will be discussed. An insight into the fracture mechanics of these two substrates clearly shows how sophisticated Nature was when these two biologic materials evolved. An understanding of enamel and dentin shows quite clearly where the limitations and short-comings are with the man-made dental materials, and may help us in developing better restorative dental materials.

2. The tooth

Nature provided animals and humans with teeth to be used for digesting food, but also as tools for hunting and self-defense. To fulfill these functions, Nature developed enamel to become the hardest biological tissue. Tooth enamel ranks 5 on Mohs hardness scale, where steel is ranked 4.5 and thus slightly softer than enamel. Its Young's modulus is 83 GPa, which falls between aluminum (69 GPa) and bronze (96-120 GPa)[1]. The enamel can be described as the whitish looking shell covering the visible part of a tooth positioned in the alveolar socket (Figure 1).

Regarding enamel and dentin, the first hard tissue to form is dentin, produced by newly differentiated odontoblasts. The first formed dentin layer is called mantle dentin and is approximately 150 μm thick and contains loosely packed coarse collagen fibrils surrounded by precipitated hydroxyapatite crystals [2]. Tiny side-branching channels oriented parallel to the dentin-enamel-junction (DEJ) and connected to the protoplasmatic extension of the odontoblasts are parts of the mantel dentin. The mantle dentin matrix is slightly less (4 vol-%) mineralized then the rest of the finally formed dentin.

As the odontoblasts move away from the DEJ, each of them leaves a cell extension protruding from the odontoblasts to the DEJ with the side-branching channels of the mantle dentin. These cell extensions may remain in contact with the DEJ during the formation of dentin as well as during the lifetime of the tooth, and they form channels through the dentin as the odontoblasts move inwards toward the pulp. The secreted collagen fibers, which are mainly oriented perpendicularly to the dentinal tubules, act as nucleisation centers for hydroxyapatite crystallites precipitating as the odontoblasts migrate inwards (Figure 2). The dentin formed by these collagen fibers represents the so called intertubular dentin (Figure 3). However, surrounding the odontoblastice processes are thin layers of collagen oriented parallel to the odontoblastic processes. These collagen layers are also mineralized and form the so called peritubular dentin, which is denser than the intertubular dentin located between the peritubular dentin tubules. An important difference between enamel and dentin is that dentin, in contrast to enamel, is a living tissue as long as the pulp is alive, while the enamel becomes a completely dead tissue as soon as the outer layer of the enamel has formed and the ameloblasts degraded.

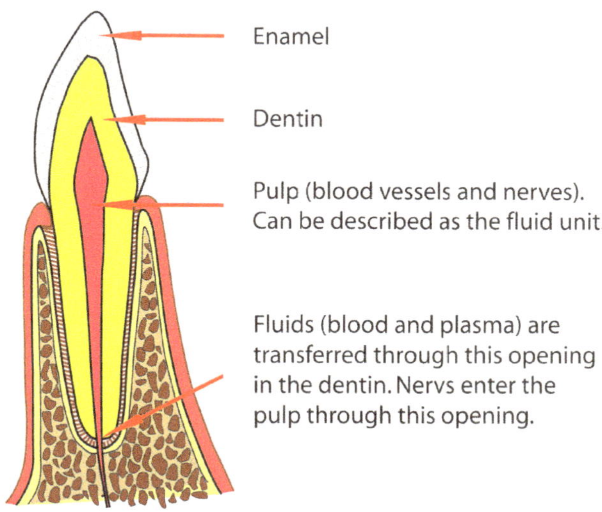

Figure 1. Drawing showing a tooth attached in its alveolar socket. The root of the tooth is attached to the alveolar socket via collagen fibers, the so called periodontal ligament. Blood vessels and nerves enter the pulp chamber via the apical opening.

The formation of mantle dentin triggers the ameloblasts to start secreting enamel proteins on the newly formed mantle dentin. The first hydroxyapatite crystals that form on the mantle dentin are randomly packed in this first formed enamel and interdigitated with the crystallites of dentin. Eventually the dentin crystallites present in the mantel dentin act as nucleation sites for the first enamel crystallites.

After the first layer of structureless enamel has formed, the ameloblasts move away from the DEJ, which permits the formation of the so called Tomes' processes, which form at the ends the ameloblasts closest to the DEJ. When the Tomes' processes are established, the enamel rods start developing (Figure 4).

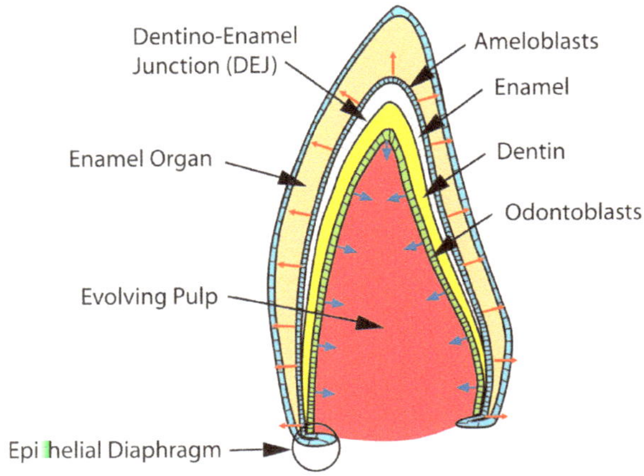

Figure 2. The hard tissues are formed by the odontoblasts (dentin) and ameloblasts (enamel). During the development of the tooth, epithelial cells have formed a bell shaped enamel organ. Inside that bell is connective tissue that shows active budding of capillaries. At a certain stage, the fibroblasts in contact with the epithelium bell become highly differentiated and develop into odontoblasts and form the first layer of dentin. That layer stimulates the epithelium cells in contact with the dentin at the DEJ to differentiate into ameloblasts and form enamel. As a consequence, the two cell types move in opposite direction as they form dentin (blue arrows) and enamel (red arrows). When the ameloblasts reach the outer cells of the enamel organ they start degrading and lose vitality. At the same time, the dentin has increased in thickness and the epithelial diaphragm with odontoblasts have grown downwards and developed the root and the pulp chamber (see Figure 1).

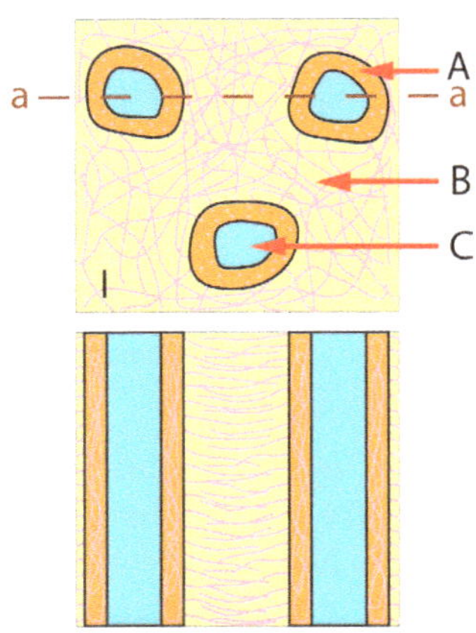

Figure 3. The top drawing represents a cross-section of dentin, perpendicular to the peritubular dentin (A). The lower drawing represents a plane parallel to the odontoblastic processes (C) and cut along a-a. In the peritubular dentin, collagen fibers represented by pink lines are present parallel to the odontoblastic processes. Hydroxyapatite precipitate along these fibers, and together they form the so called peritubular dentin (A). Collagen precipitates perpendicular to the odontoblastic processes too, and when hydroxyapatite precipitate in that matrix, the intertubular dentin (B) is formed.

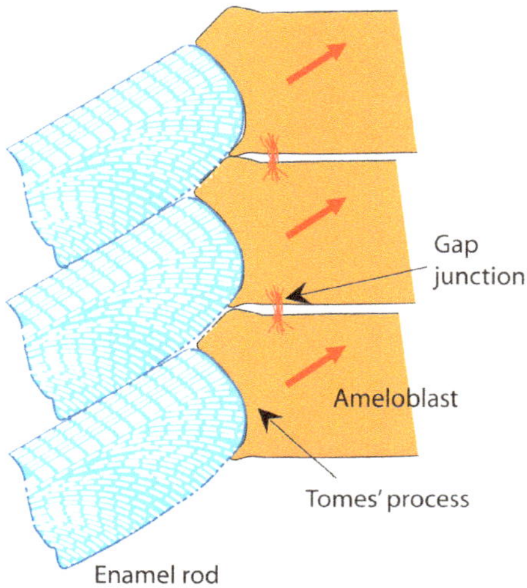

Figure 4. After the first layer of structureless enamel has formed on the mantel dentin, the ameloblast differentiate its end closest to the precipitated enamel into the so called Tomes' process. This unit can be described as a concave formation from which hydroxyapatite crystallites precipitate. The c-axis of these crystallites are perpendicular to the surface of Tomes's process, explaining the the well organized precipitation of the hydroxyapatite crystallites in each enamel rod.

The secretion from the peripheral site of the Tomes' process results in the formation of what is referred to as the enamel matrix wall. These walls enclose pits into which the Tomes' processes fit. These sites are then filled with matrix proteins acting as nucleating agents for the hydroxylapatite crystallites. The crystallites that precipitate in these two matrices (the matrix wall and the central pit) have different orientation. It is important to emphasize that the final wall and pit enamel have the same composition. The only difference is the orientation of the crystallites in these two enamel types.

A cross-section of the enamel rods reveals that the individual rods have a key-hole shaped structure (Figure 5).

As the ameloblasts move toward their final destiny, they produce enamel rods that are somewhat wavy and interwoven (Figure 6). Independent on these waves, the enamel rods form angles that are roughly perpendicular to the outer as well as inner surfaces of the enamel shell (Figure 7). The hard enamel can be described as a hard shield protecting the underlying dental tissue of the visible part of the tooth.

Enamel consists mainly of hydroxyapatite crystallites, which are oriented in very well organized larger bundles of crystallites. These larger bundles are referred to as enamel rods. Each enamel rod is made by enamel forming cells, the so called ameloblasts. The diameters of the rods range from 4-8 μm. During enamel formation, the ameloblasts secret different proteins (amelogenins and enamelins), which act as nucleating agents for the hydroxyl apatite crystallites. During enamel formation, the ameloblasts move from the dentin-enamel junction (DEJ) to the surface of the final enamel crown. When the enamel shell has reached its final shape, the ameloblasts degenerate and die, explaining why mature enamel is a non-vital tissue made up by ~85 vol-% hydroxyapatite and 15 vol-% proteins and water [2].

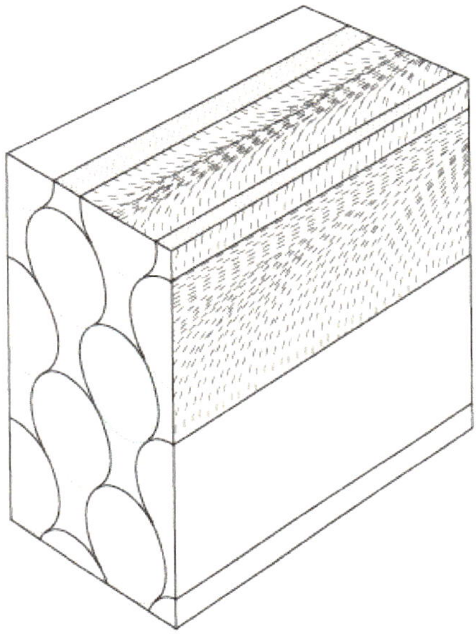

Figure 5. Cross section of enamel rods shows the key-hole structutre (blue). The longitudinal orientation of the hydroxyapatite crystallites can be explained by considering how Tomes' process controls the crystallite orientation (Figure 4). Figure redrawn after [3].

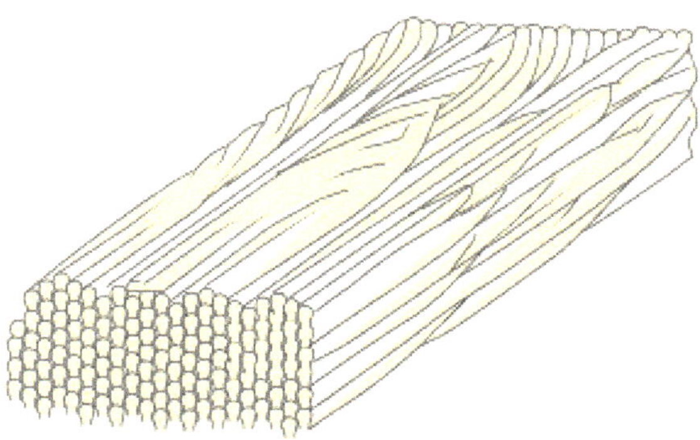

Figure 6. The keyhole shaped rods become more and more interwoven as the rods approaches the DEJ. Redrawn from [4]. The intervowen structure shown in the drawing is also characteristic for the cusp tips, where that type of enamel is called "gnarled enamel".

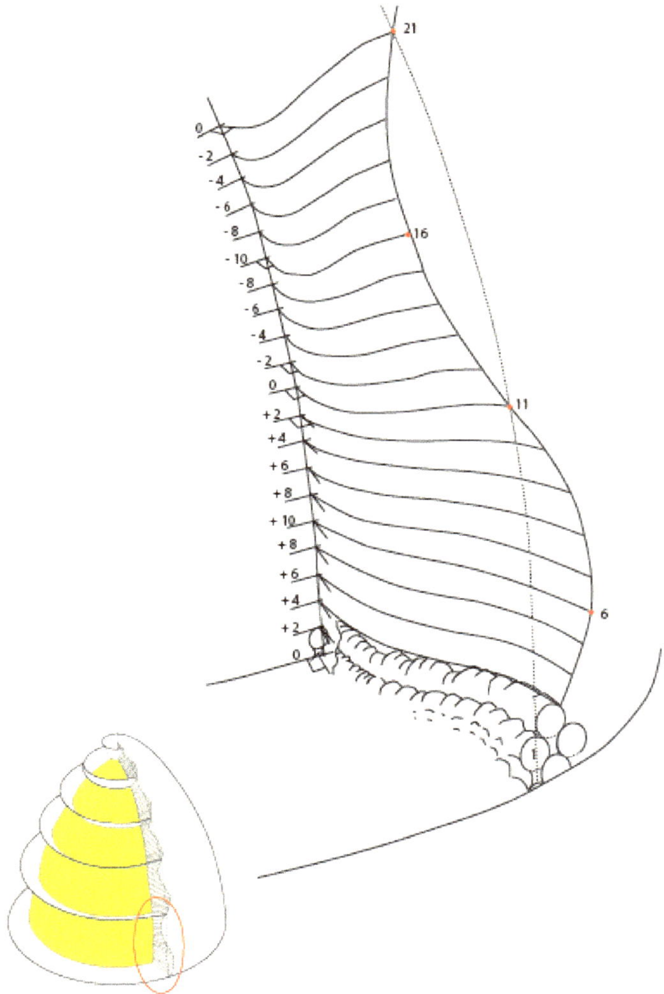

Figure 7. The bottom left drawing shows the orientation of the rods along the long axis of the tooth. As seen from both drawings, there is a continous shift in orientation resulting in the S-shaped orientation. If the line joining points 1, 11 and 21 along the S-shaped curve represents a plane forming 90 degrees to the enamel surface, it is seen from the drawing that there is a difference in rod orientation that can be described as 90 ± 10 degrees. Redrawn from [2].

The pulp chamber is a cavity inside the dentin formed by the surrounding dentin. The pulp chamber contains soft tissue, blood vessels and nerves and is lined by the odontoblasts. As the tooth grew older, the odontoblasts continue to produce dentin, causing the size of the pulp chamber to decrease with age.

As seen from the properties presented in Table 1, enamel has lower fracture toughness than dentin, but significantly higher hardness and modulus of elasticity. These properties suggest that enamel is a highly brittle material that should easily chip away from the dentin. Fortunately that is not the case. The reason can be related to a firm enamel-dentin attachment as well the sophisticated anisotropic composite structure of both enamel and dentin. If cracks propagate through the enamel, they often stop before they reach the enamel-dental interface, and if they continue propagating they usually stop when they reach the enamel-dentin interface. That explains why fractured teeth are not as common as one otherwise would expect by considering force and fatigue levels teeth have to withstand.

Hard tissue	Modulus of elasticity (GPa)	Fracture toughness (MPa $m^{1/2}$)	Hardness (GPa)
Enamel	78 ± 1 to 98 ± 4	0.44 ± 0.04 to 1.55 ± 0.29	2.83 ± 0.10 to 3.74 ± 0.48
Dentin	18 ± 1 to 22 ± 1	3.08 ± 0.33	0.53 ± 0.01 to 0.63 ± 0.03

Table 1. Highest and lowest reported values in Xu et al.'s study[5], except for the fracture toughness value of dentin which is from El Mowafy and Watts study[6]. Identified variations relate to the anisotropic nature of enamel and dentin as well as variations among teeth.

2.1. Fracture mechanical aspects of enamel

As discussed earlier, the tooth can be described as a rather complicated composite structure developed to serve the user. Nature adapted the principle that teeth must be hard and rigid in order to generate sufficiently high local stress levels. These stresses are capable of penetrating tissues during hunting and fighting, but also capable of crushing hard food. At the same time, enamel has also been designed to limit the inherent brittle nature of hydroxyapatite by dispersing propagating cracks and thereby resist some brittle failures.

By orienting the rods on the cusp tips along the axis of the tooth, a parallel model composite is formed in that region (Figure 8). At the same time, by orienting the rods more or less perpendicular to the long axis of the tooth in the remaining parts of the crown, a series model composite is formed in that part of the crown. These models are valid under the assumption the load is in an axial direction. Since the parallel model results in a stiffer combination than a series model material, the tooth has been designed so that rigidity is optimized in the chewing/biting direction and flexing in a direction perpendicular to that direction.

As the stiffness of the parallel model exceeds the stiffness of the series model, we can understand how such a design assists an animal attacking another animal. During such an attack, the canines of the attacking animal may penetrate the tissue of the attacked animal, but that bite may not necessarily result in an instant kill. During the biting action, the attacking animal benefits from the stiffness of the canines (parallel model behavior of the tip of the canines make the tooth stiff like a steel arrow). However, if the attacked animal was not killed instantaneously, most likely it will try to get loose from the attacking animal's jaws. During that attempt the risk of fracturing the canines of the attacking animal increases.

However, thanks to the rod and tubule orientations in the cervical and mid crown regions, the material characteristics of the enamel in these bendable regions are represented by the series model, thereby allowing the tooth to flex somewhat and absorb mechanical energy rather than fracture.

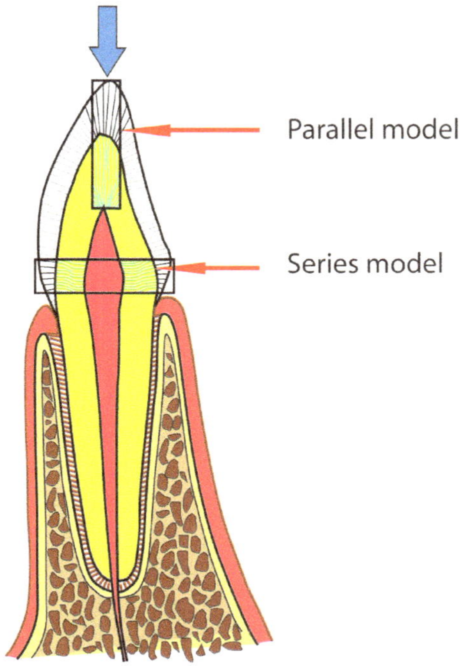

Figure 8. A tooth loaded in axial direction (blue arrow) responds in two ways. On the cusp tips, the rods and the tubules are oriented parallel with the load and resulting in a material which modulus can be predicted from a parallel model prediction. In the cervical region of the tooth, the modulus can be predicted from a series model.

Some basic science studies have been conducted to study the fracture behavior of enamel and dentin through the years. In one such study [7] the investigators studied a mandibular molar tooth restored with different Class II amalgam preparations. By use of finite element analysis, the stress distribution induced along the internal edges as a result of occlusal loading was calculated, and by use of Paris law the cyclic crack growth rate of sub-surface flaws located along the dentinal internal edges was determined. Based on the assumptions used in their calculations, they claimed that flaws located within the dentin along the buccal and lingual internal edges can reduce the fatigue life of restored teeth significantly. Sub-surface cracks as short as 25 μm were capable of promoting tooth fracture well within 25 years from the time of restoration placement. Furthermore, cracks longer than 100 μm reduced the fatigue life of the tooth to less than 5 years. Consequently, sub-surface cracks introduced during cavity preparation with conventional dental burs may serve as a principal source for premature restoration failure.

As the hardest and one of the most durable load bearing tissues of the body, enamel has attracted considerable interest from both material scientists and clinical practitioners due to its excellent mechanical properties. In a recent article [8] possible mechanisms responsible for the excellent mechanical properties of enamel were explored and summarized. What these authors emphasized was the hierarchical structure and the nanomechanical properties of the minor protein macromolecular components. The experimental and numerical results supported the made assumptions. For example, enamel showed to have lower elastic modulus, higher energy absorption ability and greater indentation creep behavior than sintered hydroxyapatite material. These findings suggest that the structural and compositional characteristics of the minor protein component significantly regulate the mechanical properties of enamel in order to better match its functional needs.

The fascinating aspect of enamel is that its structure seems to have evolved and adapted to the need of the user of the teeth. For example, in some recent publications [9, 10], these issues have been discussed. Lucas et al. [9] proposed a model based on how fracture and deformation concepts of teeth may be adapted to the mechanical demands of diet, while Constantino et al [10] used that model by examining existing data on the food mechanical properties and enamel morphology of great apes (Pan, Pongo, and Gorilla). They paid particular attention to whether the consumption of fallback foods plays a key role in influencing great ape enamel morphology. Their results suggest that so is the case, and that their findings may explain the evolution of the dentition of extinct hominins.

Along these lines, Lee et al.[11] did a comparative study of human and great ape molar tooth enamel. They used nano-indentation techniques to map profiles of elastic modulus and hardness across sections from the enamel–dentin junction to the outer tooth surface. The measured data profiles overlapped between species, suggesting a degree of commonality in material properties. Using established deformation and fracture relations, critical loads to produce function-threatening damage in the enamel of each species were calculated for characteristic tooth sizes and enamel thicknesses. The results suggest that differences in load-bearing capacity of molar teeth in primates are less a function of underlying material properties than of morphology.

From the above studies, it is quite clear that Nature has adapted the structure of enamel to resist fractures. In a study by Bajaj [4] the crack growth resistance behavior and fracture toughness of human tooth enamel was determined. The results were quantified using incremental crack growth measures and conventional fracture mechanics. The results revealed that enamel undergoes an increase in crack growth resistance (i.e. rising R-curve) with crack extension from the outer to the inner enamel, and that the rise in toughness is a function of distance from the dentin enamel junction (DEJ). The outer enamel exhibited the lowest apparent toughness (0.67± 0.12 MPa $m^{0.5}$), and the inner enamel exhibited a rise in the growth toughness from 1.13 MPa $m^{0.5}$/mm to 3.93 MPa $m^{0.5}$/mm. The maximum crack growth resistance at fracture (i.e. fracture toughness (K_C)) ranged from 1.79 to 2.37 MPa $m^{0.5}$.

Crack growth in the inner enamel was accompanied by a host of mechanisms operating from the micro- to the nano-scale. Decussation in the inner enamel promoted crack deflection and twist, resulting in a reduction of the local stress intensity at the crack tip (Figures 6 and 7). In addition, extrinsic mechanisms such as bridging by unbroken ligaments of the tissue and the organic matrix promoted crack closure. Micro-cracking due to loosening of prisms was also identified as an active source of energy dissipation. The unique microstructure of enamel in the decussated region promotes crack growth toughness that is approximately three times that of dentin and over ten times that of bone.

In addition to the micro- and nano-structure of enamel, the tooth anatomy by itself is such that it has adapted to force conditions present in the oral cavity. Anderson et al. [12] modeled what they believed drove the initial evolution of the cingulum. Recent work on physical modeling of fracture mechanics has shown that structures which approximate mammalian dentition (hard enamel shell surrounding a softer/tougher dentine interior) undergo specific fracture patterns dependent on the material properties of the food items [9, 13]. Soft materials result in fractures occurring at the base of the stiff shell away from the contact point due to heightened tensile strains. These tensile strains occur around the margin in the region where cingula develop. In Anderson et al.'s [12] study, they tested whether the presence of a cingulum structure would reduce the tensile strains seen in enamel using basic finite element models of bilayered cones. Finite element models of generic cone shaped "teeth" were created both with and without cingula of various shapes and sizes. Various forces were applied to the models to examine the relative magnitudes and directions of average maximum principal strain in the enamel. The addition of a cingulum greatly reduces tensile strains in the enamel caused by "soft-food" forces. The relative shape and size of the cingulum has a strong effect on strain magnitudes as well. Scaling issues between shapes are explored and show that the effectiveness of a given cingulum to reducing tensile strains is dependent on how the cingulum is created. Partial cingula, which only surround a portion of the tooth, are shown to be especially effective at reducing strain caused by asymmetrical loads, and shed new light on the potential early function and evolution of mammalian dentitions.

2.2. Fracture mechanical aspects of dentin

Dentin is not as brittle as enamel. However, considering that enamel rests on dentin, and that cracks may propagate through the enamel, it is important to understand the fracture mechanics of dentin.

Human dentin is known to be susceptible to failure under repetitive cyclic fatigue loading. Nalla et al. [14] addressed the paucity of fatigue data through a systematic investigation of the effects of prolonged cyclical loading on human dentin. They performed the evaluations in an environment of ambient temperature and where the dentin was kept in a Hank's balanced salt solution. The results they got were discussed in the context of possible mechanisms of fatigue damage and failure. The stiffness loss data collected were used to deduce crack velocities and the thresholds for such cracking. They concluded that the presence of small (on the order of 250 μm) incipient flaws in human dentin will not radically affect their useful life as Arola et al.[7] claimed.

Kruzic et al. [15] investigated the fracture toughness properties of dentin in terms of resistance-curve (R-curve) behavior, i.e., fracture resistance increase with crack extension. Of particular interest was the identification of relevant toughening mechanisms involved in the crack growth. Their study was conducted on elephant dentin, and they compared hydrated and dehydrated dentin. Crack bridging by uncracked ligaments, observed directly by microscopy and X-ray tomography, was identified as a major toughening mechanism. Further experimental evidence were provided by compliance-based experiments. In addition, with hydration, dentin was observed to display significant crack blunting leading to a higher overall fracture resistance than in the dehydrated material. In this paper they show how uncracked bridges remain behind the propagating crack, giving the dentin some fracture toughness.

Bajaj et al. [16] used striations resulting from fatigue crack growth in the dentin of human teeth to identify difference between young and old dentin. They used compact tension (CT) specimens obtained from the coronal dentin of molars from young ($17 \leq$ age ≤ 37 years) and senior (age ≥ 50 years) individuals, and exposed the dentin to cyclic Mode I loads. Striations evident on the fracture surfaces were examined using a scanning electron microscope and contact profilometer. Fatigue crack growth striations that developed in vivo were also examined on fracture surfaces of restored molars. The average spacing in the dentin of seniors (130 ± 23 μm) was significantly larger ($p < 0.001$) than that in young dentin (88 ± 13 μm). Fatigue striations in the restored teeth exhibited features that were consistent with those that developed in vitro and a spacing ranging from 59 to 95 μm. Unlike metals, the striations in dentin developed after a period of cyclic loading that ranged from 1×10^3 to 1×10^5 cycles. The study showed that the cracks tend to propagate perpendicular towards the orientation of the tubules, and climb along a plane tangential to the peritubular cuffs and then continue perpendicularly to the tubules.

Yan et al. [17] showed that rather than using a linear-elastic fracture mechanics (LEFM)(K_C) that ignores plastic deformation and tend to underestimate the fracture toughness, a plastic fracture mechanics (EPFM)(K_{JC}) approach was used. The presence of collagen (approximately 30% by volume) was assumed to enhance the toughening mechanisms in dentin. By comparing the values of the fracture toughness values estimated using either LEFM or EPFM, they found that the K_C and K_{JC} values of plane parallel as well as antiplane parallel specimens were different. The fracture toughness estimated based on K_{JC} was significantly greater than that estimated based on K_C (32.5% on average; $p<0.001$). In addition, K_{JC} of antiplane parallel specimens was significantly greater than that of in-plane parallel specimens. Consequently, in order to critically evaluate the fracture toughness of human dentin, EPFM should be employed rather than LEFM.

3. Man made dental materials

By considering the sophistication of the biological materials enamel and dentin, it is easy to understand why it is such a challenge to identify a man made material that can compete with the biological hard tissues. In addition to their mechanical properties, such a material should be biocompatible, aesthetic, corrosion resistant, easy to process and reasonable inexpensive, making such an identification extremely challenging. Of these properties, strength values within a group of materials are often used by manufacturers in their marketing and by the dentist when it comes to selecting a product. Unfortunately, strength by itself may not be the best parameter to choose. The reason is that strength is a conditional rather than an inherent material property [18]. Strength data alone should therefore not be used to extrapolate and predict the performance of a structure. Instead, they should be used together with the microstructure of the material, processing history, testing methodology, testing environment and failure mechanism. Structural failures are determined by additional failure probability variables in concert with strength that describe stress distributions, flaw size distributions, which can contribute to either single or multiple failure modes. Lifetime predictions require additional information about the time dependence of slow crack growth. Basic fracture mechanics principles and Weibull failure modeling are important to consider.

To make dental treatment even more challenging, just consider how dentists cut teeth and use different materials. During the cutting process, flaws of different sizes are most likely induced in the remaining tooth structure. Flaws and different defects are also most likely induced during handling and insertion of different materials. The impact of such flaws can be devastating for any material, particularly for brittle ceramic materials. To show how different surface treatments can affect the strength properties, Table 2 has been included to show how different surface treatments of glass can affect its strength [19]. A severely sandblasted glass lose as much as 67% of its original strength, while a drawn silica fiber tested in vacuum is 400 times stronger than the glass, a difference that can be related to the presence of water molecules in air.

Glass treatment	Strength (MPa)
Glass rods "as received" from factory	45
Severely sand blasted	14
Acid etched and lacquered	1725
Drawn silica fibers tested in vacuum	12000 – 16000

Table 2. Effect of surface treatments on the strength of glass

By use of Griffith's equation[20], one can show how the stress level is affected by flaw size and surface energy and explain the results presented in Table 2. That equation further shows that any processing step affecting the size, orientation or distribution of flaws will affect the measured strength of materials, particularly brittle materials. It also shows how environmental conditions may affect surface energy and thereby also the strength.

3.1. Fracture mechanics aspects of ceramics

Clinical experience suggests that all-ceramic crowns may not be as durable as their porcelain-fused-to-metal counterparts, particularly when placed on molar teeth. The reason relates to the brittleness of ceramics, making them prone for chipping and fracturing [21-27]. In the 1980s and 1990s, crowns were fabricated as enamel-like monoliths from micaceous glass-ceramics (Dicor, Dentsply/Caulk, Milford, DE) and high leucite porcelains (IPS Empress, Ivoclar, Schaan, Lichtenstein), but these ceramics showed unacceptably high failure rates and were soon replaced by improved ceramics [28, 29]. Subsequent crown design has focused on retention of porcelain as an aesthetic veneer fired to much stronger alumina-based ceramics, either glassinfiltrated (InCeram, Vita Zahnfabrik, Bad S.ackingen, Germany) or pure and dense (Procera, Nobel Biocare, Goteborg, Sweden) alumina, as supporting cores. Although alumina-based crowns continued to replace metal-based crowns, failure rates remained an issue [30]. During the past 15 years, ultra-strong core ceramics, e.g. yttria-stabilized zirconia (Y-TZP) and alumina-matrix composites (AMC)[31] have gained in popularity but have yet to be documented regarding their clinical long-term success.

Clinically, bulk fractures are the reported cause of all ceramic crown failure whether the crown is a monolith or a layered structure. According to a fractographic evaluation by Thompson et al.[32], in which they evaluated fractured and recovered Dicor and Cerestore crowns, they found that failures generally did not ensue from damage at the occlusal surface. Instead, for Dicor the cracks emerged from the internal surface, while in the case of Cerestore, the initiation occurred at the porcelain/core interface inside the core materials. In other studies it has been shown that radial cracks are initially contained within the inner core layer, but subsequently propagate to the core boundaries, ultimately causing irretrievable failure. This failure mechanism raises an interesting question: If the ceramic core materials are so strong, why do the cracks not originate in the weak outer porcelain? In the case of porcelain-fused-to-metal, porcelain failures do seem to occur preferentially in the porcelain, although there is some indication that such failures may be preceded by plasticity in the ductile metal [33] That in turn raises the question: What are the important material parameters that govern these failure modes in crown structures, and how may they be optimized? Maybe McLean's [33] suggestion from 1983 that layered all-ceramic crowns should perform well if the core fracture strength exceeded the yield strength of base metal alloys (about 400–500MPa for gold)

Before diving deeper into the fracture strength of the core, let us accept that there are several factors which can be associated with crack initiation and propagation in dental ceramic restorations. These factors include: (a) shape of the restoration; (b) micro-structural inhomogeneities; (c) size and distribution of surface flaws; (d) residual stresses and stress gradients, induced by polishing and/or thermal processing; (e) the environment in contact with the restoration; (f) ceramic/cement interfacial features; (g) thickness and thickness variation of the restoration; (h) elastic module of restoration components; and (i) magnitude and orientation of applied loads. The possible interactions among these variables complicate the interpretation of failure analysis observations, explaining why fracture behavior of all-ceramic crowns is rather tricky problem to understand.

Even though McLean's[34] suggestion that the core fracture strength exceeded the yield strength of base metal alloys (about 400–500MPa for gold) might be tempting to adopt to, it is very important to realize that ceramics, in contrast to metals, are brittle materials, and that strength is more of a "conditional" than an inherent material property, and strength data alone cannot be directly extrapolated to predict structural performance [18]. Strength data, particularly of brittle materials, are meaningful when placed into context via knowledge of material microstructure, processing history, testing methodology, testing environment and failure mechanism(s). Lifetime predictions require additional information about the time dependence of slow crack-growth. Basic fracture mechanics principles and Weibull failure modeling are key factors to consider as well as the role of interfacial stresses. Thus, in order to understand the actual clinical failure mode it is absolutely necessary to consider all the variables listed in the previous paragraph until results from in vitro strength testing can be considered to have any clinical value.

Natural teeth as well as most modern ceramic restorations can be described as layered structures. In the case of teeth the layers are enamel and dentin, while in the case of all ceramics a core ceramic and a porcelain coating. There are also unlayered ceramics in use, but since they are resting on a cement layer and dentin, even they can be described as layered structures. In a study by Jung et al. [35], they determined whether coating thickness and coating/substrate mismatch are key factors in the determination of contact induced damage in clinically relevant bilayer composites. They studied crack patterns in two bilayer systems conceived to simulate crown and tooth structures, at opposite extremes of elastic/plastic mismatch. In one case they looked at porcelain on glass-infiltrated alumina ("soft/hard"), and glass-ceramic on resin composite ("hard/soft"). Hertzian contacts were used to investigate the evolution of fracture damage in the coating layers, as functions of contact load and coating thickness. The crack patterns differed radically in the two bilayer systems: In the porcelain coatings, cone cracks initiate at the coating top surface; in the glass-ceramic coatings, cone cracks again initiate at the top surface, but additionally, upward-extending transverse cracks initiate at the internal coating/substrate interface, where the latter were dominant. This study revealed that the substrate has a profound influence on the damage evolution to ultimate failure in bilayer systems. It was also found that the cracks were highly stabilized in both systems, with wide ranges between the loads to initiate first cracking and to cause final failure, implying damage-tolerant structures. Finite element modeling was used to evaluate the tensile stresses responsible for the different crack types.

In a follow up study, Jung et al.[36] assumed that the lifetimes of dental restorations are limited by the accumulation of contact damage introduced during chewing, and that the strengths of dental ceramics are significantly lower after multi-cycle loading than after single-cycle loading. To test that hypothesis, they looked at indentation damage and associated strength degradation from multi-cycle contacts using spherical indenters in water. They evaluated four dental ceramics: "aesthetic" ceramics porcelain and micaceous glass-ceramic (MGC), and "structural" ceramics--glass-infiltrated alumina and yttria-stabilized tetragonal zirconia polycrystal (Y-TZP) They found that at large numbers of contact cycles, all materials showed an abrupt transition in damage mode, consisting of strongly enhanced damage inside the contact area and attendant initiation of radial cracks outside. This transition in damage mode is not observed in comparative static loading tests, attesting to a strong mechanical component in the fatigue mechanism. Radial cracks, once formed, lead to rapid degradation in strength properties, signaling the end of the useful lifetime of the material. Strength degradation from multi-cycle contacts were examined in the test materials, after indentation at loads from 200 to 3000 N up to 10^6 cycles. Degradation occurs in the porcelain and MGC after ~ 10^4 cycles at loads as low as 200 N; comparable degradation in the alumina and Y-TZP requires loads higher than 500 N, well above the clinically significant range.

In another study from the same year, Drummond et al. [37] evaluated the flexure strength under static and cyclic loading and determined the fracture toughness under static loading of six restorative ceramic materials. Their intent was primary to compare four leucite ($K_2O \bullet Al_2O_3 \bullet 4SiO_2$) strengthened feldspathic (pressable) porcelains to a low fusing feldspathic porcelain and an experimental lithium disilicate containing ceramic. All materials were tested as a control in air and distilled water (without aging) and after three months aging in air or distilled water to determine flexure strength and fracture toughness. A staircase approach was used to determine the cyclic flexure strength. The mean flexure strength for the controls in air and water (without aging or cyclic loading) ranged from 67 to 99 MPa, except the experimental ceramic that was twice as strong with mean flexure strength of 191–205 MPa. For the mean fracture toughness, the range was 1.1–1.9 MPa $m^{0.5}$ with the experimental ceramic being 2.7 MPa m $^{0.5}$. The effect of testing in water and aging for three months caused a moderate reduction in the mean flexure strength (6–17%), and a moderate to severe reduction in the mean fracture toughness (5–39%). The largest decrease (15–60%) in mean flexure strength was observed when the samples were subjected to cyclic loading. The conclusion they draw from the study was that the lithium disilicate containing ceramic had significantly higher flexure strength and fracture toughness when compared to the four pressable leucite strengthened ceramics and the low fusing conventional porcelain. All of the leucite containing pressable ceramics did provide an increase in mean flexure strength (17–19%) and mean fracture toughness (3–64%) over the conventional feldspathic porcelain. Further, the influence of testing environment and loading conditions implies that these ceramic materials in the oral cavity might be susceptible to cyclic fatigue, resulting in a significant decrease in the survival time of all-ceramic restorations.

The studies conducted by Jung et al. [35, 36] were followed up by Rhee et al. [38] who approached the onset of competing fracture modes in ceramic coatings on compliant substrates from Hertzian-like contacts. They paid special attention to a deleterious mode of radial cracking that initiates at the lower coating surface beneath the contact, in addition to traditional cone cracking and quasiplasticity in the near contact area. The critical load relations were expressed in terms of well-documented material parameters (elastic modulus, toughness, hardness, and strength) and geometrical parameters (coating thickness and sphere radius). Data from selected glass, Al_2O_3 and ZrO_2 coating materials on polycarbonate substrates were used to demonstrate the validity of the relations. The formulation provides a basis for designing ceramic coatings with optimum damage resistance.

Deng et al. [39] used spherical indenters on flat ceramic coating layers bonded to compliant substrates. They identified critical loads needed to produce various damage modes, cone cracking, and quasi-plasticity at the top surfaces and radial cracking at the lower (inner) surfaces are measured as a function of ceramic-layer thickness. The characteristic features of these were;

i. Cone cracks initiate from the top surface outside the contact circle, where the Hertzian tensile stress level reaches its maximum [40, 41]. The crack first grows downward as a shallow, stable surface ring, resisted by the material toughness T (K_{IC}), before popping into full cone geometry at load
$P_C = A(T^2/E)r$
with $A = 8.6x10^3$ from fits to data from monolithic ceramics with known toughness [42]
ii. Quasiplasticity initiates when the maximum shear stress in the Hertzian near field exceeds Y/2, with yield stress Y ~ H/3 determined by the material hardness H (load/projected area, Vickers indentation)[43].
The critical load is $P_Y = DH(H/E)^2r^2$
with D = 0.85 from fits to data for monolithic ceramics with known hardness [42].
iii. Radial cracks initiate spontaneously from a starting flaw at the lower ceramic surface when the maximum tensile stress in this surface equals the bulk flexure strength σ_F, at load
$P_R = B\sigma_F d^2/\log(E_C/E_S)$
with d being the ceramic layer thickness and B = 2.0 from data fits to well-characterized ceramic-based bilayer systems [38].

Thus, given basic material parameters, one can in principal make priori predictions of the critical loads for any given bilayer system. Note that P_C and P_Y are independent of layer thickness d, whereas P_R is independent of sphere radius r. These relations, within the limits of certain underlying assumptions, have been verified for model ceramic/substrate bilayer systems [38, 44]. They claimed that these damage modes, especially radial cracking, were directly relevant to the failure of all-ceramic dental crowns. The critical load data were analyzed with the use of explicit fracture-mechanics relations, expressible in terms of routinely measurable material parameters (elastic modulus, strength, toughness, hardness) and essential geometrical variables (layer thickness, contact radius).

Lawn et al. [45] conducted tests on model flat-layer specimens fabricated from various dental ceramic combinations bonded to dentin-like polymer substrates in bilayer (ceramic/polymer) and trilayer (ceramic/ceramic/polymer) configurations. The specimens were loaded at their top surfaces with spherical indenters, simulating occlusal function. The onset of fracture was observed in situ using a video camera system mounted beneath the transparent polymer substrate. Critical loads to induce fracture and deformation at the ceramic top and bottom surfaces were measured as functions of layer thickness and contact duration. Radial cracking at the ceramic undersurface occurred at relatively low loads, especially in thinner layers. Fracture mechanics relations were used to confirm the experimental data trends, and to provide explicit dependencies of critical loads in terms of key variables (material—elastic modulus, hardness, strength and toughness; geometric—layer thicknesses and contact radius). Tougher, harder and (especially) stronger materials show superior damage resistance. Critical loads depend strongly (quadratically) on crown net thickness. The analytic relations provided a seemingly sound basis for the materials design of next-generation dental crowns.

3.2. Fracture mechanics aspects of dental composites

Dental composite resins consist of ceramic filler particles, usually within a size range of 1-5 μm and mixed with nano-sized (20-40 nm) particles. These inorganic filler particles are silane coated and mixed with a curable monomer to form a viscous paste that can be inserted into a prepared cavity, whereupon it can be shaped and cured. During curing, the silane coated particles bond chemically with the polymer matrix. The filler fraction in dental composites rarely exceeds 60-65 vol-% because of problems with having higher volumes of randomized packed filler particles. Depending on filler size and filler size distribution, it is possible to make different types of dental composites. Since the total filler surface area per gram filler increases as the filler size decreases, finer particles tie up more resin, causing the viscosity of the material to increase fastest with filler fraction of smallest particles. Because of that phenomenon, composites with the finest filler particles tend to contain the lowest filler volume. The modulus of elasticity of a dental composite can roughly be estimated by determine the theoretical modulus of both the series as well as parallel models, and assume that the modulus of the composite for a certain filler fraction falls somewhere between these boundaries.

When the first modern dental composites were introduced during the 60s, it soon became clear that their wear resistance when used on load bearing surfaces was not high enough to be able to resist wear on occlusal surfaces. As a consequence, research performed during the 60s to the 80s focused on finding a solution to the wear problem as well as developing an understanding of the wear mechanism of these materials. During that era, it became clear that some of the key factors associated with composite wear were the quality of the filler matrix bond as well as the filler particle size and distribution. At a symposium supported by 3M in 1984 [46], research findings revealed that the best posterior composites at that time had reached a wear resistance of the commonly used amalgams.

During the research involving wear of composites, researchers had identified that cracks sometimes developed in regions in contact with an opposing cusp. The wear in those regions were often described as two-body wear, while the more general and less dramatic wear occurring on other surfaces were described as a three-body wear caused by abrasive particles sliding over the composite surface during chewing. When it came to the so-called two-body wear, it seemed reasonable to assume that during cusp sliding, micro-cracks could be induced. Another possible wear mechanism induced in the contact region could also be fatigue wear, triggered by a Hertzian failure [47]. In both these cases, microscopic flaws would develop, and these flaws would then contribute to an accelerated wear in these regions. In 1988, Roulet [48] claimed that fractures within the body of restorations and at the margins were a major problem regarding the failure of posterior composites.

However, during the 70s and 80s, the focus on dental composites were related to what clinicians perceived as being the major reasons for failures, which included wear, recurrent caries and discolorations. The notion that flaws were involved in the wear process led Truong and Tyas [49] to determine stable crack growth in dental composites. They did so by use of a double-torsion technique to establish the relationship between the stress intensity factor (SIF) K_I and the crack velocity (v) for commercial and experimental composites. They tested dry, water-saturated and ethanol/water (3:1 v/v) saturated specimens. At a given crack velocity, the difference between the K_I of a dry specimen and that of a water-saturated specimen was attributed solely to the change of Young's modulus caused by the plasticizing effect of water. However, microcracking occurring during immersion in an ethanol/water mixture resulted in an excessive drop of K_I values from the dry state to ethanol/water mixture saturated state for Estilux Posterior and Occlusin samples, while little effect of fluids on K_I could be observed on P10 and P30. The investigators tried to theoretically predict the wear of the composites, based on the assumption that microcracking occurs in the subsurface layer due to cyclic and impact stresses. Based on that assumption, three criteria for good wear resistance would be: (a) high fracture toughness (high critical SIF, K_{IC}) and larger threshold crack length (a_t); (b) small inherent flaw size (a_o) and (c) high crazing stress (σ_c). Based in these assumptions and the results of this study, the wear resistance of tested commercial composites should be: Occlusin > P10 > Estilux Posterior > P30 = Ful-Fil > Profile > Silux --~ Isomolar > Concept.

In a study from 1991, Higo et al. [50] used a fracture mechanics approach to investigate the fracture toughness behavior of three commercial composite resins for dental use named Clearfil photo posterior, P-50 and Occlusin. The outcome of that study was that Occlusin exhibited higher fracture toughness values than any other resin when employing a ring specimen test procedure. However, when an indentation method was used, comparable fracture toughness values for all three resins were produced.

As a fracture mechanics approaches became more popular in attempts to estimating life-times of dental restorative materials, it became important to have available data on the fatigue behavior of these materials. At the end of the 90s, efforts at estimation included several untested assumptions related to the equivalence of flaw distributions sampled by shear, tensile, and compressive stresses. However, environmental influences on material properties were so far not accounted for to any greater extent, and it was unclear if fatigue limits existed. In a study by Baran et al. [51], they characterized the shear and flexural strengths of three resins used as matrices in dental restorative composite materials by use of Weibull parameters. They found that shear strengths were lower than flexural strengths, liquid sorption had a profound effect on characteristic strengths, and the Weibull shape parameter obtained from shear data differed for some materials from that obtained in flexure. In shear and flexural fatigue, a power law relationship applied for up to 250 000 cycles; no fatigue limits were found, and the data thus implied only one flaw population is responsible for failure. Again, liquid sorption adversely affected strength levels in most materials (decreasing shear strengths and flexural strengths by factors of 2–3) and to a greater extent than did the degree of cure or material chemistry.

In a study by Manhart et al. [52], they determined some mechanical properties of three packable composites (Solitaire, Surefil, ALERT), a packable ormocer (Definite), an advanced hybrid composite (Tetric Ceram) and an ionreleasing composite (Ariston pHc) in vitro (Table 3). As seen from that table, the properties of these composites differed significantly, which could be related to differences in filler particle size and shape distributions among the different materials. Their study suggested that fracture and wear behavior of the composite resins would be highly influenced by the filler system. They found that ALERT had the highest fracture toughness value, but also the highest wear rate, which they related to the fiber like particles used in that material. Overall, Surefil demonstrated good fracture mechanics parameters and low wear rate, which they suggested could be related to their more particle shaped filler particles. This study suggested that fracture and wear behavior of the composite resins are highly influenced by the filler system.

Composite material	Flexural strength (MPa)	Flexural modulus (GPa)	Fracture toughness K_{IC} (MN $m^{-1/2}$)	Mean wear rate (μm^3 $cycle^{-1}$)
Solitare	81.6 (10.0)	4.4 (0.3)	1.4 (0.2)	1591
Definite	103.0 (19.9)	6.3 (0.9)	1.6 (0.3)	2763
Surefil	132.0 (14.3)	9.3 (0.9)	2.0 (0.2)	3028
ALERT	124.7 (22.1)	12.5 (2.1)	2.0 (0.2)	8275
Tetric Ceram	107.6 (11.4)	6.8 (0.5)	2.0 (0.1)	5417
Ariston pHc	118.1 (10.5)	7.3 (0.8)	1.9 (0.2)	7194

Table 3. Some properties of six dental composite materials [52].

Considering the importance of being able to perform life-time predictions of dental composites, McCool et al [53] , continued their research from 1998 [51], by comparing the lifetime predictions resulting from two methods of fatigue testing: dynamic and cyclic fatigue. To do so they made model composites, in which one variable was the presence of a silanizing agent. They tested their specimens in 4-point flexure, using a cyclic fatigue frequency of 5 Hz, while their dynamic fatigue testing spanned seven decades of stress rate application. Data were reduced and the crack propagation parameters for each material were calculated from both sets of fatigue data. These parameters were then used to calculate an equivalent static tensile stress for a 5-year survival time. The 5-year survival stresses predicted by dynamic fatigue data were approximately twice those predicted by cyclic fatigue data. In the absence of filler particle silanization, the survival stress was reduced by half. Aging in a water-ethanol solution reduced the survival stresses by a factor of four to five. One of the conclusions drawn from this study is that cyclic fatigue is a more conservative means of predicting lifetimes of resin-based composites.

The notion that there is a correlation between wear resistance and fracture toughness was to some degree rejected by Ruddel et al.[54]. In their study they produced pre-polymerized fused-fiber filler modified composite particles and determined their effectiveness by incorporating these fibers into composites. The results revealed that these particles decreased both flexural strength and fracture toughness, but improved wear performance. The SEM evaluations did not suggest that porosities had been incorporated during particle incorporation. Instead, fractures were transgranular through the reinforcing particles. Microscopic flaws observed in the new particles most likely explain the lower strength and toughness values. This study is important, because it shows that a composite with improved wear resistance could also suffer from an increase in fracture risk.

During the past 10 years, it has become clear that fracture is a major reason for clinical failure of dental composites. Many clinical fractures are likely to be preceded by slow sub-critical crack propagation. To study the slow sub-critical crack propagation, Loughran [55] used notched composite (Z100, 3M ESPE) specimens and fatigued them in a four-point bending test using a load cycle at 5 Hz between 25 and 230 N until failure. Displacement and load were recorded during the fatigue tests and used to derive crack propagation based on beam-compliance. What they found was that the number of cycles until failure ranged between 34 and 82,481. In the last 1500 cycles prior to final fracture, the beam compliance increased consistently, indicating sub-critical crack propagation. From the compliance change they calculated that the crack length increased 8% (77 ± 14 μm) before final failure. The crack growth rate during sub-critical crack propagation was determined as a function of the stress intensity for the last 1500 cycles before fracture. The importance of this study was that they found that the fatigue lifetime varied widely, and that stable crack growth existed prior to fracture consistently. This consistency allowed formulation of stress-based crack propagation relationships that can be used in concert with numerical simulations to predict composite restoration performance. The large variation found for specimen lifetime was attributed to the initiation process that precedes sub-critical crack propagation.

As mentioned earlier, during the early 80s, dentists regarded poor wear resistance tendency to be associated with recurrent caries and restoration discolorations as the key shortcomings with dental composites. Today, that perception has changed quite considerable. By improved filler technology and silanization methods, the poor wear resistance is no longer a major clinical problem. Improved adhesives, now making it possible to bond composites to both enamel and dentin, have decreased the risk for recurrent caries. The use of more stable chemicals and smoother composite surfaces caused by the use of finer filler particles has decreased the magnitude of restoration discolorations. In other words, what were regarded as major shortcomings with posterior composites are no longer regarded as major weaknesses. Of course, these shortcomings have not yet been completely eliminated, so there is still room for improvements. However, as the composites have been improved, another shortcoming has been identified as now being the biggest problem, namely fractures[48]. In a recently published clinical study [56], in which two composites were evaluated over a 22-year period, the authors claimed that the most common reason for failures of posterior composites were fractures. That study suggests that further understanding of the fracture mechanical behavior of dental composites is needed.

3.3. Fracture mechanics aspects of cements and adhesives

In order to attach restorations such as composite fillings, inlays/onlays, crowns and bridges, different cements/adhesives have been used in dentistry through the years. The oldest but still used cement is the zincphosphate cement, which was introduced about 150 year ago and consists mainly of a zincoxide powder mixed with phosphoric acid. During setting, that cement goes through an acid-base reaction during which a salt and water is formed. The way this cement works is simply by etching the surfaces of the tooth and the surface of the restoration the cement is in contact with, a process that occurs as the cement sets, whereupon zincphosphate crystallites precipitate into the etched surface regularities as the cement sets. With that mechanism a mechanical interlocking is established, explaining the retention of the cemented restoration.

In addition to the zinc phosphate cement, other cements such as silicate, zincsilico phosphate, polycarboxylate and glass ionomer cements have been used. In the case of the silicate and zincsilico phosphate cements, phosphoric acid is used in both cases, while the powders of these two cements are either a silicate glass powder or a mixture of that powder with a zinc phosphate powder. When it comes to the polycarboxylate and the glass ionomer cements, the powders are either the zinc oxide powder or the glass powder used in the silicate cement, while the acid has been replaced with a polyalceonic acid. The polyalkeoinic acid, often polyacrylic acid, is capable of reacting with the powder through an acid-base reaction, but also with the dentin or enamel surface. During that reaction the –COO^- of the polyalkeonic acid can interact with ions such as the Ca^{2+} present in the tooth surface and form some ionic interaction. Compared to the zinc phosphate and silicate cements, the polycarboxylate and glass ionomer cements were introduced to dentistry during the 60s and the 70s. Regarding the ability to bond to hard tooth tissues, it is generally assumed that zinc phosphate, zincsilico phosphate, and silicate cements only bond via micromechanical retention, while polycarboxylate and glass ionomer cements bond both via a micromechanical retention as well ionic surface interaction.

The idea to develop some kind of chemical bond to dental hard tissues was however introduced before the zincpolycarboxylate and glass ionomer cements had been invented. The first idea to use some kind of chemical interaction to form a bond to the hard dental tissues was introduced during the late 40s when Hagger [57] suggested that a molecule that had a phosphate group capable of interaction with Ca^{2+} at the tooth surface and a methylmethacrylate group capable of forming a covalent bond to a curing methacrylate based filling materials could form such a bond. Unfortunately, the molecule Hagger used to achieve such a bond did not show to be very efficient. However, when Buonocore in 1955 [58] explored the possibility to first etch the enamel surface with a phosphoric acid, then rinse and dry and coat the acid roughened surface with a curable resin, it became possible to achieve a predictable bond to enamel.

Buonocore's idea was not widely accepted initially, because dentists feared that the phosphoric acid, particularly if it came in contact with exposed dentin surface, would cause pulp irritation and eventually pulp death. Such pulp reactions were known to occur, particularly when the more slow setting silicate cement was used. As a consequence it would take several years until Buonocore's acid-etch approach took off. A major contributor for teaching dentists how to use enamel etching and composite resins was 3M, who during the 60s had expanded their products to dentistry.

To spread the usage and the acceptance of enamel etching and resin bonding as well as their composite resin, 3M sponsored a symposium entitled "The Acid Etch Technique" in 1975. The presentations presented at that symposium were published in a book [59] that was then widely distributed by representatives for the company. By having prominent researchers presenting papers related to the acid etch technique, a lot of misperceptions could be eliminated and the enamel etch technique became generally accepted [60]. When it came to testing enamel bonding, most in vitro studies relied on morphology achieved by use of SEM and different strength tests of which shear bond testing soon became the most popular.

Even though enamel bonding was a major advance in dentistry, the ability to bond to dentin was not resolved when enamel bonding took off. Because most surfaces exposed during tooth preparations of cavities and crowns consist of dentin, a reliable dentin bonding was still needed in order to truly bond different restorative materials to dentin. However, dentin bonding was much more complicated to achieve than enamel bonding. In contrast to enamel, dentin is a living tissue and therefore much more demanding than enamel when it came to biocompatibility of chosen materials. Besides, dentin contains much more water, making it difficult to adapt more or less hydrophobic materials to the dentin surface.

Parallel to these events, Bowen had already during the 60s initiated research to develop resin systems capable of bonding to cut dentin surfaces [61]. The basic principle behind his ideas was that the adhesives should contain a reactive group capable of reacting with Ca-ions present on the tooth surface, and then react with the resin when the resin cured. When these adhesives, often referred to as the first generation of dentin adhesives were explored, it soon became clear that a cut dentin surface was coated with a so called smear layer. That smear layer consisted of a few microns thick layer of smeared collagen in which fractured hydroxyapatite crystallites were embedded. It was soon clear that the first generation of adhesives developed a weak bond to the tooth surface, and that the bond was weak and worked for a short time period only, mainly because the bonds formed to the smear layer, or the bonds between the smear layer and the dentin were too weak to resist loading.

During the 70s, the dental community discussed the effect the smear layer had on bonding and whether or not it should remain on the dentin surface. Some researchers viewed it as beneficial, since the vital dentin channels were sealed off, decreasing the risk of pulp irritations caused by the restorative material. As a consequence, somewhat more acidic adhesive systems were developed, capable of removing some of the smear layer but retaining some smear serving as protective layer. The adhesives that fell into this class are often referred to as the 2nd generation adhesives.

At the end of the 70th, a major break-through occurred. That break-through consisted of a clinical study performed by Fusayama at al. [62], in which they claimed that by etching both enamel and dentin, they were able to bond composites to dentin without having any problem with pulps responding to the etching procedure. There is no doubt that Fusayama et al.'s finding was looked upon with enormous skepticism. Their explanation that resin infiltrated the tubules and thereby formed resin tags that contributed to the retention was also questioned. It was first when Nakabaiashi [63] came out with his hybridization explanation, suggesting that the resin infiltrated the etched dentin surface and formed a hybrid layer consisting of partly dissolved dentin, as dentin etching started to become accepted .

These two studies [62, 63] opened the door for more aggressive dentin etching resulting in the 3rd generation adhesives. Etching dentin with phosphoric acid was still not the general trend. Instead, weaker conditioners such as EDTA and citric acids were used[64]. However, at the end of the 80th, some bonding systems had occurred on the market that used the same etchant for both enamel and dentin. The success of these adhesives, the so called 4th generation adhesives, took of during the early 90s, when both Kanca [65] as well as Gwinnett [66] in two independent studies claimed that by leaving the dentin moist before priming, they could better infiltrate the collagen layer with the primer and thereby achieve better bonding to dentin.

Simultaneous with these trends related to bonded composite, it had also been noticed that by etching the surface of ceramic restorations located at the dentin surface with hydrofluoric acid and then silane coat the etched surface, it was possible to bond ceramic restorations to tooth surfaces. Such an approach resulted in a significantly lower risk of ceramic fracture than compared to the use of more traditional cements, including polycarboxylate and glass ionomer cements. By use of the information presented under the ceramic section in this chapter, it is quite easy to explain why resin bonded ceramics performed so well by considering fracture mechanics. In the case of the more traditional cements, they can be described as having brittle properties with limited ability to form strong bonds to the ceramic surface. In the case of the phosphoric acid based cements they did not form any strong bonds to the tooth surface neither. By realizing that even a ceramic restoration can flex during chewing, one can visualize the development of shear stresses at the ceramic-cement interface, and that these stresses can trigger a crack growth along the ceramic-cement interface. In the case of the resin bonded ceramics, the shrinkage of the resin cement initially induced some compressive stress in the ceramic surface adjacent to the resin cement. If a crack propagates to the resin interface in such a case, the more ductile nature of the resin cement will not as easily allow the crack to propagate along the ceramic-cement interface. Besides, after the load has been removed, the resin will because of its polymerization shrinkage, try to force the fractured ceramic in contact with its fractured surfaces. Thus, in this case, a ceramic fracture may occur, but in contrast to a fracture in a ceramic cemented with more traditional cement, one may not end up with a detectable catastrophic failure.

From a fracture mechanics point of view, there is no doubt that the adhesive joint is the most challenging region. The reason relates simply to practical problems such as minimizing the incorporation of defects in this region. In addition, the fact that the adhesive shrinks and induces shrinkage stresses between the tooth and the adhesive, as well as between the adhesive and the restorative material, does not make the situation manageable, which is further complicated by differences in mechanical/physical properties of the different materials forming a joint. In the following section we will approach the adhesive joint in an attempt to identify different challenges associated with this region.

When it comes to the failure mechanism at the dentin resin interface, there are certain questions that need to be addressed. These questions include: (1) does failure at the human dentin-resin interface occur by a cohesive or an adhesive mechanism? (2) is the failure mechanism accompanied by a plastic deformation, and if so how important is it? To address these questions, Lin and Douglas [67] performed a computational analysis and fractography of two different bonding systems: Scotchbond- (SB2) and Scotchbond-Multipurpose (SBM). The difference between these two systems is that SB2 consists of a mixture of primer and a so called bonding resin, while SBM uses the same primer and bonding resin, but in contrast to SB2 they are placed as separate systems on the dentin surface. According to their estimates, the dentin-resin interracial fracture toughness (G_{IC}), for the SB2 and for the SBM were 30.22 ± 5.61 and 49.56 ± 7.65 J m^{-2}, respectively, which were significantly different ($p < 0.01$). Both SB2 and SBM interfaces with dentin displayed significant degrees of plasticity (0.15 and 0.19) which were beneficial to crack resistance. Thus, correcting for the plasticity, the G_{IC} for SB2 and for SBM increased to 42.83 ± 7.75 and 74.97 ± 10.47 J m^{-2}, respectively. The fractography of the two systems reflected these numeric differences. SB2 showed largely interfacial adhesive failure, while SBM showed adhesive-cohesive failure with occasional dentin adhesions attached to the composite interface and vice versa.

In another study, Toparli [68] determined the reliability and validity of the adhesive bond toughness of dentin/composite resin interfaces from the standpoint of fracture mechanics. The fracture toughness (K_{IC}) and fracture energy (J_{IC}) values of two different composite resins (Brilliant Dentin and P50) were determined. The fracture toughness and energy values obtained experimentally for Brilliant Dentin were found to be higher than those for P50. It was seen that calculated J values (J_{adh} and J_{res}) changed with the crack length; but the effective fracture energy (J_{eff}) was independent of the crack length, as expected. The applied fracture energy (J_{appl}) and effective fracture energy (J_{eff}) are considerably smaller than the experimentally determined J_{IC} values of composite resins. The important finding was that the bonded interface tends to produce microscopic flaws which could act as critical stress risers promoting interfacial failures. The initiation and propagation of such flaws under the mastication forces can be followed by fracture toughness (K_{IC}) or fracture energy (J_{IC}) in linear elastic fracture mechanics (LEFM).

The effect of crack growth at a resin bonded metal interface after storage in water was studied by Moulin [69], who found that the adherence energy dramatically decreased with time in water. The slope of the regression straight line appeared to be a good criterion for evaluating the durability of the alloy/adhesive interface. The study revealed the importance of silica coating the metal surface and, especially, the effectiveness of the Rocatec system upon the degree of hydrolytic degradation. The study showed how the development of cracks depends upon surface treatment.

Adhesion at the titanium–porcelain interface using a fracture mechanics approach has also been used to investigate the bonding mechanism of such systems [70]. In that study they used specimens of five different titanium–porcelain and one base metal–porcelain bonding systems on which they performed a four-point bending interfacial delaminating test. The pre-cracked specimen was subjected to load and the strain energy release rate (G) was calculated from the critical load to induce stable crack extension in each system. The strain energy release rate of titanium–porcelain with a Gold Bonder interface layer was highest among the five different systems. No attempt was made to explain the experimental findings.

In two studies by Ichim et al. [71, 72] they looked at a typical non-carious cervical lesion, a so called abfraction, treated with a glass ionome or a combination of glassionomer and composite. The approach they used was that they used a nonlinear fracture mechanical approach simulated by use of FEA. They used a novel Rankine and rotating crack model to trace the fracture failure process of the cervical restorations. The approach involves an automatic insertion of an initial crack, mesh updating for crack propagation and self contact at the cracked interface. The results were in good agreement with published clinical data, in terms of the location of the fracture failure of the simulated restoration and the inadequacy of the dental restoratives for abfraction lesions.

In their second study [72] they investigated the influence of the elastic modulus (E) on the failure of cervical restorative materials and tried to identify an E value that would minimize mechanical failure under clinically realistic loading conditions. What they found was that the restorative materials currently used in non-carious cervical lesions are largely unsuitable in terms of resistance to fracture of the restoration. They suggested that the elastic modulus of such a material should be in the range of 1 GPa rather than several GPa that is usually the case.

Despite an obvious advantage to approach adhesives and their performance from a fracture mechanics point of view, traditional bond studies usually focus on bond strength values. By comparing such strength values, it is noticed that large variations exist among different reports. These variations are due to differences among operators, but also on the day a certain tester performed a test. The standard deviation is 25-50 % of the mean value, which suggests that defects present in the adhesive region may be of a bigger concern than the true adhesive strength

In an attempt to resolve the questions related to the large variability in strength values and their clinical meaning, The Academy of Dental Materials at their annual meeting in 2010, focused that meeting on the value of bond strength measurements. In one presentation, Scherrer et al. [73 presented a literature search based on all dentin bond strength data obtained for six adhesives evaluated with four tests (shear, microshear, tensile and microtensile) and critically analyzed the results with respect to average bond strength, coefficient of variation, mode of failure and product ranking. The PubMed search was carried out for the years between 1998 and 2009. The six adhesive resins that were selected included three step systems (OptiBond FL, Scotch Bond Multi-Purpose Plus), two-step (Prime & Bond NT, Single Bond, Clearfil SE Bond) and one step (Adper Prompt L Pop). By pooling the results from the 147 references, it was revealed an ongoing high scatter in the bond strength data regardless which adhesive and which bond test was used. Coefficients of variation remained high (20–50%) even with the microbond test. The reported modes of failure for all tests still included a high number of cohesive failures. The ranking of the adhesives seemed to be dependent on the test method being used. The scatter in dentin bond strength data, independent of used test, confirmed Finite Element Analysis predicting non-uniform stress distributions due to a number of geometrical, loading, material properties and specimens preparation variables. The study reopened the question whether an interfacial fracture mechanics approach to analyze the dentin–adhesive bond would not be more appropriate for obtaining better agreement among dentin bond related papers.

In another paper presented at that meeting, Soderholm [74] emphasized the benefits of using fracture mechanics approaches when it comes to studying dental adhesives. In his review, different general aspects of fracture mechanics and adhesive joints were reviewed, serving as a foundation for a review of fracture toughness studies performed on dental adhesives. The dental adhesive studies were identified through a MEDLINE search using "dental adhesion testing AND enamel OR dentin AND fracture toughness" as search strategy. The outcome of the review revealed that fracture toughness studies performed on dental adhesives are complex, both regarding technical performance as well as achieving good discriminating ability between different adhesives. The review also suggested that most fracture toughness tests of adhesives performed in dentistry are not totally reliable because they usually did not consider the complex stress pattern at the adhesive interface. However, despite these limitations, the review strongly supports the notion that the proper way of studying dental adhesion is by use a fracture mechanics aproach.

In a study by Howard and Soderholm [75] they used a fracture mechanics approach previously described by Pilliar and Tam [76-80] to test the hypothesis that a self-etching adhesive is more likely to fail at the dentin adhesive interface than an etch-and-rinse adhesive. What they found was that the fracture toughness values (K_{IC}) of the two adhesives were not significantly different. The rather high frequency of mixed failures did not support the hypothesis that the dentin-adhesive interface is clearly less resistant to fracture than the adhesive–composite interface. The finding that cracks occurred in 6–8% in the composite suggests that defects within the composite or at the adhesive–composite interface are important variables to consider in adhesion testing.

In a recently published study by Ausello et al. [81], they used FEA and fatigue mechanic laws to estimate the fatigue damage of a restored molar. The simulated restoration consisted of an indirect class II MOD cavity preparation restored with a composite. Fatigue simulation was performed by combining a preliminary static FEA simulation with classical fatigue mechanical laws. It was found that regions with the shortest fatigue-life were located around the fillets of the class II MOD cavity, where the static stress was highest.

From the above papers, it becomes clear that adhesion tests utilized in dentistry are unable to separate the effects of adhesive composition, substrate properties, joint geometry and type of loading on the measured bond strength. This makes it difficult for the clinician to identify the most suitable adhesive for a given procedure and for the adhesive manufacturer to optimize its composition. To come to grip with these challenges, Jancar [82] proposed an adhesion test protocol based on the fracture mechanics to generate data for which separation of the effect of composition from that of the joint geometry on the shear (τ_a) and tensile (σ_a) bond strengths was possible for five commercial dental adhesives. The adhesive thicknesses (h) used varied from 15 to 500 μm, and the commercial adhesives had fracture toughness values (K_{IC}) ranging from 0.3 to 1.6 MPa $m^{1/2}$. They used double lap joint (DLJ) and modified compact tension (MCT) specimens which were conditioned for 24 h in 37°C distilled water, then dried in a vacuum oven at 37°C for 24 h prior to testing. Both τ_a and σ_a increased with increasing adhesive thickness, exhibiting a maximum bond strength at the optimum thickness (h_{opt}). For $h < h_{opt}$, both τ_a and σ_a were proportional to h, and, above h_{opt}, both τ_a and σ_a decreased with $h^{-4/10}$ in agreement with the fracture mechanics predictions. Hence, two geometry-independent material parameters, Ψ and (H_c/Q), were found to characterize τ_a and σ_a over the entire thickness interval. The results seem important, because it suggests that the adhesion tests currently used in dentistry provide the geometry dependent bond strength, and such data cannot be used either for prediction of clinical reliability of commercial dental adhesives or for development of new ones. Instead, the proposed test protocol allowed one to determine two composition-only dependent parameters determining τ_a and σ_a. A simple proposed procedure can then be used to estimate the weakest point in clinically relevant joints always exhibiting varying adhesive thickness and, thus, to predict the locus of failure initiation. Moreover, this approach can also be used to analyze the clinical relevance of the fatigue tests of adhesive joints.

In a recent paper by Kotousove [83] a conceptual framework utilizing interfacial fracture mechanics and Toya's solution for a partially delaminated circular inclusion in an elastic matrix was used , which can be applied (with caution) to approximate polymer curing induced cracking about composite resins for Class I cavity restorations. The findings indicated that: (I) most traditional shear tests are not appropriate for the analysis of the interfacial failure initiation; (II) material properties of the restorative and tooth material have a strong influence on the energy release rate; (III) there is a strong size effect; and (IV) interfacial failure once initiated is characterized by unstable propagation along the interface almost completely encircling the composite. The importance of this study is that it analyses the reliability of composite Class I restorations and provides an adequate interpretation of recent adhesion debonding experimental results utilizing tubular geometry of specimens. The approach clearly identifies the critical parameters including; curing strain, material module, size and interfacial strain energy release rate for reliable development of advanced restorative materials.

In a similar approach, Yamamoto [84] calculated stresses produced by polymerization contraction in regions surrounding a dental resin composite restoration. Initial cracks were made with a Vickers indenter at various distances from the edge of a cylindrical hole in a soda-lime glass disk. Indentation crack lengths were measured parallel to tangents to the hole edge. Resin composites (three brands) were placed in the hole and polymerized (two light irradiation protocols) at equal radiation exposures. The crack lengths were remeasured at 2 and 10 min after irradiation. Radial tensile stresses due to polymerization contraction at the location of the cracks (σ-crack) were calculated from the incremental crack lengths and the fracture toughness K_C of the glass. Contraction stresses at the composite–glass bonded interface (σ-interface) were calculated from σ-crack on the basis of the simple mechanics of an internally pressurized thick-walled cylinder. The greater the distance or the shorter the time following polymerization, the smaller was σ-crack. Distance, material, irradiation protocol and time significantly affected σ-crack. Two-step irradiation resulted in a significant reduction in the magnitude of σ-interface for all resin composites. The contraction stress in soda-lime glass propagated indentation cracks at various distances from the cavity, enabling calculation of the contraction stresses.

4. Conclusion

By reviewing enamel, dentin and their interfacial bond, it is obvious that the tooth evolved in such a way it would be able to function in an optimal way without fracturing. With the sophisticated structure of both enamel and dentin, it becomes quite clear that existing man made restorative materials are far from optimal in comparison to the biological hard tissues. The crack growth risk in ceramics needs to be reduced, something that can be achieved by use of fracture tough ceramics such as alumina and zirconia. Unfortunately, as shown in Lawn et al.'s[45] paper, rather extensive removal of existing tooth structure needs to be performed in order to minimize future failures. Such an approach, though, does not make sense if one considers that a more sophisticated material is removed in order to replace it with an inferior material.

When it comes to dental composites, we have now reached a point when fractures of composites are being judged as the most common reason for composite failures [48, 56]. To come to grip with that problem, our understanding of the fracture mechanics of dental composites needs to be improved. The same is true regarding cements/adhesives. The particulate filled resins we are now using are rather primitive when compared to both enamel and dentin. However, it seems as this group of materials has the highest chance to evolve and approach the properties of enamel and dentin.

By looking at dentistry from a fracture mechanics point of view, it becomes quite clear that traditional dentistry suffer from some major processing problems. The first problem is that restorations are individual units that differ in shape and size. These different sizes and shapes result in different levels and locations of localized stresses. The second problem is that restorations are placed by individual dentists working under different conditions and introducing different amounts and types of flaws during the different dental procedures. Considering that the theoretical strength is several magnitudes stronger than the real strength values due to the presence of defects in materials suggest that processing defects, located in a material or at an interface is a significant dental problem. The third problem is partly self-inflicted. During dental education, students learn to copy the anatomy of natural teeth. The pits and fissures present in natural teeth act naturally as stress concentrators, but because of the sophisticated structure of a substrate such as enamel, such pits and fissures may in fact act as crack stoppers. Take for example a crack that might propagate along a cusp toward the central fissure. When that crack reaches the bottom of that fissure, the thickness of the enamel decreases and one can assume that the crack will not continue to propagate up along the other cusp with increasing enamel thickness. In the case of a man-made crown or filling, the fissure will not serve the same protective purpose. However, because dental students are trained to reproduce the sharp anatomic details, sharp anatomic fissures are often regarded as a sign of good competence, while in reality such details will rather facilitate crack growth.

Based on the information provided in this chapter it seems reasonable to suggest that future dental students should receive more training in fracture mechanics in order to better understand how handling and design may affect the final outcome of a restorative procedure Besides, with such a knowledge they would be able to communicate better with other scientists and thereby facilitate the development of better restorative materials.

Author details

Karl-Johan Söderholm
College of Dentistry, University of Florida, Gainesville, Florida, USA

5. References

[1] Staines M, Robinson WH, Hood JAA (1981) Spherical Indentation of Tooth Enamel. J. mater. sci. 16:2551-2556.
[2] Ten Cate AR. Oral histology : Development, Structure, and Function. 5th ed. St. Louis, Mo.: Mosby; 1998. xi, 497 p.
[3] Meckel AH, Griebstein WJ, Neal RJ (1965) Structure of Mature Human Dental Enamel as Observed by Electron Microscopy. Arch. oral biol. 10:775-783. Epub 1965/09/01.
[4] Bajaj D, Arola DD (2009) On the R-curve Behavior of Human Tooth Enamel. Biomater. 30:4037-4046.
[5] Xu HH, Smith DT, Jahanmir S, Romberg E, Kelly JR, Thompson VP, et al. (1998) Indentation Damage and Mechanical Properties of Human Enamel and Dentin. J. dent. res. 77:472-480 Epub 1998/03/13.
[6] El Mowafy OM, Watts DC (1986) Fracture Toughness of Human Dentin. J. dent. res. 65:677-681. Epub 1986/05/01.
[7] Arola D, Huang MP, Sultan MB (1999) The Failure of Amalgam Dental Restorations due to Cyclic Fatigue Crack Growth. J. mater. sci.-mater. med. 10:319-327.
[8] He LH, Swain MV (2008) Understanding the Mechanical Behaviour of Human Enamel from its Structural and Compositional Characteristics. J. mech. behav. biomed. mater. 1:18-29. Epub 2008/01/01.
[9] Lucas P, Constantino P, Wood B, Lawn B (2008) Dental Enamel as a Dietary Indicator in Mammals. Bioessays. 30:374-385.
[10] Constantino PJ, Lucas PW, Lee JJW, Lawn BR (2009) The Influence of Fallback Foods on Great Ape Tooth Enamel. Amer. j. phys. anthrop. 140:653-660.

[11] Lee JJW, Morris D, Constantino PJ, Lucas PW, Smith TM, Lawn BR (2010) Properties of Tooth Enamel in Great Apes. Acta biomater. 6:4560-4565.
[12] Anderson PSL, Gill PG, Rayfield EJ (2011) Modeling the Effects of Cingula Structure on Strain Patterns and Potential Fracture in Tooth Enamel. J. morph. 272:50-65.
[13] Lawn BR, Lee JJW, Constantino PJ, Lucas PW (2009) Predicting Failure in Mammalian Enamel. J. mech. behav. biomed. mater. 2:33-42.
[14] Nalla RK, Imbeni V, Kinney JH, Staninec M, Marshall SJ, Ritchie RO (2003) In vitro Fatigue Behavior of Human Dentin with Implications for Life Prediction. J. biomed. mater. res. A. 66:10-20. Epub 2003/07/02.
[15] Kruzic JJ, Nalla RK, Kinney JH, Ritchie RO (2003) Crack Blunting, Crack Bridging and Resistance-Curve Fracture Mechanics in Dentin: Effect of Hydration. Biomater. 24:5209-5221. Epub 2003/10/22.
[16] Bajaj D, Sundaram N, Arola D (2008) An Examination of Fatigue Striations in Human Dentin: In vitro and In vivo. J. biomed. mater. res. part B-appl. biomater. 85B:149-159.
[17] Yan JH, Taskonak B, Platt JA, Mecholsky JJ (2008) Evaluation of Fracture Toughness of Human Dentin using Elastic-Plastic Fracture Mechanics. J. biomech. 41:1253-1259.
[18] Kelly JR (1995) Perspectives on Strength. Dent. mater. 11:103-110. Epub 1995/03/01.
[19] Phillips CJ (1965) Strength and Weakness of Brittle Materials. Amer. scientist. 53:20-&.
[20] Griffith AA (1968) Phenomena of Rupture and Flow in Solids. Asm. trans. quarterly. 61:871-&.
[21] Kelsey WP, 3rd, Cavel T, Blankenau RJ, Barkmeier WW, Wilwerding TM, Latta MA (1995) 4-Year Clinical Study of Castable Ceramic Crowns. Am. j. dent. 8:259-262. Epub 1995/10/01.
[22] Fradeani M, Aquilano A (1997) Clinical Experience with Empress Crowns. Inter. j. prosthet. 10:241-247.
[23] Sjogren G, Lantto R, Granberg A, Sundstrom BO, Tillberg A (1999) Clinical Examination of Leucite-Reinforced Glass-Ceramic Crowns (Empress) in General Practice: A Retrospective Study. Inter. j. prosthet. 12:122-128. Epub 1999/06/18.
[24] Sjogren G, Lantto R, Tillberg A (1999) Clinical Evaluation of All-Ceramic Crowns (Dicor) in General Practice. J. prosthet. dent. 81:277-284. Epub 1999/03/02.
[25] Malament KA, Socransky SS (1999) Survival of Dicor Glass-Ceramic Dental Restorations over 14 Years. Part II: Effect of Thickness of Dicor Material and Design of Tooth Preparation. J. prosthet. dent. 81:662-667. Epub 1999/05/29.
[26] Malament KA, Socransky SS (1999) Survival of Dicor Glass-Ceramic Dental Restorations over 14 Years: Part I. Survival of Dicor Complete Coverage Restorations and Effect of Internal Surface Acid Etching, Tooth Position, Gender, and Age. J. prosthet. dent. 81:23-32. Epub 1999/01/08.
[27] Malament KA, Socransky SS (2001) Survival of Dicor Glass-Ceramic Dental Restorations over 16 Years. Part III: Effect of Luting Agent and Tooth or Tooth-Substitute Core Structure. J. prosthet. dent. 86:511-519. Epub 2001/11/29.
[28] McLaren EA, White SN (2000) Survival of In-Ceram Crowns in a Private Practice: A Prospective Clinical Trial. J. prosthet. dent. 83:216-222. Epub 2000/02/11.
[29] Fradeani M, Redemagni M (2002) An 11-Year Clinical Evaluation of Leucite-Reinforced Glass-Ceramic Crowns: A Retrospective Study. Quintessence inter. 33:503-510.

[30] Oden A, Andersson M, Krystek-Ondracek I, Magnusson D (1998) Five-Year Clinical Evaluation of Procera AllCeram Crowns. J. prosthet. dent. 80:450-456. Epub 1998/10/29.

[31] Willmann G (2001) Improving Bearing Surfaces of Artificial Joints. Adv. eng. mater. 3:135-141.

[32] Thompson JY, Anusavice KJ, Naman A, Morris HF (1994) Fracture Surface Characterization of Clinically Failed All-Ceramic Crowns. J. dent. res. 73:1824-1832. Epub 1994/12/01.

[33] Zhao H, Hu XZ, Bush MB, Lawn BR (2001) Cracking of Porcelain Coatings Bonded to Metal Substrates of Different Modulus and Hardness. J. mater. res. 16:1471-1478.

[34] McLean JW. Dental Ceramics : Proceedings of the First International Symposium on Ceramics. Chicago: Quintessence Pub. Co.; 1983. 541 p.

[35] Jung YG, Wuttiphan S, Peterson IM, Lawn BR (1999) Damage Modes in Dental Layer Structures. J. dent. res. 78:887-897. Epub 1999/05/18.

[36] Jung YG, Peterson IM, Kim DK, Lawn BR (2000) Lifetime-Limiting Strength Degradation from Contact Fatigue in Dental Ceramics. J. dent. res. 79:722-731. Epub 2000/03/23.

[37] Drummond JL, King TJ, Bapna MS, Koperski RD (2000) Mechanical Property Evaluation of Pressable Restorative Ceramics. Dent. mater. 16:226-233.

[38] Rhee YW, Kim HW, Deng Y, Lawn BR (2001) Contact-Induced Damage in Ceramic Coatings on Compliant Substrates: Fracture Mechanics and Design. J. ameri. ceram. soc. 84:1066-1072.

[39] Deng Y, Lawn BR, Lloyd IK (2002) Characterization of Damage Modes in Dental Ceramic Bilayer Structures. J. biomed. mater. res. 63:137-145.

[40] Frank FC LB (1967) On the Theory of Hertzian Fracture. Proc . r. soc. London Ser A. 299:291-306.

[41] Lawn B, Wilshaw R (1975) Indentation Fracture - Principles and Applications. J. mater. sci. 10:1049-1081.

[42] Rhee YW, Kim HW, Deng Y, Lawn BR (2001) Brittle Fracture versus Quasi Plasticity in Ceramics: A Simple Predictive Index. J. amer. cer. soc. 84:561-565.

[43] Tabor D. The Hardness of Metals. Oxford: Oxford University Press; 2000. ix, 175 p. p.

[44] Lawn BR, Lee KS, Chai H, Pajares A, Kim DK, Wuttiphan S, et al. (2000) Damage-Resistant Brittle Coatings. Adv. eng. mater. 2:745-748.

[45] Lawn BR, Pajares A, Zhang Y, Deng Y, Polack MA, Lloyd IK, et al. (2004) Materials Design in the Performance of All-Ceramic Crowns. Biomater. 25:2885-2892.

[46] Vanherle G, Smith DC, Minnesota Mining and Manufacturing Company. Dental Products Division. International symposium on posterior composite resin dental restorative materials. St. Paul, MN.: Minnesota Mining & Mfg. Co.; 1985. 558 p. p.

[47] Prasad SV, Calvert PD (1980) Abrasive Wear of Particle-Filled Polymers. J. mater. sci. 15:1746-1754.

[48] Roulet JF (1988) The Problems Associated with Substituting Composite Resins for Amalgam: A Status Report on Posterior Composites. J. dent. 16:101-113. Epub 1988/06/01.

[49] Truong VT, Tyas MJ (1988) Prediction of in vivo Wear in Posterior Composite Resins: A Fracture Mechanics Approach. Dent. mater. 4:318-327. Epub 1988/12/01.

[50] Higo Y, Damri D, Nunomura S, Kumada K, Sawa N, Hanaoka K, et al. (1991) The Fracture Toughness Characteristics of Three Dental Composite Resins. Biomed. mater. eng. 1:223-231. Epub 1991/01/01.

[51] Baran GR, McCool JI, Paul D, Boberick K, Wunder S (1998) Weibull Models of Fracture Strengths and Fatigue Behavior of Dental Resins in Flexure and Shear. J. biomed. mater. res. 43:226-233.

[52] Manhart J, Kunzelmann KH, Chen HY, Hickel R (2000) Mechanical Properties and Wear Behavior of Light-Cured Packable Composite Resins. Dent. mater. 16:33-40. Epub 2001/02/24.

[53] McCool JI, Boberick KG, Baran GR (2001) Lifetime Predictions for Resin-Based Composites using Cyclic and Dynamic Fatigue. J. biomed. mater. res. 58:247-253. Epub 2001/04/25.

[54] Ruddell DE, Maloney MM, Thompson JY (2002) Effect of Novel Filler Particles on the Mechanical and Wear Properties of Dental Composites. Dent. mater. 18:72-80. Epub 2001/12/13.

[55] Loughran GM, Versluis A, Douglas WH (2005) Evaluation of Sub-Critical Fatigue Crack Propagation in a Restorative Composite. Dent. mater. 21:252-261. Epub 2005/02/12.

[56] Da Rosa Rodolpho PA, Donassollo TA, Cenci MS, Loguercio AD, Moraes RR, Bronkhorst EM, et al. (2011) 22-Year Clinical Evaluation of the Performance of Two Posterior Composites with Different Filler Characteristics. Dent. mater. 27:955-963. Epub 2011/07/19.

[57] Hagger O; British patent 687299 and Swiss patent 2789461951.

[58] Buonocore MG (1955) A Simple Method of Increasing the Adhesion of Acrylic Filling Materials to Enamel Surfaces. J. dent. res. 34:849-853. Epub 1955/12/01.

[59] Silverstone LM, Dogon IL. Proceedings of an International Symposium on the Acid Etch Technique / edited by Leon M. Silverstone and I. Leon Dogon. St. Paul, Minn.: North Central Pub. Co.; 1975. xvii, 293 p.

[60] Dogon IL. Studies demonstrating the need for an intermediate resin of low viscosity for the acid etch technique. In: Proceedings of an international symposium on the acid etch technique. St Paul Minn: North Central Publishing Company; 1975.

[61] Bowen RL, Marjenhoff WA (1992) Development of an Adhesive Bonding System. Oper. dent. suppl. 5:75-80. Epub 1992/01/01.

[62] Fusayama T, Nakamura M, Kurosaki N, Iwaku M (1979) Non-Pressure Adhesion of a New Adhesive Restorative Resin. J. dent. res. 58:1364-1370. Epub 1979/04/01.

[63] Nakabayashi N, Kojima K, Masuhara E (1982) The Promotion of Adhesion by the Infiltration of Monomers into Tooth Substrates. J. biomed. mater. res. 16:265-273. Epub 1982/05/01.

[64] Asmussen ME. Adhesion of Restorative Resins to Dental Tissues - In: Posterior composite resin dental restorative materials. Vanherle G SD, editor. The Netherlands: Peter Szule Publishing CO; 1985.

[65] Kanca J, 3rd (1992) Improving Bond Strength through Acid Etching of Dentin and Bonding to Wet Dentin Surfaces. J. amer. dent. assoc. 123:35-43. Epub 1992/09/01.

[66] Gwinnett AJ (1992) Moist versus Dry Dentin: Its Effect on Shear Bond Strength. Am. j. Dent. 5:127-129. Epub 1992/06/01.

[67] Lin CP, Douglas WH (1994) Failure Mechanisms at the Human Dentin-Resin Interface - a Fracture-Mechanics Approach. J. biomech. 27:1037-1047.

[68] Toparli M, Aksoy T (1998) Fracture Toughness Determination of Composite Resin and Dentin/Composite Resin Adhesive Interfaces by Laboratory Testing and Finite Element Models. Dent. mater. 14:287-293. Epub 1999/06/24.

[69] Moulin P, Picard B, Degrange M (1999) Water Resistance of Resin-Bonded Joints with Time Related to Alloy Surface Treatment. J. dent. 27:79-87. Epub 1999/01/29.

[70] Tholey MJ, Waddell JN, Swain MV (2007) Influence of the Bonder on the Adhesion of Porcelain to Machined Titanium as Determined by the Strain Energy Release Rate. Dent. mater. 23:822-828. Epub 2006/08/16.

[71] Ichim I, Li Q, Loughran J, Swain MV, Kieser J (2007) Restoration of Non-Carious Cervical Lesions Part I. Modelling of Restorative Fracture. Dent. mater. 23:1553-1561. Epub 2007/03/30.

[72] Ichim IP, Schmidlin PR, Li Q, Kieser JA, Swain MV (2007) Restoration of Non-Carious Cervical Lesions Part II. Restorative Material Selection to Minimise Fracture. Dent. mater. 23:1562-1569. Epub 2007/03/30.

[73] Scherrer SS, Cesar PF, Swain MV (2010) Direct Comparison of the Bond Strength Results of the Different Test Methods: A Critical Literature Review. Dent. mater. 26:e78-93. Epub 2010/01/12.

[74] Soderholm KJ (2010) Review of the Fracture Toughness Approach. Dent. mater. 26:e63-77. Epub 2010/01/05.

[75] Howard K, Soderholm KJ (2010) Fracture Toughness of Two Dentin Adhesives. Dent. mater. 26:1185-1192. Epub 2010/10/12.

[76] Pilliar RM, Vowles R, Williams DF (1987) Fracture-Toughness Testing of Biomaterials Using a Mini-Short Rod Specimen Design. J. biomed. mater. res. 21:145-154.

[77] Tam LE, Dev S, Pilliar RM (1994) Fracture-Toughness of Dentin-Glass Ionomer Interfaces. J. dent. res. 73:183-183.

[78] Tam LE, Pilliar RM (1993) Fracture-Toughness of Dentin Resin-Composite Adhesive Interfaces. J. dent res. 72:953-959.

[79] Tam LE, Pilliar RM (1994) Effects of Dentin Surface Treatments on the Fracture-Toughness and Tensile Bond Strength of a Dentin-Composite Adhesive Interface. J. dent. res. 73:1530-1538.

[80] Tam LE, Pilliar RM (1994) Fracture Surface Characterization of Dentin-Bonded Interfacial Fracture-Toughness Specimens. J. dent. res. 73:607-619.

[81] Ausiello P, Franciosa P, Martorelli M, Watts DC (2011) Numerical Fatigue 3D-FE Modeling of Indirect Composite-Restored Posterior Teeth. Dent. mater. 27:423-430. Epub 2011/01/14.

[82] Jancar J (2011) Bond Strength of Five Dental Adhesives using a Fracture Mechanics Approach. J. mech. behav. biomed. mater. 4:245-254. Epub 2011/02/15.

[83] Kotousov A, Kahler B, Swain M (2011) Analysis of Interfacial Fracture in Dental Restorations. Dent. mater. 27:1094-1101. Epub 2011/08/10.

[84] Yamamoto T, Nishide A, Swain MV, Ferracane JL, Sakaguchi RL, Mornoi Y (2011) Contraction Stresses in Dental Composites Adjacent to and at the Bonded Interface as Measured by Crack Analysis. Acta biomater. 7:417-423.

Section 3

Fracture Mechanics Based Models of Fatigue

Chapter 5

Good Practice for Fatigue Crack Growth Curves Description

Sylwester Kłysz and Andrzej Leski

Additional information is available at the end of the chapter

1. Introduction

Fatigue life estimation and crack propagation description are the most important components in the analysis of life span of structural components but it may require time and expense to investigate it experimentally. For fatigue crack propagation studying in cases when it is difficult to obtain detailed results by direct experimentation computer simulation is especially useful. Hence, to be efficient, the crack propagation and durability of construction or structural component software should estimate the remaining life both experimentally and by simulation. The critical size of the crack or critical component load can be calculated using material constants which have been derived experimentally and from the constant amplitude crack propagation curve, crack size-life data and curve using crack propagation software. Many works in the field of fracture mechanics prove significant development in the numerical analysis of test data from fatigue crack propagation tests.

A simple stochastic crack growth analysis method is the maximum likelihood and the second moment approximation method, where the crack growth rate is considered as a random variable. A deterministic differential equation is used for the crack growth rate, while it is assumed that parameters in this equation are random variables. The analytical methods are implemented into engineering practice and are use to estimate of the statistics of the crack growth behavior (Elber, 1970; Forman et al., 1967; Smith, 1986).

Though many models have been developed, none of them enjoys universal acceptance. Due to the number and complexity of mechanisms involved in this problem, there are probably as many equations as there are researchers in the field. Each model can only account for one or several phenomenological factors - the applicability of each varies from case to case, there is no general agreement among the researchers to select any fatigue crack growth model in relation to the concept of fatigue crack behavior (Kłysz, 2001; Paris & Erdogan, 1963; Wheeler, 1972; Willenborg et al., 1971). Mathematical models proposed e.g. by Paris,

Forman, and further modifications thereof describe crack propagation with account taken of such factors as: material properties, geometry of a test specimen/structural component, the acting loads and the sequence of these loads (AFGROW, 2002; Kłysz et al., 2010a; NASGRO®, 2006; Newman, (1992); Skorupa, 1996). Application of the NASGRO equation, derived by Forman and Newman from NASA, de Koning from NLR and Henriksen from ESA, of the general form (AFGROW, 2002; NASGRO, 2006):

$$\frac{da}{dN} = C \cdot \left[\frac{(1-f)}{(1-R)} \cdot \Delta K \right]^n \cdot \frac{\left(1 - \frac{\Delta K_{th}}{\Delta K}\right)^p}{\left(1 - \frac{K_{\max}}{K_c}\right)^q} \tag{1}$$

has significantly extended possibilities of describing the crack growth rate tested according to the standard (ASTM E647). The coefficients stand for:

a – crack length [mm],
N – number of load cycles,
C, n, p, q – empirical coefficients,
R – stress ratio,
ΔK – the stress-intensity-factor (SIF) range that depends on the size of the specimen, applied loads, crack length, $\Delta K = K_{max} - K_{min}$ $[MPa\sqrt{m}]$,
ΔK_{th} – the SIF threshold, i.e. minimum value of ΔK, from which the crack starts to propagate:

$$\Delta K_{th} = \left(\Delta K_1 \cdot \left(\frac{a}{a+a_0} \right)^{\frac{1}{2}} \right) \cdot \frac{\left[\frac{1-R}{1-f} \right]^{(1+R \cdot C_{th})}}{(1-A_0)^{(1-R) \cdot C_{th}}} \tag{2}$$

or

$$\Delta K_{th} = \left(\Delta K_0 \cdot \left(\frac{a}{a+a_0} \right)^{\frac{1}{2}} \right) / \left(\frac{1-f}{(1-A_0)(1-R)} \right)^{(1+C_{th} R)} \tag{2a}$$

where: a_0 – structural crack length that depends on the material grain size [mm],

ΔK_0 – threshold SIF at $R \to 0$,
ΔK_1 – threshold SIF at $R \to 1$,
C_{th} – curve control coefficient for different values of R; equals 0 for negative R, equals 1 for $R \geq 0$, for some materials it can be found in the NASGRO database,
K_{max} – the SIF for maximum loading force in the cycle,
K_c – critical value of SIF,
f – Newman's function that describes the crack closure:

$$f = \begin{cases} \max\left(R, A_0 + A_1 R + A_2 R^2 + A_3 R^3\right) & for \;\; R \geq 0 \\ A_0 + A_1 R & for \;\; -2 \leq R < 0 \end{cases} \tag{3}$$

where A_0, A_1, A_2, A_3 coefficients are equal:

$$A_0 = (0.825 - 0.34 \cdot \alpha + 0.05 \cdot \alpha^2) \cdot \left[\cos\left(\frac{\pi}{2} \cdot \frac{S_{\max}}{\sigma_0}\right)\right]^{\frac{1}{\alpha}}, \tag{4}$$

$$A_1 = (0.415 - 0.071 \cdot \alpha) \cdot \frac{S_{\max}}{\sigma_0}, \tag{5}$$

$$A_2 = 1 - A_0 - A_1 - A_3, \tag{6}$$

$$A_3 = 2 \cdot A_0 + A_1 - 1. \tag{7}$$

α, S_{max}/σ_0 – Newman's empirical coefficients.

Determination of the above coefficients for equation that correctly approximates test data is difficult and causes some singularities described below, when the Least Squares Method (LSM) is used.

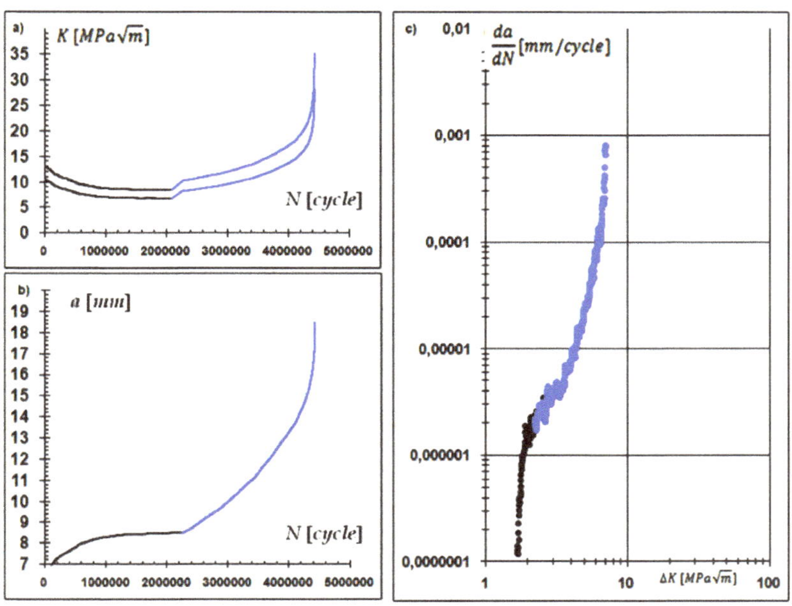

Figure 1. Fatigue crack propagation graphs: a) *K=f(N)*; b) *a=f(N)* and c) *da/dN=f(ΔK)*

The fatigue crack growth test results provide an illustration of relations such as: specimen stress intensity vs. number of cycles (*K=f(N)*), crack growth vs. number of cycles (*a=f(N)*); crack growth rate vs. stress intensity factor range (*da/dN=f(ΔK)*). These experimental curves can be presented, for example, in the graphical form shown in Fig. 1 (for a single specimen, two-stage test: stage I - decreasing ΔK test, black curve; stage II – constant amplitude test, blue curve).

Specifically, the *da/dN=f(ΔK)* plots can be obtained directly from the material test machine control software (e.g. by employing the compliance method and by using a clip gauge) or can be obtained by differentiating the *a=f(N)* curve after correlating it with *K=f(N)*. These plots, for single specimen tests, as well as for tests with multiple specimens under different load conditions (e.g. various stress ratio R values), can be successfully described analytically when appropriate mathematical models and equations are employed.

2. Test data

Fatigue tests for structural components durability analysis can be conducted with the RCT (*Round Compact Tension*) (Fig. 2a) or with SEN (*Single Edge Notch*) (Fig. 2b) or other specimens according to the corresponding ASTM E647 standard.

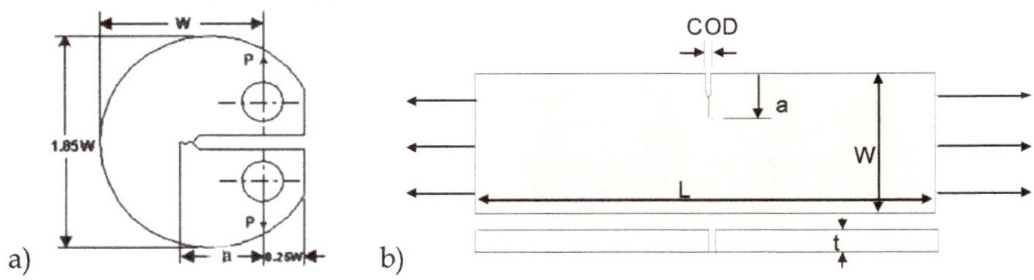

Figure 2. RCT & SEN specimens for fatigue crack propagation tests

The general formula that describes the stress intensity factor is as follows:

$$K_I = \frac{P}{B\sqrt{W}}Y, \tag{8}$$

where: P – applied force,
B, W – the specimen's thickness and width,
Y – the specimen's shape function (ASTM E647, Fuchs & Stephens 1980, Murakami 1987):
for the RCT specimen:

$$Y = \frac{\left(2+\frac{a}{W}\right)}{\left(1-\frac{a}{W}\right)^{\frac{1}{2}}} \cdot \left(0.886 + 4.64 \cdot \frac{a}{W} - 13.32 \cdot \left(\frac{a}{W}\right)^2 + 14.72 \cdot \left(\frac{a}{W}\right)^3 - 5.56 \cdot \left(\frac{a}{W}\right)^4\right) \tag{9}$$

for the SEN specimen:

$$Y = 1,12 - 0,231\left(\frac{a}{W}\right) + 10,55\left(\frac{a}{W}\right)^2 - 21,72\left(\frac{a}{W}\right)^3 + 30,39\left(\frac{a}{W}\right)^4 \tag{10}$$

where a/W is a non-dimensional crack length.

The compliance function to compute the crack length in the RCT specimen has the form:

$$\frac{a}{W} = 1 - 4.459\,u + 2.066\,u^2 - 13.041\,u^3 + 167.627\,u^4 - 481.4\,u^5 \tag{11}$$

for the SEN specimen (Bukowski & Kłysz 2003):

$$\frac{a}{W} = 1.407 - 4.132\,u + 3.928\,u^2 - 1.364\,u^3 \tag{12}$$

where: u – compliance described by the following formula:

$$u = \frac{1}{1 + \left(\frac{E \cdot B \cdot COD}{F}\right)^{0.5}}, \tag{13}$$

E – Young's modulus,
COD – Crack Opening Displacement.

An example of the *F-COD* relationship has been plotted in Fig. 3. The plot has been gained from the fatigue crack growth test conducted for the SEN specimen made from constructional steel subjected to constant amplitude loading with overloads (Bukowski & Kłysz, 2003). The records were taken in the course of statically applied 2-cycle overloads of 40% order at subsequent stages of crack propagation. Load base level and overload level were gradually reduced as the crack growth kept increasing and after non-linearity (hysteresis loop) had occurred in the *F-COD* plot. The objective was to avoid failure of the specimen in a subsequent overload cycle to be able then to continue the crack-propagation test. Any change in the angle of inclination of the rectilinear segment of each of the hysteresis loops (i.e. the *F/COD* proportion from formula (13)) is a measure of the specimen's compliance u and proves the crack length in the specimen under examination keeps growing.

Results presented below come from the examination of the 2024 aluminum alloy taken from the helicopter rotor blades (Kłysz & Lisiecki, 2009) or from the aircraft ORLIK's fuselage skins (Kłysz et al., 2010b) and are obtained for three values of test stress ratio $R = 0.1$; 0.5; 0.8, under laboratory conditions, with loading frequency 15 Hz. The crack length was measured with the *COD clip gauge* using the compliance method. The crack growth rate was determined using the polynomial method.

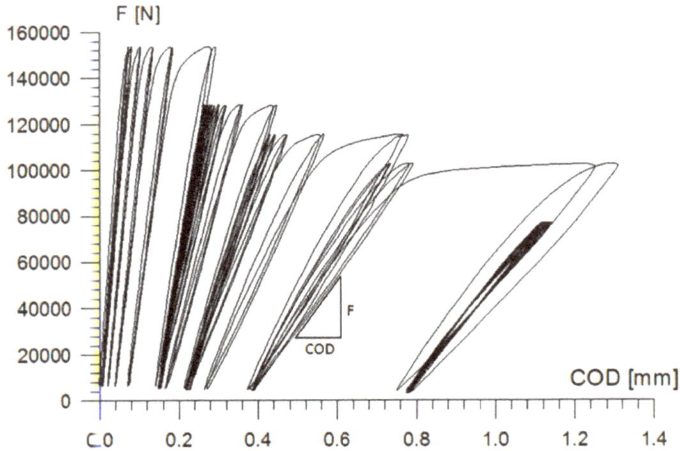

Figure 3. Relationship of *F-COD* recorded in subsequent overload cycles of fatigue crack growth test

3. Data analysis

Results of fatigue crack growth rate tests for 3 specimens (for *R* = *0.1; 0.5; 0.8*) are presented in Fig. 4 (Lisiecki & Łysz, 2007).

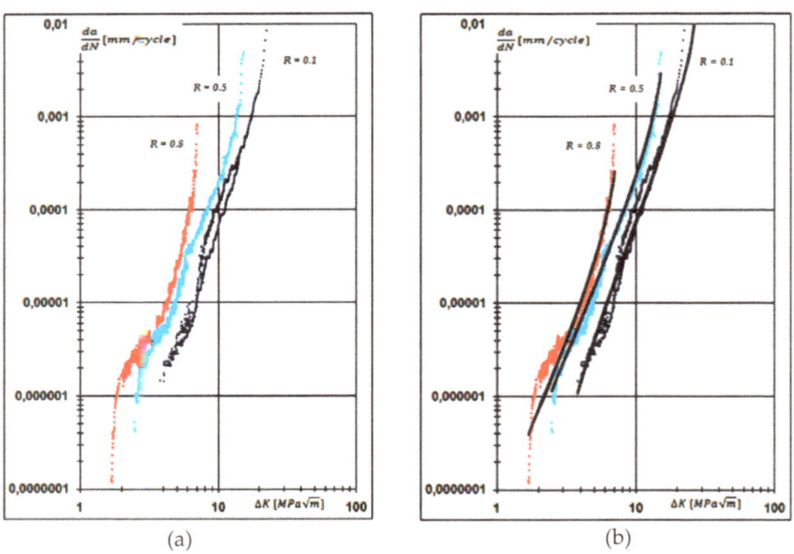

Figure 4. Fatigue crack growth rates in 3 specimens for different *R* values – a) test data, b) results of approximation

The NASMAT curve fitting algorithms use the least-squares error minimization routines in the log-log domain to obtain the corresponding constants using the NASMAT module contained within the NASGRO suite of software (NASGRO, 2006). The constants C and n, i.e. the main fit parameters, are determined through the minimization of the sum of squares of errors, where the error term corresponding to the *i-th* data pair $(\Delta K, da/dN)_i$ is (Forman et al., 2005):

$$e_i = \log\left(\frac{da}{dN}\right)_i - \log C - n\log\left[\frac{(1-f)}{(1-R)}\cdot\Delta K\right]_i - \log\cdot\left[\frac{\left(1-\frac{\Delta K_{th}}{\Delta K}\right)^p}{\left(1-\frac{K_{\max}}{K_c}\right)^q}\right]_i . \quad (14)$$

Values of da/dN are determined using the method of differentiating the dependence a-N with the secant or the polynomial method applied (AFGROW, 2002; ASTM 647; NASGRO, 2006).

Generally the curve fitting of crack growth data is an iterate process that consists in using established values of various constants (other than C and n), specifying the data sets that typify the material, applying the least-squares algorithm to compute C and n, and plotting the data for various R values with the curve fit of each stress ratio. The process is continued by making slight modifications in the entered values until the best fit to the test data is obtained. In general fitting the NASGRO equation is really a multi-step process involving:

- fitting or defining the threshold region;
- fitting or defining the critical stress intensity or toughness to be used at the instability asymptote;
- making initial assumptions on key parameters such as p and q;
- performing the least squares fit to obtain C and n; and finally;
- using engineering judgment to adjust the results for consistency and/or a desired level of conservatism.

For the LSM approximation of test data, analytical description thereof, and determination of coefficients of approximation equations, according to which the criterion used in the analysis is the minimum of the square sum:

$$S = \sum_{i=1}^{n}\left(\overline{y_i} - y_i\right)^2 \quad (15)$$

of deviations between values of the test data y_i and those of the approximated function $\bar{y}_i$. This method of approximation is characterized with the following properties that in some cases may be considered as disadvantages (Forman et al., 2005; White et al., 2005; Huang et al., 2005; Taheri et al., 2003):

- in respect of the order of magnitude, value of the sum S increases as magnitudes of approximated values increase, e.g. if values of test data are of the order of magnitude

10, 1000, 100000, with the scatter of 10%, the summed differences are of the order of magnitude 1, 100, 100000, and hence, dynamic changes in the total value of the sum S depend on values of differences – as a quadratic function it is characterized by a linear function of the derivative, which also means that for differences close to zero (e.g. 10^{-5}, 10^{-8}, etc.) this dynamic change is much smaller than for differences of higher magnitudes, which influences the „flexibility" of the performed approximation;

- if the test data significantly differ from each other in magnitude (e.g. from 1 to 100000 or from 10^{-8} to 10^{-2}), the approximated values near the lower threshold contribute much less to the total sum S than approximated values near the upper threshold; this means that, e.g. tens or hundreds of test data with differences in magnitude of 100% from value 1 are less significant in performing the approximation than one or a few data points which differ by 1% from value 100000.

According to the above stated example, the approximation is "asymmetric" since better approximation will be achieved for higher values of test data, neglecting differences around smaller values – an example of such approximation is shown in Fig. 4b, where one can see a good fit of theoretical description of 3 curves for large values of *da/dN* (over 10^{-4} mm/cycle) while there is an evident misfit for smallest values (below 10^{-5} mm/cycle). The presented approximation has been achieved by satisfying the LSM criterion, i.e. the minimum value of the sum S. When the test data are within a wide range of values, e.g. 5 orders of magnitude, i.e. from 10^{-2} to 10^{-7} mm/cycle, then differences between the highest values and the approximating function will have the largest effect on the square sum S of deviations while differences for small values, sometimes of 2-3 orders of magnitude, do not contribute much to the total sum S.

Hence, the misfit of the approximating function for low values of *da/dN*, practically for values lower by only 1-2 orders of magnitude than the maximum values of *da/dN*. Within this range the theoretical description is rather random and has rather no effect on the value of the sum S, which indicates that this criterion is rather useless for this type of analysis.

It seems reasonable to use one of the following criterion modifications, which will allow to remove the above stated problems:

- changing the form of the criterion, or
- using logarithmic values of *da/dN*,

$$S=\sum_{i=1}^{n}\left(\overline{\log y_i}-\log y_i\right)^2 \qquad or \qquad S=\sum_{i=1}^{n}\left(\log \overline{y_i}-\log y_i\right)^2. \tag{16}$$

In the present study the first variant has been examined (see section 3.3) due to the fact that it is more general since it does not limit itself only to positive values of predicted y_i, which is a requirement in the second variant. In the case of crack propagation test data all the *da/dN* values are positive; therefore the second variant could also be used.

Since the criterion for fitting the theoretical description to the test data in the form of equation (15) or (16), or any other, is closely connected with the number of approximated points (in the case under discussion, coordinates in the graph (da/dN_i,ΔK_i)), the quality of fit has to depend on:

- the distribution of the number of test points among particular curves,
- the distribution of test points on particular curves,

not to mention

- the scatter of test points and accuracy of finding them.

If the distribution of points among particular curves is not uniform, the approximation will show better fit *to the curves* with a larger number of points than to those with a smaller number of points – the contribution thereof to the pooled error included in the approximation criterion will be greater; the minimization thereof will occur around the larger data cluster. Similar situation occurs while fitting the description *to a given* experimentally gained *curve* – where the data concentration is larger, the approximation will be better than where there is less data, or where the data are only individual points. Therefore, essential to the analysis of test data and to description thereof is the regular distribution of the test data over the whole range to be subject to approximation. Since it is sometimes beyond the reach of researchers while recording the test data directly during the testing work, some modification or recalculation of the test data set may prove indispensable.

3.1. Data set modification

As clearly seen in Fig. 4, the number of points in the threshold and critical areas of the scope of the stress intensity factor ΔK is very small, which results from the specific nature of the performed test and data recording.

For crack growth rates lower than 10^{-6} mm/cycle the increment by 1 mm occurs after approx. 1 million cycles, i.e. the process is a long-lasting one, and the recording of the crack-length increment for instance every 0.01 mm gives 100 points of test data only (while in the case of taking records every 0.005 mm, the number of points will be 200). The testing work for even lower crack growth rates is still more time- and energy-consuming. With as little crack-length increments as these there is practically no chance that in single load cycles any random jump will occur in values of recorded data of the order of 0.01 or at least 0.001 mm (i.e. by approximately 3 – 4 orders of magnitude higher than the crack growth rate under examination). This provides relatively regular recording of crack lengths in the course of the testing work, i.e. for subsequent increments 0.01, 0.02, 0.03, ... mm, etc. (even if measurements are taken for crack-length increments by only a fraction of a millimetre, i.e. in a shorter time, which means for the number of cycles lower than the above-mentioned 1 million).

In the range of critical crack propagation, at the crack growth rate higher than 10^{-3} mm/cycle, the recordings of the crack length increments every 0.01 mm (as above) take place more frequently than every 10 cycles. For load-applying frequencies of 10 – 20 Hz this means 1 s long data-recording intervals in the course of the testing work. The final several millimetres' crack-length increment occurs as fast as over only several minutes of the testing work, with crack-length increments significantly increasing every cycle. Hence, at the testing rate getting as high, the number of test points remains relatively low and, because of these ever-growing increments, lower than the above-mentioned 100 or 200 points per every 1 mm of the crack length.

In the intermediate area of the graph (10^{-6} through 10^{-3} mm/cycle, i.e. covering 3 orders of magnitude of the *da/dN* value) the above-mentioned exemplary crack-length increments every 0.01 mm take place on a regular basis, however, with random fluctuations typical of the phenomenon under examination – there are no identical data recordings after 0.01, 0.02, 0.03, 0.04, … mm of crack-length increment, since instantaneous readings (variations) from the measuring sensors may cause that the data recording during the test, with the same recording criterion assumed, can occur for increments of, e.g. 0.01, 0.028, 0.038, 0.057, … mm, disturbing at the same time the regular basis of increments in the number of cycles between particular measurements. Fig. 5a illustrates the non-uniformity of such data-recording practice; the arrows point to where such disturbances have occurred, and after which the subsequent record is taken after the higher number of cycles. This, in turn, affects the crack growth rate. Calculation of the *da/dN* derivative based on the in this way recorded data must also be burdened with a random scatter, Fig. 5b, larger than that resulting from the properties of the material under examination.

To eliminate these incidental disturbances, the experimentally recorded time function may become smoothed by means of interpolation of results on the basis of any linear regression function (with either a straight line or a polynomial). Fig. 5c shows an example of such smoothening: presented with a full line is result of the 7-point regression, i.e. after having interpolated each point $(a_i;N_i)$, with account taken of 6 adjacent points: 3 points in front of and 3 points behind a given point $(a_i;N_i)$. It is evident that this smoothed curve represents in a reliable (or even better, in a more reliable way) the experimentally recorded dependence between measured quantities. On the other hand, the above-discussed disturbances have been removed from particular measurements.

Calculation of the da/dN_i derivative for any point of the plot $(a_i;N_i)$ can be carried out on the basis of linear or polynomial regression for e.g. 5, 7, or 9 adjacent points around a given *i*-th point. Fig. 5d shows result of the 5-point linear regression (2 points in front of the $(a_i;N_i)$ point, the $(a_i;N_i)$ point, 2 points behind the $(a_i;N_i)$ point), of calculations of the *da/dN* derivative against the unsmoothed plot *a-N*. What in this case is arrived at from the equation for the line of regression $y_i = m_i x_i + n_i$ (and more exactly, $a_i = m_i N_i + n_i$) is:

$$da/dN_i = (y_i)' = m_i. \tag{17}$$

In the case of linear regression with polynomials of the 2nd ($y_i = m_i x_i^2 + n_i x_i + l_i$) or 3rd ($y_i = m_i x_i^3 + n_i x_i^2 + l_i x_i - k_i$) order, the crack growth rate is calculated from the formulae, respectively:

$$da/dN_i = (y_i)' = 2m_i x_i + n_i, \tag{18}$$

$$da / dN_i = \left(y_i\right)' = 3m_i x_i^2 + 2n_i x_i + l_i. \quad (19)$$

What becomes evident is a considerable scatter of calculated values of the crack growth rate *da/dN*, and for points indicated with arrows it can be stated that:

- any measurement disturbance results in that the resulting (calculated) value of *da/dN* at one or two subsequent points is always lower than that for the point in question,
- the measurement disturbance is not expected to reflect the accelerated crack propagation, even though in the form of a local maximum, which all the more confirms the correctness of treating this disturbance as a random effect,
- where the disturbance occurs in the local-maximum area, it magnifies its value; however, the scale of this increase may prove too large as compared to the actual crack growth rate.

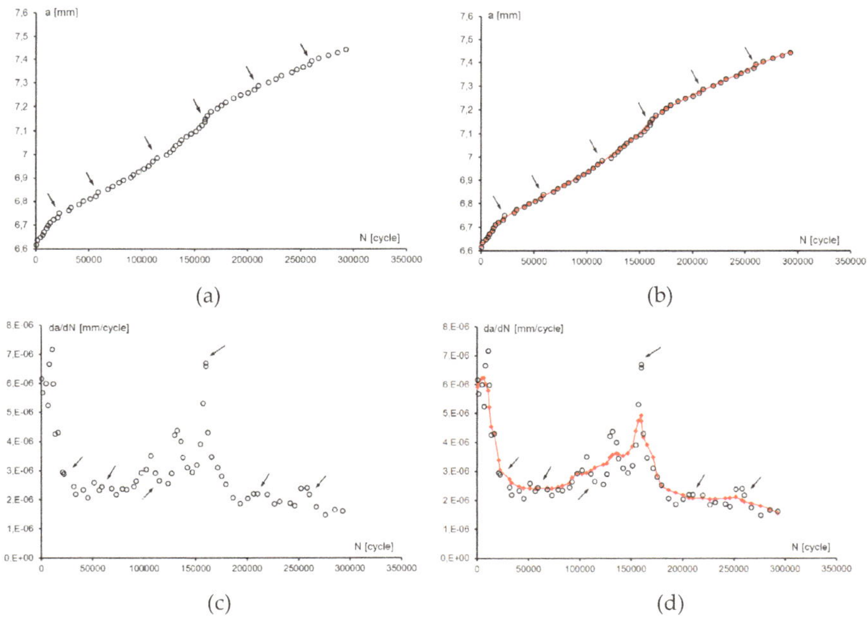

Figure 5. Calculated values of *da/dN* with corresponding test data *a-N*: 5-point linear regression, a) and b) – output data; c) and d) – smoothed data

If calculations are carried out for the smoothed curve *a-N* (Fig. 5c), the resulting *da/dN* curve presented in Fig. 5d takes the form of a solid line. The scatter of values of the crack growth rate over the whole range of calculations is much smaller. The local extremes have been maintained, however, slightly scaled down than in Fig. 5b.

In the case the regression used to calculate the *da/dN* derivative is carried out for a greater number of points adjacent to a given computational point, the corresponding curves look like in Fig. 5a (for 7-point regressions: 3 points in front of the $(a_i;N_i)$, the point in question $(a_i;N_i)$ and 3 points behind the $(a_i;N_i)$) and Fig. 5b (for 9-point regressions: 4 points in front of the $(a_i;N_i)$, the point in question $(a_i;N_i)$ and 4 behind the $(a_i;N_i)$). In all the cases the derivative of da/dN has been found from equation (17).

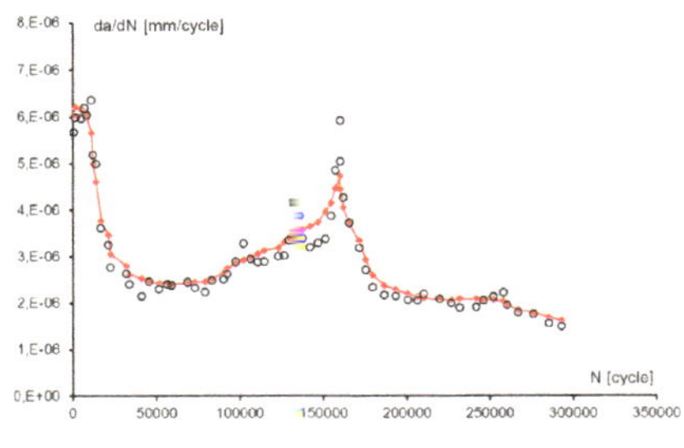

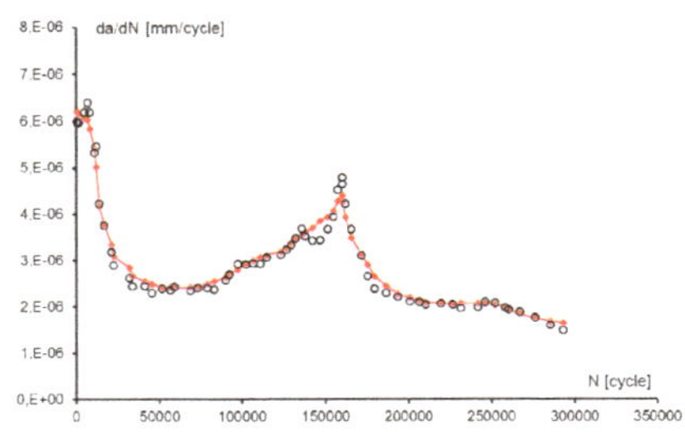

Figure 6. Calculated values of *da/dN* together with corresponding experimental data *a-N*: a) 7-point linear regression, b) 9 point linear regression

It is obvious that as the number of points taken into account in the regression analysis increases, the scatter of computational results gets reduced and the curve plotted for unsmoothed data (circles in Fig. 6) ever more resembles the curve plotted for smoothed data (solid line in Fig. 6). It is effected by the fact that the greater number of data accepted for regression brings the result closer to that of regression for smoothed data. There is of course some disadvantage: the greater number of data taken into account in regression analyses, the more reduced number of details referring to, e.g. local changes in value of *da/dN* are to be seen on the plotted curves. In the extreme, if all the points are subject to regression at once, the smoothed curve *a-N* would be a straight line and the *da/dN* curve would run horizontally. Another extreme consists in that the whole curve *a-N* would be described with only one regression equation, which in turn would provide the reliable mapping of the whole *a-N* curve; the *da/dN* derivative could be calculated by means of differentiating this equation. However feasible, it seems unpractical, work-consuming, more of the 'art for art's sake' category. Results presented in Figs 5 and 6 could be considered optimal: they provide good mapping of local changes in the approximated curves and do not require any complicated mathematical apparatus.

Characteristic of these plots (for both the unsmoothed and smoothed data) is that the calculated rates *da/dN* may be the same for different numbers of cycles *N* (hence, for different crack lengths *a* and different values of ΔK). This is the effect of more common, for this range of crack growth rate *da/dN*, occurrences of changes in the monotonicity of curves *a-N* than in threshold or critical ranges of *da/dN*-ΔK. Curves plotted in Figs 5 and 6 correspond to approx. 1-millimetre increment in the crack length (6.6 through 7.5 mm) and cover crack growth rates of $2 \div 4 \cdot 10^{-6}$ mm/cycle. At the further stage of the crack growth as the crack length increases, the crack growth rate increases as well, and before the crack reaches the critical growth range the calculated values of *da/dN* from the range 10^{-6} through 10^{-3} mm/cycle will repeatedly appear in the calculations. Hence, the number of measuring points recorded throughout the testing work for this range of *da/dN* will be higher than for threshold or critical ranges of *da/dN*-ΔK, what is to be seen also in Fig. 4.

Moreover, in practice, the plotting of a complete crack propagation curve *da/dN*-ΔK, i.e. starting from critical crack growth rates of 10^{-8} mm/cycle up to critical ones of 10^{-2} mm/cycle, is not performed in the course of one test only. This is closely related with difference in levels of ΔK for the stage of the specimen's precracking and the threshold range typical of the rates of 10^{-8} mm/cycle. The precracking usually finishes at higher values of ΔK, since it cannot proceed with the threshold growth rate. The reason is that it would take much more time than the test itself. Therefore, the test started after the specimen's precracking stage from the threshold values of the crack growth (change in the loading level from high to lower), would be connected with the crack growth retardation effect, which - in turn - would disturb test results in this area, i.e. it would not allow the researchers to gain the correct curve *da/dN*-ΔK. Such tests are usually conducted as a two-stage effort – see Fig. 1:

I stage – with exponentially decreasing ΔK ($\Delta K=\Delta K_0 e^{-ga}$), with constant relative gradient, i.e. $\frac{1}{\Delta K}\frac{\partial \Delta K}{\partial a} = -g$, up to having the left side of the plot within the threshold range. The test starts from the level of loads higher than those at the already completed stage of the specimen's precracking, so as to eliminate the crack growth retardation effect that appears as if the test is started at loads lower than those at the termination of the specimen's precracking.

The decreasing ΔK, starting from some suitably high value, and the crack length both cause that the crack growth rate becomes reduced to reach then the threshold range of the plot. At this stage, the *a-N* curve asymptotically approaches the horizontal line as the testing time increases. The testing time depends on the scientifically and economically justified needs of the researcher, although in practice this time much more depends on sensitivity of applied sensors, since both the level of applied loads and the crack opening size decrease for this range to values comparable to electric noise of the testing machine, which usually results in the test being automatically interrupted and the testing machine being stopped for crack growth rates lower than 10^{-8} mm/cycle. The at this stage obtained curve *a-N* and the propagation-curve section *da/dN*-ΔK may look like e.g. those presented in Fig. 7 (to be also seen in Fig. 1b).

II stage – at constant amplitude load (*CA test, constant amplitude test*) up to the acquisition of the right side within the critical range. The test is carried out at the level of loads higher than the level at which the stage I was completed; as the crack length increases, there is a systematic increase in the ΔK, up to the moment the critical value is reached, at which the specimen fails. The at this stage obtained curve *a-N* and the propagation-curve section *da/dN*-ΔK may look like e.g. those presented in Fig. 7 (to be also seen in Fig. 4).

The total result of both the stages has been presented in Fig. 8 – both the curves from Figs 6 and 7 complement one another to full propagation-curve plot *a-N* and *da/dN*-ΔK: experimentally found points in the form of circles, curves smoothed in the form of full lines.

It is quite clear that the mid section (range) of the *da/dN*-Δ*K* curve contains much more experimentally gained points despite the same criterion for data recording in the course of testing work for all three ranges, and also, independently of the fact that both the curves overlap over some specific section common to both of them. Furthermore, the plot presents the above-discussed changes in the monotonicity of how they run, independently of whether the calculations of the *da/dN* derivative have been conducted for unsmoothed or smoothed data – Fig. 8.

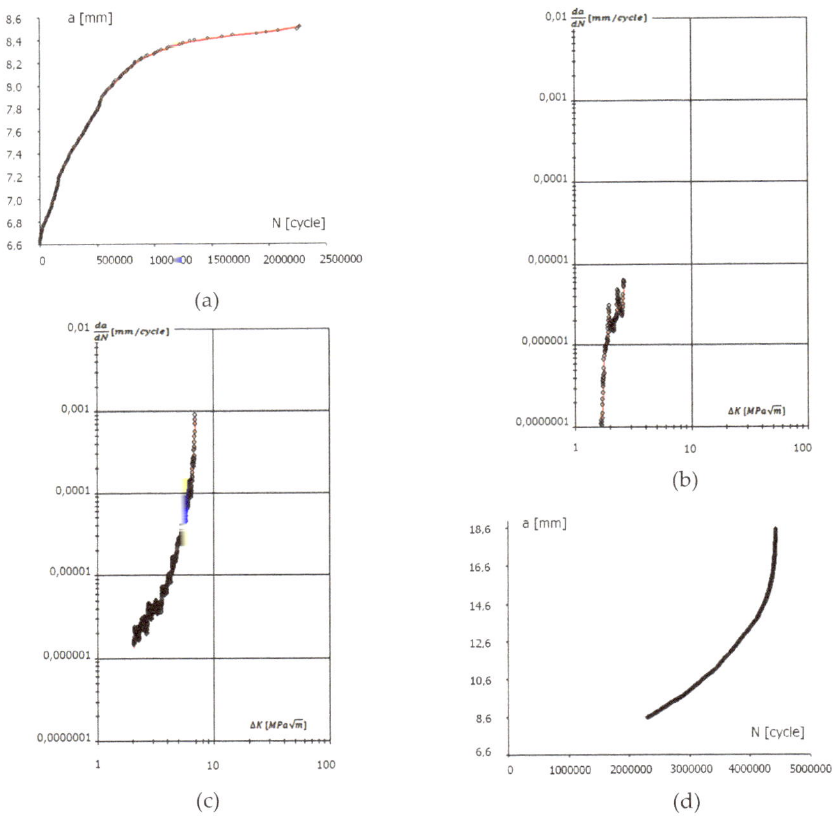

Figure 7. Curves a-N and da/dN-ΔK for the first (I) – a), b), and the second (II) test stages – c), d)

3.2. A Method of Regular Curves Mapping (MRCM)

Disturbances in the run, monotonicity of curves *da/dN*-Δ*K* as well as different measuring-data density in particular areas of the graph do not serve well any attempts to theoretically describe these curves. As mentioned earlier, the least squares methods better fit regression curves to areas where there is more approximated points, in the case given consideration, in the middle ranges of the *da/dN-ΔK* curves. To eliminate this effect, application of the Authors' Method of Regular Curves Mapping (MRCM) to approximate the *da/dN-ΔK* curves is advisable.

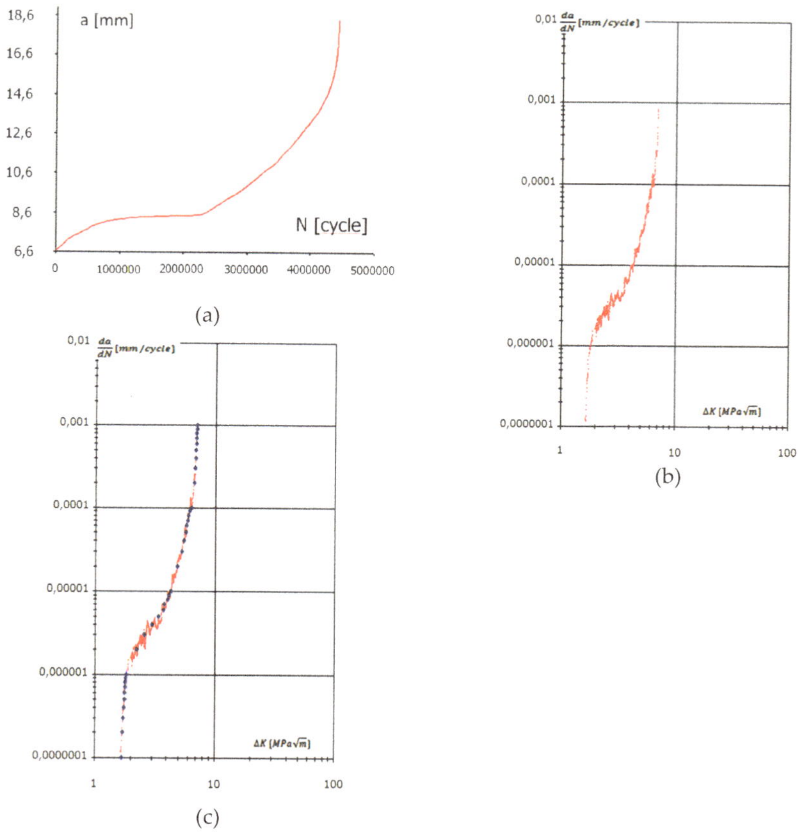

Figure 8. Curves *a-N* and *da/dN-ΔK* and how they run at the I and II test stages: – results for data after the curve has been smoothed - a), b), and effect of having applied the MRCM – c)

The MRCM technique of mapping test data consists in fixing, at regular intervals (along axes x or y), the *k* number of representative points in the data set under analysis (upon the experimentally gained curve). The following actions are to be taken:

a. determined are selected values of coordinates x_i (or y_i), for which the above-mentioned points will be fixed ($i = 1,2, \ldots,k$),
b. from the curve under analysis, point x'_i (or y'_i) is fixed, of coordinate value closest to the assumed value of x_i (or y_i), and $2m$ of adjacent points – by assumption, in half these are

points of values lower than x_i (or y_i) and in half - of higher values, (m is equal to, e.g. 2, 3, 4 or 5),

c. a set of in this way gained data $2m+1$, (x'_{i-m}, y_{i-m}) through (x'_{i+m}, y_{i+m}) (or (x_{i-m}, y'_{i-m}) through (x_{i+m}, y'_{i+m})) – grouped around some selected value of x_i (or y_i) is subject to regression with any function to determine the approximated value of y_i^* (or x_i^*) corresponding to the selected value of x_i (or y_i),

d. the point of coordinates (x_i, y_i^*) (or (x_i^*, y_i)) is mapped on the curve under analysis – as the i-th representative data item found on the basis of the assumed criteria,

e. steps b) through d) are repeated for subsequent k number of values determined in a), until a set of k number of points that represent (map) the curve is obtained.

The effect of the in this way performed mapping of values of da/dN, regularly distributed within particular intervals (orders of magnitude), in selected $k = 37$ points, for the curve shown in Fig. 8b, is presented in Fig. 8c. The points in question:

- well represent (map) the curve under analysis,
- are equidense distributed within the whole range of da/dN variability,
- do not show any more or less significant fluctuations/scatter of values resulting from, e.g. random measuring-data dispersions.

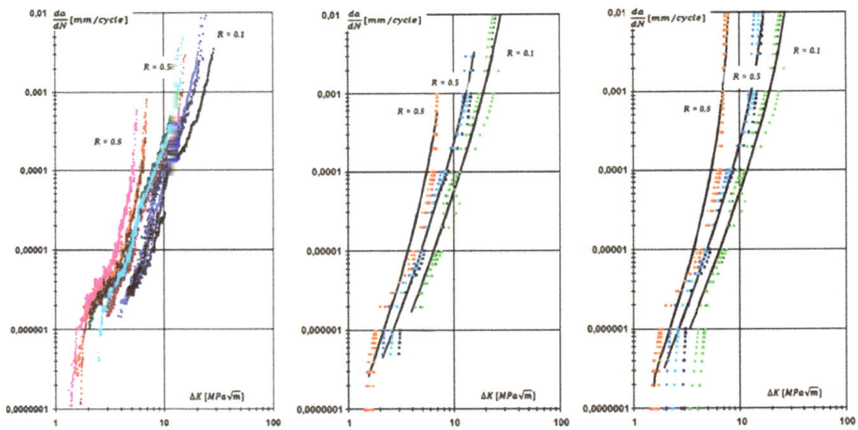

(a) (b) (c)

Figure 9. Experimentally gained curves da/dN-ΔK for 9 specimens tested at different stress ratios R a) and the same curves having been mapped with points using the MRCM, b) and with extrapolated points that 'perform' the mapping according to the MRCM, c) – together with approximation thereof with the NASGRO equation

The set of points that map the curve seems to give good basis, owing to the above described features, for analyses of theoretical description of a given, experimentally gained curve. In the case of nine (9) curves that correspond to tests with three (3) values of the stress ratio R –

Fig. 9a, the result of experimentally gained data modification with the MRCM applied (points in the graph) is presented in Fig. 9b, together with the data approximation by means of the NASGRO equation, with the LSM criterion used, according to formula (15).

The MRCM technique also enables, if need be and with scientific correctness maintained, the extrapolation of the mapping points *beyond the range* of recorded test data, i.e. into the area of crack growth rates lower (the threshold range) or higher (the critical range) than those recorded experimentally, with their tendency to change which is peculiar to those areas, on the basis of regression at boundary (in the graph – lower or upper) points of experimentally gained curves – the extrapolation result has been shown in Fig. 9c. Application of extrapolation to prepare data for the analytical modelling may prove advantageous in the case the particular experimentally gained curves show different ranges of values, and hence, different numbers of mapping points. After correctly performed extrapolation one can arrive at the situation when they are equalized, which means the same 'power' of each of the with the regression method approximated curves.

3.3. Modifications of the LSM method criterion

In order to eliminate the approximation misfit as shown in Fig. 1 and to improve the quality of approximation, modification of formula (15) takes the following form:

$$S^* = \sum_{i=1}^{n}\left(\frac{\overline{y_i} - y_i}{y_i}\right)^2 = \sum_{i=1}^{n}\left(\frac{\overline{y_i}}{y_i} - 1\right)^2 . \tag{20}$$

The fraction in brackets in formula (13), as a relative error, is a measure of deviation independent of the order of magnitude of compared values (approximated and approximating ones), so that the contribution of all the test data is equally "strong" to the total error *S*, which should have good effect on the approximation within the whole range, since:

- each value among test data y_i has equal contribution to the sum S^*, independent of its magnitude 10^{-7}, 10^{-2}, 1 or 100000 (i.e. it fits in any magnitude range) – always a deviation of e.g. 10-, 50-, 200-percent of approximating value will give a component of the sum S^* equal to 0.01, 0.25, 4, respectively;
- the criterion assures that the achieved approximation is "symmetric", i.e. the degree of approximation around lower and higher values is the same;
- disadvantages of the criterion described with formula (15) are no longer valid.

The criterion described with formula (20) has also some specific property: if the approximating value equals zero (i.e. for the approximation smaller by 100%) or it is twice as big as the approximated value (i.e. for approximation larger by 100%), then independently of the approximated value the component of the sum S^* will equal 1.

In order to carry out the approximation of test data it is necessary to calculate coefficients of the approximating equation used to determine $\bar{y}_i$. Equation (1), after applying logarithms, takes the form:

$$\log\left(\frac{da}{dN}\right) = \log(C) + n \cdot \log\left[\frac{(1-f)}{(1-R)} \cdot \Delta K\right] + p \cdot \log\left(1 - \frac{\Delta K_{th}}{\Delta K}\right) - q \cdot \log\left(1 - \frac{K_{\max}}{K_c}\right), \quad (21)$$

and can be presented in the following general way:

$$\bar{y} = b_0 + b_1 \cdot f_1 + b_2 \cdot f_2 + b_3 \cdot f_3. \quad (22)$$

Coefficients b_i are directly connected with C, n, p and q ($b_0 = log(C)$, $b_1 = n$, $b_2 = p$, $b_3 = -q$), whereas functions f_i depend on ΔK and R and include all the remaining coefficients of the NASGRO equation. Coefficients b_i of the approximating equation are calculated from the minimum condition of the equation (20), i.e.:

$$\frac{\partial S}{\partial b_k} = \frac{\partial\left(\sum_{i=1}^{n}\left(\frac{b_0 + b_1 \cdot f_{1,i} + b_2 \cdot f_{2,i} + b_3 \cdot f_{3,i} - y_i}{y_i}\right)^2\right)}{\partial b_k} = 0 \quad (23)$$

$$k = 1,2,3,4.$$

This leads to the following system of equations:

$$\begin{aligned}
\frac{\partial S}{\partial b_0} &= 2 \cdot \sum_{i=1}^{n}\left[\left(\frac{b_0 + b_1 \cdot f_{1,i} + b_2 \cdot f_{2,i} + b_3 \cdot f_{3,i} - y_i}{y_i}\right) \cdot \frac{1}{y_i}\right] = 0,\\
\frac{\partial S}{\partial b_1} &= 2 \cdot \sum_{i=1}^{n}\left[\left(\frac{b_0 + b_1 \cdot f_{1,i} + b_2 \cdot f_{2,i} + b_3 \cdot f_{3,i} - y_i}{y_i}\right) \cdot \frac{f_1}{y_i}\right] = 0,\\
\frac{\partial S}{\partial b_2} &= 2 \cdot \sum_{i=1}^{n}\left[\left(\frac{b_0 + b_1 \cdot f_{1,i} + b_2 \cdot f_{2,i} + b_3 \cdot f_{3,i} - y_i}{y_i}\right) \cdot \frac{f_2}{y_i}\right] = 0,\\
\frac{\partial S}{\partial b_3} &= 2 \cdot \sum_{i=1}^{n}\left[\left(\frac{b_0 + b_1 \cdot f_{1,i} + b_2 \cdot f_{2,i} + b_3 \cdot f_{3,i} - y_i}{y_i}\right) \cdot \frac{f_3}{y_i}\right] = 0.
\end{aligned} \quad (24)$$

It is a system of 4 linear equations with 4 unknowns b_i, which after transformation takes a form:

$$\begin{aligned}
&n\sum_{i=1}^{n}\frac{1}{y_i} - b_0\sum_{i=1}^{n}\frac{1}{y_i^2} - b_1\sum_{i=1}^{n}\frac{f_{1,i}}{y_i^2} - b_2\sum_{i=1}^{n}\frac{f_{2,i}}{y_i^2} - b_3\sum_{i=1}^{n}\frac{f_{3,i}}{y_i^2} = 0,\\
&n\sum_{i=1}^{n}\frac{f_{1,i}}{y_i} - b_0\sum_{i=1}^{n}\frac{f_{1,i}}{y_i^2} - b_1\sum_{i=1}^{n}\frac{f_{1,i}^2}{y_i^2} - b_2\sum_{i=1}^{n}\frac{f_{1,i}f_{2,i}}{y_i^2} - b_3\sum_{i=1}^{n}\frac{f_{1,i}f_{3,i}}{y_i^2} = 0,\\
&n\sum_{i=1}^{n}\frac{f_{2,i}}{y_i} - b_0\sum_{i=1}^{n}\frac{f_{2,i}}{y_i^2} - b_1\sum_{i=1}^{n}\frac{f_{1,i}f_{2,i}}{y_i^2} - b_2\sum_{i=1}^{n}\frac{f_{2,i}^2}{y_i^2} - b_3\sum_{i=1}^{n}\frac{f_{2,i}f_{3,i}}{y_i^2} = 0,\\
&n\sum_{i=1}^{n}\frac{f_{3,i}}{y_i} - b_0\sum_{i=1}^{n}\frac{f_{3,i}}{y_i^2} - b_1\sum_{i=1}^{n}\frac{f_{1,i}f_{3,i}}{y_i^2} - b_2\sum_{i=1}^{n}\frac{f_{2,i}f_{3,i}}{y_i^2} - b_3\sum_{i=1}^{n}\frac{f_{3,i}^2}{y_i^2} = 0.
\end{aligned} \quad (25)$$

and is easily solved by subtracting in the following steps:

- eliminating b_0

$$
\begin{aligned}
&\frac{n\sum_{i=1}^{n}\frac{1}{y_i}}{\sum_{i=1}^{n}\frac{1}{y_i^2}}-\frac{n\sum_{i=1}^{n}\frac{f_{1,i}}{y_i^2}}{\sum_{i=1}^{n}\frac{f_{1,i}}{y_i^2}}-b_1\left(\frac{\sum_{i=1}^{n}\frac{f_{1,i}}{y_i^2}}{\sum_{i=1}^{n}\frac{1}{y_i^2}}-\frac{\sum_{i=1}^{n}\frac{f_{1,i}^2}{y_i^2}}{\sum_{i=1}^{n}\frac{f_{1,i}}{y_i^2}}\right)-b_2\left(\frac{\sum_{i=1}^{n}\frac{f_{2,i}}{y_i^2}}{\sum_{i=1}^{n}\frac{1}{y_i^2}}-\frac{\sum_{i=1}^{n}\frac{f_{1,i}f_{2,i}}{y_i^2}}{\sum_{i=1}^{n}\frac{f_{1,i}}{y_i^2}}\right)-\\
&-b_3\left(\frac{\sum_{i=1}^{n}\frac{f_{3,i}}{y_i^2}}{\sum_{i=1}^{n}\frac{1}{y_i^2}}-\frac{\sum_{i=1}^{n}\frac{f_{1,i}f_{3,i}}{y_i^2}}{\sum_{i=1}^{n}\frac{f_{1,i}}{y_i^2}}\right)=0,\\
&\frac{n\sum_{i=1}^{n}\frac{1}{y_i}}{\sum_{i=1}^{n}\frac{1}{y_i^2}}-\frac{n\sum_{i=1}^{n}\frac{f_{2,i}}{y_i^2}}{\sum_{i=1}^{n}\frac{f_{2,i}}{y_i^2}}-b_1\left(\frac{\sum_{i=1}^{n}\frac{f_{1,i}}{y_i^2}}{\sum_{i=1}^{n}\frac{1}{y_i^2}}-\frac{\sum_{i=1}^{n}\frac{f_{1,i}f_{2,i}}{y_i^2}}{\sum_{i=1}^{n}\frac{f_{2,i}}{y_i^2}}\right)-b_2\left(\frac{\sum_{i=1}^{n}\frac{f_{2,i}}{y_i^2}}{\sum_{i=1}^{n}\frac{1}{y_i^2}}-\frac{\sum_{i=1}^{n}\frac{f_{2,i}^2}{y_i^2}}{\sum_{i=1}^{n}\frac{f_{2,i}}{y_i^2}}\right)-\\
&-b_3\left(\frac{\sum_{i=1}^{n}\frac{f_{3,i}}{y_i^2}}{\sum_{i=1}^{n}\frac{1}{y_i^2}}-\frac{\sum_{i=1}^{n}\frac{f_{2,i}f_{3,i}}{y_i^2}}{\sum_{i=1}^{n}\frac{f_{2,i}}{y_i^2}}\right)=0,\\
&\frac{n\sum_{i=1}^{n}\frac{1}{y_i}}{\sum_{i=1}^{n}\frac{1}{y_i^2}}-\frac{n\sum_{i=1}^{n}\frac{f_{3,i}}{y_i^2}}{\sum_{i=1}^{n}\frac{f_{3,i}}{y_i^2}}-b_1\left(\frac{\sum_{i=1}^{n}\frac{f_{1,i}}{y_i^2}}{\sum_{i=1}^{n}\frac{1}{y_i^2}}-\frac{\sum_{i=1}^{n}\frac{f_{1,i}f_{3,i}}{y_i^2}}{\sum_{i=1}^{n}\frac{f_{3,i}}{y_i^2}}\right)-b_2\left(\frac{\sum_{i=1}^{n}\frac{f_{2,i}}{y_i^2}}{\sum_{i=1}^{n}\frac{1}{y_i^2}}-\frac{\sum_{i=1}^{n}\frac{f_{2,i}f_{3,i}}{y_i^2}}{\sum_{i=1}^{n}\frac{f_{3,i}}{y_i^2}}\right)-\\
&-b_3\left(\frac{\sum_{i=1}^{n}\frac{f_{3,i}}{y_i^2}}{\sum_{i=1}^{n}\frac{1}{y_i^2}}-\frac{\sum_{i=1}^{n}\frac{f_{3,i}^2}{y_i^2}}{\sum_{i=1}^{n}\frac{f_{3,i}}{y_i^2}}\right)=0
\end{aligned}
\tag{26}
$$

what gives 3 equations of the general form:

$$
\begin{aligned}
&B_k-b_1B_{1,k}-b_2B_{2,k}-b_3B_{3,k}=0\\
&k=\ 1,2,3.
\end{aligned}
\tag{27}
$$

- eliminating b_1

$$\frac{B_1}{B_{1,1}} - \frac{B_2}{B_{2,1}} - b_2\left(\frac{B_{1,2}}{B_{1,1}} - \frac{B_{2,2}}{B_{2,1}}\right) - b_3\left(\frac{B_{1,3}}{B_{1,1}} - \frac{B_{2,3}}{B_{2,1}}\right) = 0,$$
$$\frac{B_1}{B_{1,1}} - \frac{B_3}{B_{3,1}} - b_2\left(\frac{B_{1,2}}{B_{1,1}} - \frac{B_{3,2}}{B_{3,1}}\right) - b_3\left(\frac{B_{1,3}}{B_{1,1}} - \frac{B_{3,3}}{B_{3,1}}\right) = 0. \quad (28)$$

what gives 2 equations of the general form:

$$C_k - b_2 C_{2,k} - b_3 C_{3,k} = 0$$
$$k = 2,3. \quad (29)$$

- eliminating b_2

$$\frac{C_2}{C_{2,2}} - \frac{C_3}{C_{2,3}} - b_3\left(\frac{C_{3,2}}{C_{2,2}} - \frac{C_{3,3}}{C_{2,3}}\right) = 0, \quad (30)$$

hence:

$$b_3 = \frac{\dfrac{C_2}{C_{2,2}} - \dfrac{C_3}{C_{2,3}}}{\dfrac{C_{3,2}}{C_{2,2}} - \dfrac{C_{3,3}}{C_{2,3}}} = \frac{\dfrac{\left(\frac{B_1}{B_{1,1}} - \frac{B_2}{B_{2,1}}\right)}{\left(\frac{B_{1,2}}{B_{1,1}} - \frac{B_{2,2}}{B_{2,1}}\right)} - \dfrac{\left(\frac{B_1}{B_{1,1}} - \frac{B_3}{B_{3,1}}\right)}{\left(\frac{B_{1,2}}{B_{1,1}} - \frac{B_{3,2}}{B_{3,1}}\right)}}{\dfrac{\left(\frac{B_{1,3}}{B_{1,1}} - \frac{B_{2,3}}{B_{2,1}}\right)}{\left(\frac{B_{1,2}}{B_{1,1}} - \frac{B_{2,2}}{B_{2,1}}\right)} - \dfrac{\left(\frac{B_{1,3}}{B_{1,1}} - \frac{B_{3,3}}{B_{3,1}}\right)}{\left(\frac{B_{1,2}}{B_{1,1}} - \frac{B_{3,2}}{B_{3,1}}\right)}}. \quad (31)$$

Hence, coefficient b_2 can be calculated from one of the formulae (29); secondly, coefficient b_1 from one of equations (27), and finally, coefficient b_0 from one of equations (25).

The in this way found coefficients of the NASGRO equation enable approximation of curves *da/dN-ΔK* from Fig. 9 to the form shown in Fig. 10a. Considerable improvement in the theoretical (analytical) description for the whole range of plotted curves is evident.

Both criteria (15) and (20) have also some disadvantage consisting in that if the approximating value $\bar{y}_i$ is much smaller than the approximated value y_i (i.e. by 3, 5, 7 orders of magnitude) or simply close to zero then the component of the sum *S* and *S** is close to the squared value y_i (in case of (15)) or to 1 (in case of (20)), independently of how these two values differ from each other.

Obviously, it is important whether the approximation and behavior of the approximating curve near value y_i at the level of e.g. 10^{-6} and lower (i.e. for strongly decreasing values within the "threshold" range of the graph) take place at the level of 10^{-8}, 10^{-12} or 10^{-20} (what is not hard to achieve for curves showing strong vertical courses on graphs plotted with the logarithmic scale applied); it is much better when the possible difference between values $\bar{y}_i$ and y_i is not too large.

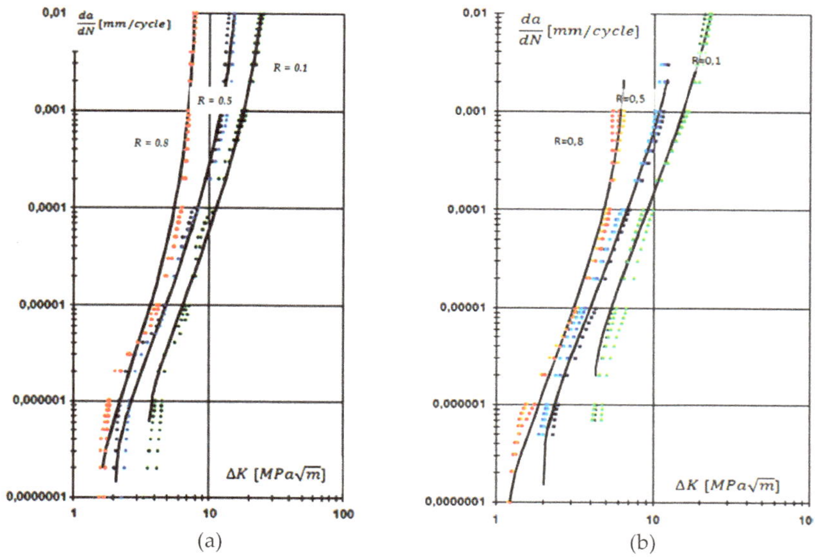

Figure 10. Result of approximation of curves from Fig. 9 with the NASGRO equation with the LSM criterion applied: a) by equation (20), b) by equation (32)

Due to dynamic changes around value $\bar{y}_i$ equal to zero (completely monotonic, as for the second-degree polynomial), functions (15) and (20) are practically insensitive to that the approximated value equals e.g. 0.01, 10^{-5}, 10^{-8} or 10^{-20}. Hence, it is most preferable if the LSM approximating criterion takes such cases into account.

Therefore, a modification is proposed to transform the criterion into the following form:

$$S^{**} = \sum_{i=1}^{n} \left(\frac{\bar{y}_i}{y_i} - 1 \right) \left(1 - \frac{y_i}{\bar{y}_i} \right). \tag{32}$$

Owing to this for both large values $\bar{y}_i$ (much different from the approximated value y_i) and small values (approaching zero) with respect to value y_i, the components of the sum take significant values, i.e. in both cases they give a significant (although - as it can be seen - diverse/unsymmetrical for each of the cases) contribution to the total approximation error – as shown in Fig. 11. In order to make the S_i components of the sum (32) and the total sum S^{**} as an approximation criterion reaches the minimum (not the maximum, as in Fig. 11) and also, when the reversal of sign takes place between the approximated value y_i and the approximating value $\bar{y}_i$), the following form would be better:

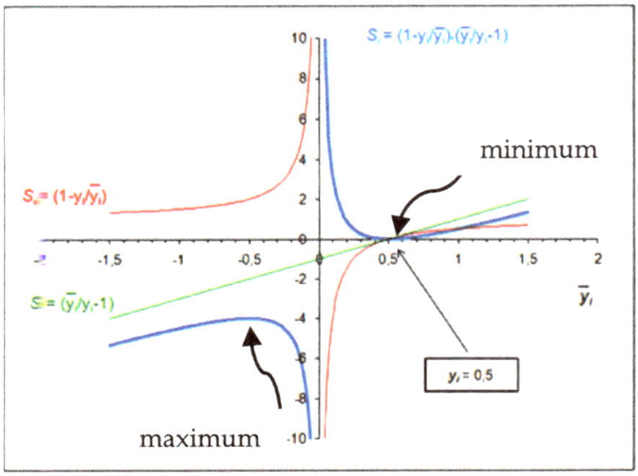

Figure 11. Component of the sum for the approximation criterion (32)

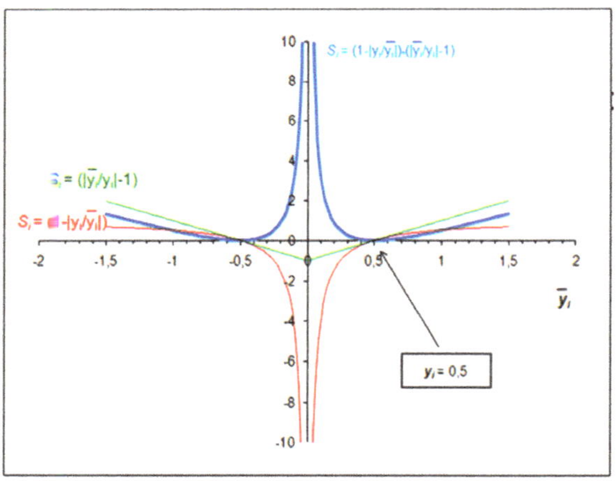

Figure 12. Component of the sum for the approximation criterion (32a)

$$S^{**} = \sum_{i=1}^{n} \left(\left| \frac{\overline{y_i}}{y_i} \right| - 1 \right) \left(1 - \left| \frac{y_i}{\overline{\overline{y_i}}} \right| \right). \tag{32a}$$

Both extremes of the S function for both positive and negative values of $\bar{y}_i$ are the minima shown in Fig. 12. Approximation criterion functions for (15), (20) and (32) (and their components) as related to approximating values $\bar{y}_i$, for:

- different approximated values y_i equal to 5; 2; 1; 0.25; 0.01; 0.00001,
- the same range of variability of $\bar{y}_i$, i.e. ($-3y_i$, $3y_i$), in order to show the $\bar{y}_i \to 0$ effect, are shown in Fig. 13.

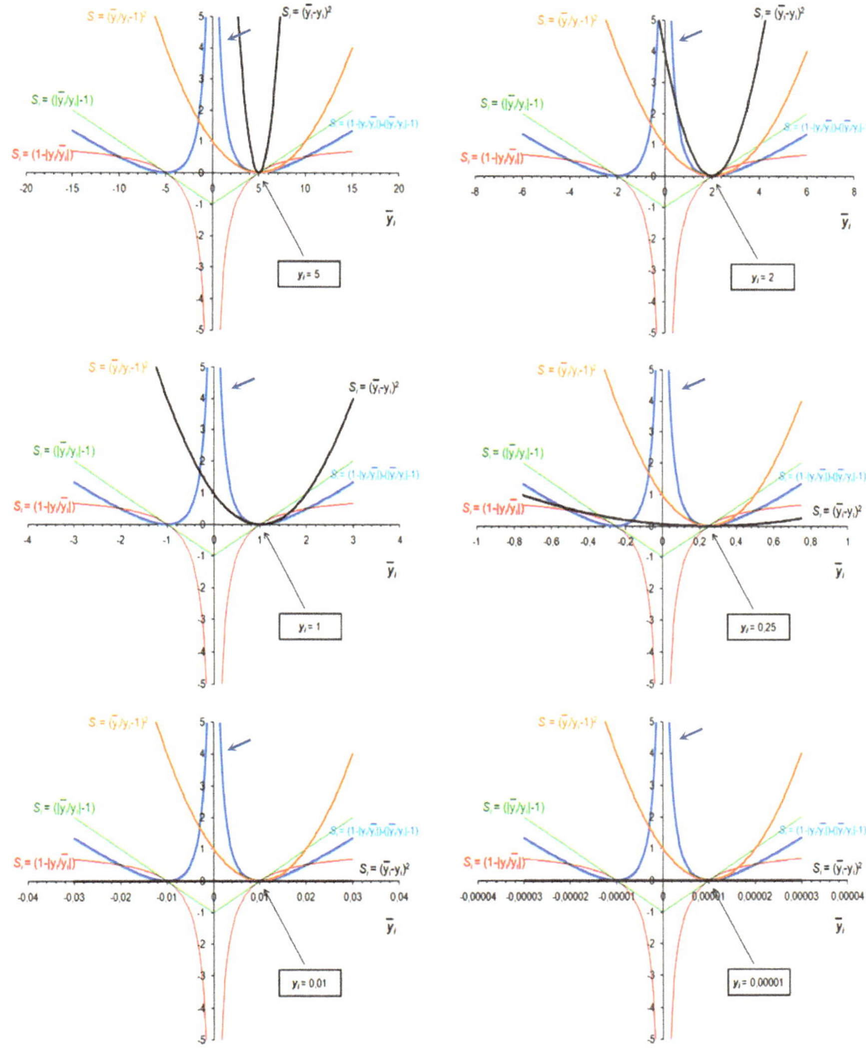

Figure 13. Approximation criterion functions for different approximated values y_i for the ($-3y_i$, $3y_i$) interval – arrow for equation (32a)

All advantages and disadvantages of the above presented LSM approximation criteria can be seen on the graphs above, in particular:

- significant dependence of values of components of the sum S (formula (15)) on the approximated value y_i;

- invariability of values of components of sums S^* (formula (20)) and S^{**} (formulae (32) and (32a)) on all the graphs, i.e. for any approximated value y_i;
- no response of the components of sums S and S^* to the $\bar{y}_i \rightarrow 0$ effect and dynamic change in the components of the sum S^{**} near value $\bar{y}_i = 0$.

The only curve that changes in the graphs presented in Fig. 13 is the plot for components of the sum S graph, i.e. for the standard form of the LSM.

Result of approximation with criterion (32a) applied is shown in Fig. 10b – for data sets with no extrapolation points. The same approximation for only 1 specimen tested at different R is shown in Fig. 14a, and for only 2 specimens tested at different R - in Fig. 14b.

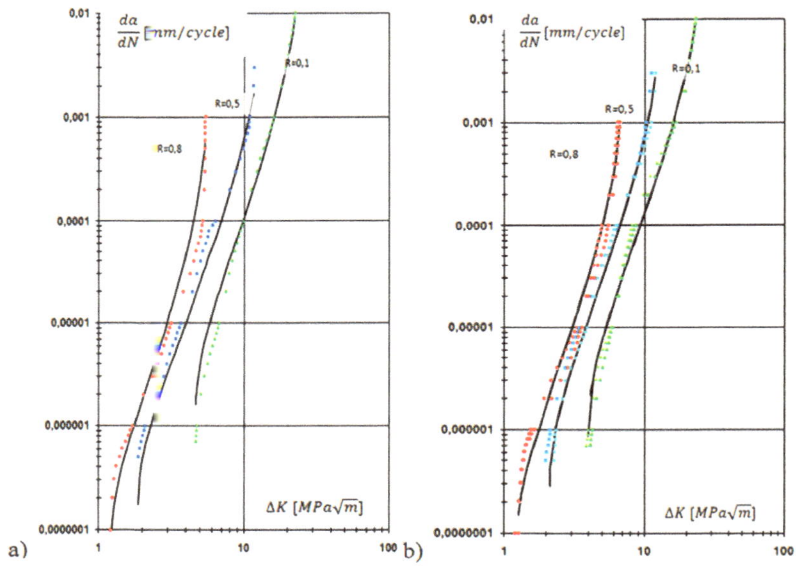

Figure 14. Approximation of $da/dN=f(\Delta K)$ data in different variants with the NASGRO equation, LSM modified according to formula (32a)

Exemplary results of approximation for different test data (with slightly smaller scatter between individual da/dN-$f(\Delta K)$ curves) is presented in Fig. 14.

Favorable effects of the approximation (in comparison with results showed in Fig. 1) after implementation of the modified LSM criterion can easily be seen. They tend to represent all the test data, within the whole range of data variability, independently of their absolute values, independently of the number of described curves – 3 (Fig. 14a), 6 (Fig. 14b), 9 (Fig. 10b). This effect has been achieved only by modifying the LSM criterion, since the idea underlying the approximation method for all the presented graphs is identical – the minimum of the sum of squared deviations between the approximated test data and the approximating values.

3.4. Regression of dependences in the NASGRO equation

Approximation of curves da/dN-ΔK substantially depends on preset values of parameters K_c and K_{th}. Hence, it is very important whether they can be determined on the grounds of the test data only (if they cover the whole range of the curve, i.e. 10^{-7} through 10^{-2} mm/cycle, which is not always easy to reach), or whether they need any other method/way to be determined, e.g. formula (2), functional dependences of the type $\Delta K_{th} = f(R)$ and $K_c = f(R)$, or the above-mentioned extrapolation. The above-discussed results of extrapolation correspond to the case when both the parameters show constant values for all the approximated curves. The plots for the test data show, however, that they depend on the stress ratio R – for each of nine experimentally gained curves parameters $\Delta K_{th,i}$ and $K_{c,i}$ can be estimated and the data gained can then be used to determine dependences $\Delta K_{th} = f(R)$ and $K_c = f(R)$, including coefficients for equation (2).

Formulae (2) and (2a) are special cases of a general formula of the following form:

$$\Delta K_{th} = \Delta K_0 \cdot \left(\frac{a}{a+a_0}\right)^{\frac{1}{2}} \left(\frac{(1-R)(1-A_0)}{(1-f)}\right)^{(C+C_{th}R)} . \tag{33}$$

Having re-arranged this formula, the following is arrived at:

$$\log(\Delta K_{th}) = \log\left(\Delta K_0 \cdot \left(\frac{a}{a+a_0}\right)^{\frac{1}{2}}\right) + C\log\left(\frac{(1-R)(1-A_0)}{(1-f)}\right) + C_{th}R\log\left(\frac{(1-R)(1-A_0)}{(1-f)}\right) \tag{34}$$

and then:

$$\begin{aligned}\log(\Delta K_{th}) - \tfrac{1}{2}\log\left(\frac{a}{a+a_0}\right) &= \log(\Delta K_0) + C\log\left(\frac{(1-R)(1-A_0)}{(1-f)}\right) + \\ &+ C_{th}R\log\left(\frac{(1-R)(1-A_0)}{(1-f)}\right)\end{aligned} \tag{35}$$

which can be described with the linear-regression equation as:

$$y = m_0 + m_1 F(R) + m_2 R F(R), \tag{36}$$

where $F(R) = \log\left(\frac{(1-R)(1-A_0)}{(1-f)}\right)$,

and the corrected value of the threshold range of the stress intensity factor is:

$$y = \log(\Delta K_{th}) - \tfrac{1}{2}\log\left(\frac{a}{a+a_0}\right).$$

Having found coefficients m_0, m_1, m_2 of the regression equation (36) we can calculate coefficients of equation (33):

$$C_{th} = m_2, \quad C = m_1 \quad \text{and} \quad \Delta K_0 = 10^{m_0} \tag{37}$$

at the same time, value of the ΔK_{th} function is calculated from the regression equation by formula:

$$\Delta K_{th} = 10^{y} \cdot \left(\frac{a}{a + a_0} \right)^{\frac{1}{2}}. \tag{38}$$

So, if we have data sets (R_i, $\Delta K_{th,i}$, a_i) – in the case under analysis there are 9 such sets – we automatically can find coefficients by formula (37), thus reducing the number of coefficients of the NASGRO equation to approximate the test data, which we are looking for.

Since there is no similar dependence for the K_c, parameter, the relationship $K_c = f(R)$ can be found in the same way (i.e. using the test data) from the ordinary linear regression $K_c = m_0 + m_1 R$ and also use it to describe 9 experimentally gained curves.

Functions $\Delta K_{th} = f(R)$ and $K_c = f(R)$ found in this way with the test data applied are shown in Fig. 15, whereas Fig. 16 illustrates effect of approximating curves da/dN-ΔK in the case given consideration.

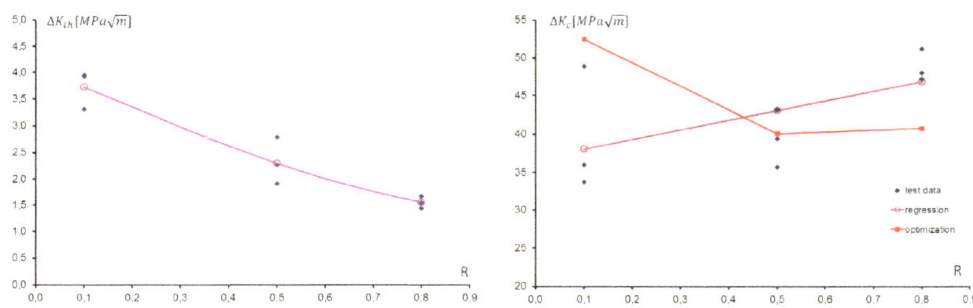

Figure 15. Functions a) $\Delta K_{th} = f(R)$ by formula (33) and b) $K_c = f(R)$ - regression

Evident is good fit of analytical description in both the critical and threshold ranges, whereas worse - in the middle section. The *ad hoc* accepted linear regression for the experimentally found relationship $K_c = f(R)$ not too precisely describes this relationship (straight line in Fig. 15, correlation coefficient reaches in this case the 0.3 level). Optimisation of values of the K_c coefficient for $R = 0.1$; 0.5 and 0.8 (here denoted as K_c*) with the LSM method to reach the minimum deviation error (32) results in the da/dN-ΔK curves approximating courses as in Fig. 16b. The curve illustrating the $K_c^* = f(R)$ dependence is in this case a broken line shown in Fig. 15b, which – easy to see – considerably strays away from the linear dependence. This proves that, among other things, one cannot *ad hoc* impose any form upon it. It can be assumed that for a larger number of experimentally gained curves, including the wider scope of values of *R*, the suggested method of determining the relationship $K_c = f(R)$ will offer better results that better correspond to the actual dependence and will remain useful for approximating the *da/dN-ΔK* curves. The broken-line curve, as that resulting from the optimisation process, may be described with, e.g. a straight line or a quadratic equation (as in Fig. 17) and used as a component of the theoretical (analytical) description of the test data with the NASGRO equation. In the case of a straight line, the correlation coefficient increases up to approx. 0.78 for the polynomial. Obviously, with three points K_c^* the correlation is complete, but if the scope of values of the asymmetry coefficient was greater, i.e. there would be more experimentally gained curves of different values of *R* (then the number of these points would increase), one should also expect high correlation for the relationship $K_c = f(R)$.

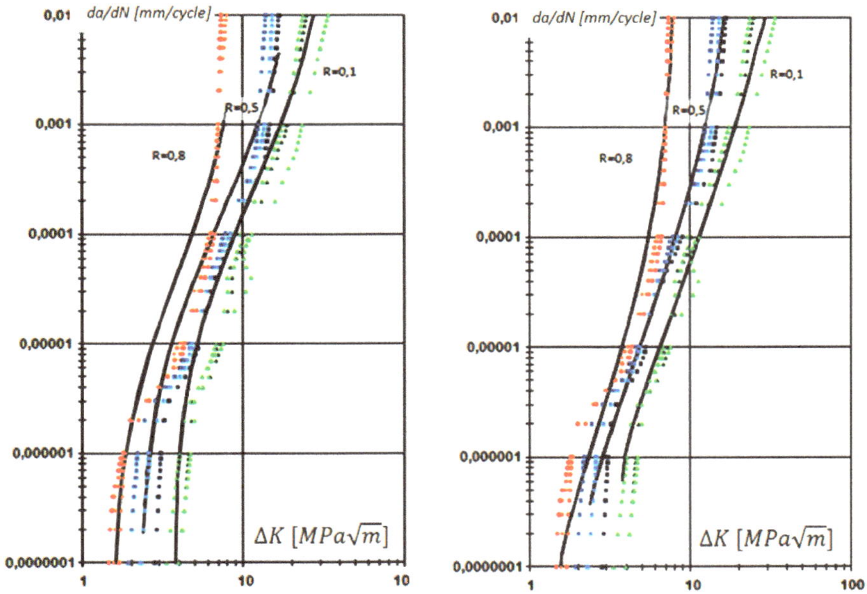

Figure 16. Approximation of curves *da/dN-ΔK* with the NASGRO equation, modified by formula (32) LSM, with extrapolated mapping points according to the MRCM: a) with regression applied as in Fig. 15, b) with optimisation for values of coefficients K_c^*

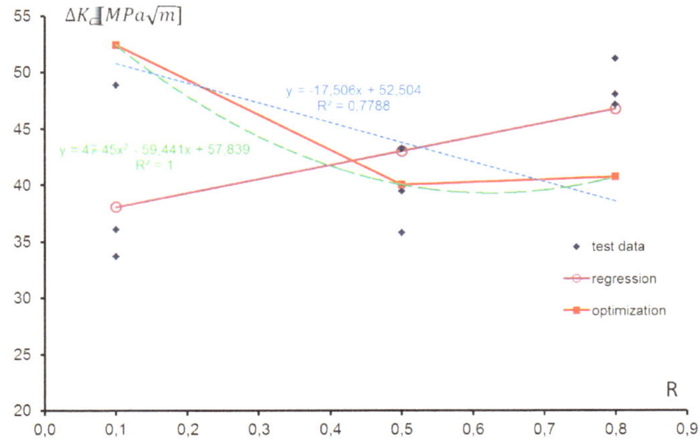

Figure 17. Linear $K_c = f(R)$ and polynomial $K_c^* = f(R)$ functions to optimise theoretical (analytical) description of curves da/dN-ΔK

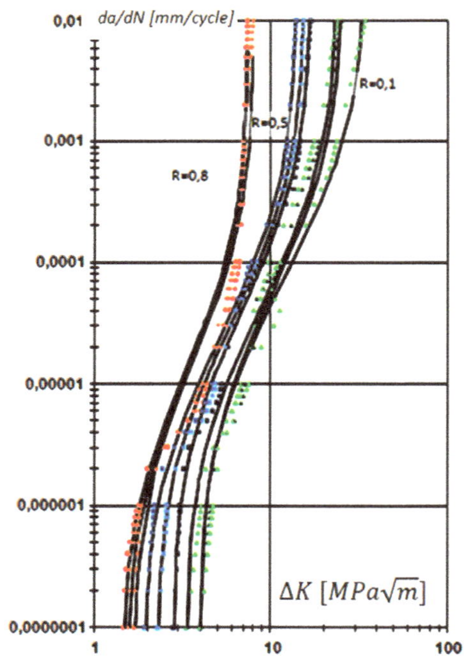

Figure 18. Curves da/dN-ΔK with coefficients ΔK_{th} and K_c individually fitted to each experimentally gained curve

If we use values of ΔK_{th} i K_c coefficients in forms determined not with the above-mentioned regression and optimisation methods, but as ones individually found for each of the experimentally gained curves ($\Delta K_{th,ind}$ i $K_{c,ind}$), the theoretical (analytical) description by means of the NASGRO equation - with the above-described methodology of finding other coefficients applied - should give even better result, see Fig. 18.

This variant of the theoretical (analytical) description is of only little practical importance, however, it shows that both the above-described methodology of analysis and the way of finding coefficients of the NASGRO equation result in correct description of experimentally found curves of fatigue-crack propagation and may be applied to this and similar categories of research issues.

4. Conclusion

Application of the Least Square Method in its classical form to determine coefficients of the NASGRO equation that describes fatigue crack propagation curve is ineffective, since data of the approximated function $da/dN=f(\Delta K)$ take values from the range of a few orders of magnitude, measuring points of the curves are irregular and in different numbers distributed in the graph (in threshold, stable-increase, and critical ranges), and subject to approximation are also several curves grouped in several sets (for different values of R).

The paper offers some techniques to modify the LSM criterion to significantly improve approximation results. These include:

- modification of the approximation-method criterion,
- smoothing of the experimentally gained curves to eliminate slight random disturbances resulting from, e.g. data recording process,
- different variants of calculating the derivative da/dN,
- regular mapping of the experimentally gained curves in the form of selected points,
- regression for points that represent (map) the experimentally gained curves to find coefficients of the crack growth equation,
- regression or optimisation of the description of partial dependences of the NASGRO equation as based on experimental data.

Values of parameters to be found as well as quantitative and qualitative results of performed approximations and theoretical (analytical) description are affected by, among other things, the number of tests that produce experimental data, and configurations thereof.

They provide a wider or narrower range of variability of parameters of significance that affect the courses of curves da/dN-ΔK, and also enable determination of accuracy and repeatability of obtained results. Reliability of the theoretical (analytical) description increases and the description itself better characterises properties of the material under examination if there are tens of curves gained experimentally from tests conducted for many (e.g. 5, 7, or 9) levels of the stress ratio R, for a wider range thereof, e.g. 0.2 through 0.9.

The proposed modification of the LSM criterion offers better fit of results of the test data approximation (unachievable with the classical LSM method). These effects are as follows:

- the provision of equal "weights" of each of the test data points in the total sum that determines this criterion (i.e. the sum of differences between approximated and approximating values) – independently of the magnitude of difference between values of data subject to approximation and that of difference between the approximated and approximating values,
- high effectiveness while approximating single, several, as well as a great number of sets/curves of test data,
- it becomes even more precise as the test data from the same (research-testing) groups show smaller scatter,
- may be used in other analyses of the same type related with test data regression, since it offers an all-purpose approach not related to propagation curves da/dN-ΔK.

Author details

Sylwester Kłysz
Air Force Institute of Technology, Warsaw, Poland
University of Warmia and Mazury in Olsztyn, Poland

Andrzej Leski
Air Force Institute of Technology, Warsaw, Poland

5. References

AFGROW Users Guide And Technical Manual. (2002). *AFRL-VA-WP-TR-2002-XXXX,* Version 4.0005.12.10, James A. Harter, Air Vehicles Directorate, Air Force Research Laboratory, WPAFB OH 45433-7542

ASTM E647 Standard test method for measurement of fatigue crack growth rates

Bukowski, L. & Kłysz, S. (2003). Compliance curve for single-edge notch specimen. *Zagadnienia Eksploatacji Maszyn,* No. 2(134), pp. 95-104

Elber, W. (1970). Fatigue crack closure under cyclic tension, *Eng. Fracture Mechs.,* Vol.2, pp. 37-45

Forman, R.G.; Kearney, V.E. & Engle, R.M. (1967). Numerical analysis of crack propagation in cyclic loaded structures, *J. Bas. Engng,* Vol.89, pp. 459-464

Forman, R.G.; Shivakumar, V.; Cardinal, J.W.; Williams, L.C. & McKeighan, P.C. (2005). Fatigue crack growth database for damage tolerance analysis, *DOT/FAA/AR-05/15,* U.S. Dep. of Transportation, FAA Office of Aviation Research, Washington, DC 20591

Fuchs, H.O. & Stephens, R.I. (1980). Metal fatigue in engineering, A Willey-Interscience Publication

Huang, X.P.; Cui, W.C. & Leng, J.X. (2005). A model of fatigue crack growth under various load spectra, *Proc. of 7th Int. Conf. of MESO,* Montreal, Canada, pp. 303–308

Kłysz, S. (2001). Fatigue crack growth in aircraft materials and constructional steel with overload consideration, (in Polish), *Air Force Institute of Technology Publication*, ISSN 1234-3544, Warsaw, Poland

Kłysz, S. & Lisiecki, J. (2009). Report of fatigue crack growth rate testing, (in Polish), *Laboratory for Material Strength Testing Report*, No. 1d/09, Air Force Institute of Technology, Warsaw, Poland

Kłysz, S.; Lisiecki, J. & Bąkowski, T. (2010a). Modification of the east-squares method for description of fatigue crack propagation, *Research Works of Air Force Institute of Technology*, ISSN 1234-3544, No.27, pp. 85-91, Warsaw, Poland

Kłysz, S.; Lisiecki, J. & Klimaszewski, S. (2010b). Analysis of crack propagation curve description for ORLIK aircraft fuselage skin material, (in Polish), *Technical News*, No 1(31), 2(32), pp.148-151, Lvov, Ukraine

Kłysz, S.; Lisiecki, J.; Leski, A. & Bąkowski, T. (2012). Least squares method modification applied to the NASGRO equation, *J. of Applied and Theoretical Mechanics*, Warsaw, Poland (in printing)

Kłysz, S.; Lisiecki, J.; Leski, A. & Bąkowski, T. (2012). Modification of the LSM criteria to approximate test data for fatigue crack growth rate, *J. of Applied and Theoretical Mechanics*, Warsaw, Poland (in printing)

Lisiecki, J. & Kłysz, S. (2007). Report of fatigue crack growth rate testing, (in Polish), *Laboratory for Material Strength Testing Report*, No. 8c/07, Air Force Institute of Technology, Warsaw, Poland

Murakami, Y. (Ed.). (1987). Stress Intensity Factors Handbook, *The Society of Materials Science*, Pergamon Press, Japan

NASGRO® Fracture Mechanics and Fatigue Crack Growth Analysis Software. (2006). v5.0, NASA-JSC and Southwest Research Institute

Newman, J.C. jr. (1992). Fastran II – A fatigue crack growth structures analysis program, *NASA TN 104159*, Langley Res. Centre, Hampton, VA

Paris, P.C. & Erdogan, A. (1963). A critical analysis of crack propagation laws, *J. Bas. Engng*, No.85, pp. 528-534

Skorupa, M. (1996). Empirical trends and prediction models for fatigue crack growth under variable amplitude loading, *ECN-R-96-007*, Netherlands Energy Research Foundation

Smith, R.A. (Ed.). (1986). *Fatigue Crack Growth. 30 Years of Progress*, Pergamon Press, Cambridge, UK

Taheri, F.; Trask, D. & Pegg, N. (2003). Experimental and analytical investigation of fatigue characteristics of 350WT steel under constant and variable amplitude loadings, *J. of Marine Structure*, No.16, pp. 69-91

Wheeler, O.E. (1972). Spectrum loading and crack growth, *Trans. ASME, J. Basic Eng.*, 94, 181

White, P.; Molent, L. & Barter, S. (2005). Interpreting fatigue test results using a probabilistic fracture approach, *Int. J. of Fatigue*, No.27, pp. 752-767

Willenborg, J.; Engle, R.M. & Wood, H.A. (1971). A crack growth retardation model using an effective stress concept, AFFDL-TR-71-1

Zhao, T. & Jiang Y. (2008). Fatigue of 7075-T651 aluminum alloy, *Int. J. of Fatigue*, No.30, pp. 834-849

Chapter 6

Fracture Mechanics Based Models of Structural and Contact Fatigue

Ilya I. Kudish

Additional information is available at the end of the chapter

1. Introduction

The subsurface initiated contact fatigue failure is one of the dominating mechanisms of failure of moving machine parts involved in cyclic motion. Structural fatigue failure may be of surface or subsurface origin. The analysis of a significant amount of accumulated experimental data obtained from field exploitation and laboratory testing provides undisputable evidence of the most important factors affecting contact fatigue [1]. It is clear that the factors affecting contact fatigue the most are as follows (a) acting cyclic normal stress and frictional stress (detailed lubrication conditions, surface roughness, etc.) which in part are determined by the part geometry, (b) distribution of residual stress versus depth, (c) initial statistical defect/crack distribution versus defect size, and location, (d) material elastic and fatigue parameters as functions of materials hardness, etc., (e) material fracture toughness, (f) material hardness versus depth, (g) machining and finishing operations, (h) abrasive contamination of lubricant and residual surface contamination, (i) non-steady cyclic loading regimes, etc. In case of structural fatigue the list of the most important parameters affecting fatigue performance is similar. None of the existing contact or structural fatigue models developed for prediction of contact fatigue life of bearings and gears as well as other structures takes into account all of the above operational and material conditions. Moreover, at best, most of the existing contact fatigue models are only partially based on the fundamental physical and mechanical mechanisms governing the fatigue phenomenon. Most of these models are of empirical nature and are based on assumptions some of which are not supported by experimental data or are controversial as it is in the case of fatigue models for bearings and gears [1]. Some models involve a number of approximations that usually do not reflect the actual processes occurring in material.

Therefore, a comprehensive mathematical models of contact and structural fatigue failure should be based on clearly stated mechanical principles following from the theory of elasticity, lubrication theory of elastohydrodynamic contact interactions, and fracture mechanics. Such models should take into consideration all the parameters described in items (a)-(e) and beyond. The advantage of such comprehensive models would be that the effect of variables such as steel cleanliness, externally applied stresses, residual stresses, etc. on contact and structural fatigue life could be examined as single or composite entities.

The goal of this chapter is to provided fracture mechanics based models of contact and structural fatigue. Historically, one of the most significant problems in realization of such an approach is the availability of simple but sufficiently precise solutions for the crack stress intensity factors. To overcome this difficulty some problems of fracture mechanics will be analyzed and their solutions will be represented in an analytical form acceptable for the further usage in modeling of contact and structural fatigue. In particular, a problem for an elastic half-plain weakened with a number of subsurface cracks and loaded with contact normal and frictional stresses as well with residual stress will be formulated and its asymptotic analytical solution will be presented. The latter solutions for the crack stress intensity factors are expressed in terms of certain integrals of known functions. These solutions for the stress intensity factors will be used in formulation of a two-dimensional contact fatigue model. In addition to that, a three-dimensional model applicable to both structural and contact fatigue will be formulated and some examples will be given. The above models take into account the parameters indicated in items (a)-(e) and are open for inclusion of the other parameters significant for fatigue.

In particular, these fatigue models take into account the statistical distribution of inclusions/cracks over the volume of the material versus their size and the resultant stress acting at the location of every inclusion/crack. Some of the main assumptions of the models are that (1) fatigue process in any machine part or structural unit runs in a similar manner and it is a direct reflection of the acting cyclic stresses and material properties and (2) the main part of fatigue life corresponds to the fatigue crack growth period, i.e. fatigue crack initiation period can be neglected. Furthermore, the variation in the distribution of cracks over time due to their fatigue growth is accounted for which is absent in all other existing fatigue models. The result of the above fatigue modeling is a simple relationship between fatigue life and cyclic loading, material mechanical parameters and its cleanliness as well as part geometry.

2. Three-dimensional model of contact and structural fatigue

The approach presented in this section provides a unified model of contact and structural fatigue of materials [1, 2]. The model development is based on a block approach, i.e. each block of the model describes a certain process related to fatigue and can be easily replaced by another block describing the same process differently. For example, accumulation of new more advanced knowledge of the process of fatigue crack growth may provide an opportunity to replaced the proposed here block dealing with fatigue crack growth with an improved one. Two examples of the application of this model to structural fatigue are provided.

2.1. Initial statistical defect distribution

It is assumed that material defects are far from each other and practically do not interact. However, in some cases clusters of nonmetallic inclusions located very close to each other are observed. In such cases these defect clusters can be represented by single defects of approximately the same size. Suppose there is a characteristic size L_σ in material that is determined by the typical variations of the material stresses, grain and surface geometry. It is also assumed that there is a size L_f in material such that $L_d \ll L_f \ll L_\sigma$, where L_d is the typical distance between the material defects. In other words, it is assumed that the defect population in any such volume L_f^3 is large enough to ensure an adequate statistical representation. It means that any parameter variations on the scale of L_f are indistinguishable for the fatigue analysis purposes and that in the further analysis any volume L_f^3 can be represented by its center point (x, y, z).

Therefore, there is an initial statistical defect distribution in the material such that each defect can be replaced by a subsurface penny-shaped or a surface semi-circular crack with a radius approximately equal to the half of the defect diameter. The usage of penny-shaped subsurface and semi-circular surface cracks is advantageous to the analysis because such fatigue cracks maintain their shape and their size is characterized by just one parameter. The orientation of these crack propagation will be considered later. The initial statistical distribution is described by a probabilistic density function $f(0, x, y, z, l_0)$, such that $f(0, x, y, z, l_0)dl_0dxdydz$ is the number of defects with the radii between l_0 and $l_0 + dl_0$ in the material volume $dxdydz$ centered about point (x, y, z). The material defect distribution is a local characteristic of material defectiveness. The model can be developed for any specific initial distribution $f(0, x, y, z, l_0)$. Some experimental data [3] suggest a log-normal initial defect distribution $f(0, x, y, z, l_0)$ versus the defect initial radius l_0

$$f(0, x, y, z, l_0) = 0 \, if \; l_0 \leq 0,$$
$$f(0, x, y, z, l_0) = \frac{\rho(0,x,y,z)}{\sqrt{2\pi}\sigma_{ln} l_0} \exp\left[-\frac{1}{2}\left(\frac{\ln(l_0) - \mu_{ln}}{\sigma_{ln}}\right)^2\right] if \; l_0 > 0, \tag{1}$$

where μ_{ln} and σ_{ln} are the mean value and standard deviation of the crack radii, respectively.

2.2. Direction of fatigue crack propagation

It is assumed that the duration of the crack initiation period is negligibly small in comparison with the duration of the crack propagation period. It is also assumed that linear elastic fracture mechanics is applicable to small fatigue cracks. The details of the substantiation of these assumptions can be found in [1]. Based on these assumptions in the vicinity of a crack the stress intensity factors completely characterize the material stress state. The normal k_1 and shear k_2 and k_3 stress intensity factors at the edge of a single crack of radius l can be represented in the form [4]

$$k_1 = F_1(x, y, z, \alpha, \beta)\sigma_1\sqrt{\pi l}, \; k_2 = F_2(x, y, z, \alpha, \beta)\sigma_1\sqrt{\pi l},$$
$$k_3 = F_3(x, y, z, \alpha, \beta)\sigma_1\sqrt{\pi l}, \tag{2}$$

where σ_1 is the maximum of the local tensile principal stress, F_1, F_2, and F_3 are certain functions of the point coordinates (x, y, z) and the crack orientation angles α and β with respect to the coordinate planes. The coordinate system is introduced in such a way that the x- and

y-axes are directed along the material surface while the z-axis is directed perpendicular to the material surface.

The resultant stress field in an elastic material is formed by stresses $\sigma_x(x,y,z)$, $\sigma_y(x,y,z)$, $\sigma_z(x,y,z)$, $\tau_{xz}(x,y,z)$, $\tau_{xy}(x,y,z)$, and $\tau_{zy}(x,y,z)$. Some regions of material are subjected to tensile stress while other regions are subjected to compressive stress. Conceptually, there is no difference between the phenomena of structural and contact fatigue as the local material response to the same stress in both cases is the same and these cases differ in their stress fields only. As long as the stress levels do not exceed the limits of applicability of the quasi-brittle linear fracture mechanics when plastic zones at crack edges are small the rest of the material behaves like an elastic solid. The actual stress distributions in cases of structural and contact fatigue are taken in the proper account. In contact interactions where compressive stress is usually dominant there are still zones in material subjected to tensile stress caused by contact frictional and/or tensile residual stress [1, 5].

Experimental and theoretical studies show [1] that after initiation fatigue cracks propagate in the direction determined by the local stress field, namely, perpendicular to the local maximum tensile principle stress. Therefore, it is assumed that fatigue is caused by propagation of penny-shaped subsurface or semi-circular surface cracks under the action of principal maximum tensile stresses. Only high cycle fatigue phenomenon is considered here. On a plane perpendicular to a principal stress the shear stresses are equal to zero, i.e. the shear stress intensity factors $k_2 = k_3 = 0$. To find the plane of fatigue crack propagation (i.e. the orientation angles α and β), which is perpendicular to the maximum principal tensile stress, it is necessary to find the directions of these principal stresses. The latter is equivalent to solving the equations

$$k_2(N,\alpha,\beta,l,x,y,z) = 0,\ k_3(N,\alpha,\beta,l,x,y,z) = 0. \tag{3}$$

Usually, there are more than one solution sets to these equations at any point (x,y,z). To get the right angles α and β one has to chose the solution set that corresponds to the maximum tensile principal stress, i.e. maximum of the normal stress intensity factor $k_1(N,l,x,y,z)$. That guarantees that fatigue cracks propagate in the direction perpendicular to the maximum tensile principal stress if α and β are chosen that way.

For steady cyclic loading for small cracks $k_{20} = k_2/\sqrt{l}$ and $k_{30} = k_3/\sqrt{l}$ are independent from the number of cycles N and crack radius l together with equation (3) lead to the conclusion that for cyclic loading with constant amplitude the angles α and β characterizing the plane of fatigue crack growth are independent from N and l. Thus, angles α and β are functions of only crack location, i.e. $\alpha = \alpha(x,y,z)$ and $\beta = \beta(x,y,z)$. For the most part of their lives fatigue cracks created and/or existed near material defects remain small. Therefore, penny-shaped subsurface cracks conserve their shape but increase in size.

Even for the case of an elastic half-space it is a very difficult task to come up with sufficiently precise analytical solutions for the stress intensity factors at the edges of penny-shaped subsurface or semi-circular surface crack of arbitrary orientation at an arbitrary location (x,y,z). However, due to the fact that practically all the time fatigue cracks remain small and exsert little influence on the material general stress state the angles α and β can be determined in the process of calculation of the maximum tensile principle stress σ. The latter is equivalent to solution of equations (3). As soon as σ is determined for a subsurface crack its normal stress intensity factor k_1 can be approximated by the normal stress intensity factor for the case of a single crack of radius l in an infinite space subjected to the uniform tensile stress σ, i.e. by $k_1 = 2\sigma\sqrt{l/\pi}$.

2.3. Fatigue crack propagation

Fatigue cracks propagate at every point of the material stressed volume V at which $\max_T(k_1) > k_{th}$, where the maximum is taken over the duration of the loading cycle T and k_{th} is the material stress intensity threshold. There are three distinct stages of crack development: (a) growth of small cracks, (b) propagation of well–developed cracks, and (c) explosive and, usually, unstable growth of large cracks. The stage of small crack growth is the slowest one and it represents the main part of the entire crack propagation period. This situation usually causes confusion about the duration of the stages of crack initiation and propagation of small cracks. The next stage, propagation of well–developed cracks, usually takes significantly less time than the stage of small crack growth. And, finally, the explosive crack growth takes almost no time.

A relatively large number of fatigue crack propagation equations are collected and analyzed in [6]. Any one of these equations can be used in the model to describe propagation of fatigue cracks. However, the simplest of them which allows to take into account the residual stress and, at the same time, to avoid the usage of such an unstable characteristic as the stress intensity threshold k_{th} is Paris's equation

$$\frac{dl}{dN} = g_0\Big(\max_{-\infty<x<\infty} \triangle k_1\Big)^n,\ l\mid_{N=0}= l_0, \tag{4}$$

where g_0 and n are the parameters of material fatigue resistance and l_0 is the crack initial radius. Notice, that in cases of loading and relaxation such as in contact fatigue $\triangle k_1 = k_1$.

Fatigue cracks propagate until they reach their critical size with radius l_c for which $k_1 = K_f$ (K_f is the material fracture toughness, i.e. to the radius of $l_c = (K_f/k_{10})^2$). After that their growth becomes unstable and very fast. Usually, the stage of explosive crack growth takes just few loading cycles.

It can be shown that the number of loading cycles needed for a crack to reach its critical radius is almost independent from the material fracture toughness K_f. This conclusion is supported by direct numerical simulations. For the further analysis, it is necessary to determine for a crack its initial radius l_{0c}, which after N loading cycles reaches the critical size of l_c. Solving the initial-value problem (4) one obtains the formula

$$l_{0c} = \{l_c^{\frac{2-n}{2}} + N(\tfrac{n}{2} - 1)g_0[\max_{-\infty<x<\infty} \triangle k_{10}]^n\}^{\frac{2}{2-n}}, \tag{5}$$

where l_{0c} depends on N, x, y, and z. Obviously, for $n > 2$ and fixed x, y, and z the value of l_{0c} is a decreasing function of N. It is important to keep in mind that $l_{0c}(N, x, y, z)$ is minimal where $k_{10}(x, y, z)$ is maximal, which, in turn, happens where the material tensile stress reaches its maximum.

2.4. Crack propagation statistics

To describe crack statistics after the crack initiation stage is over it is necessary to make certain assumptions. The simplest assumptions of this kind are: the existing cracks do not heal and new cracks are not created. In other words, the number of cracks in any material volume remains constant in time. Based on a practically correct assumption that the defect distribution is initially scarce, the coalescence of cracks and changes in the general stress field are possible only when cracks have already reached relatively large sizes. However, this may happen only during the last stage of crack growth the duration of which is insignificant for calculation of fatigue life. Therefore, it can be assumed that over almost all life span of fatigue cracks their orientations do not change. This leads to the equation for the density of crack distribution $f(N,x,y,z,l)$ as a function of crack radius l after N loading cycles in a small parallelepiped $dxdydz$ with the center at the point with coordinates (x,y,z)

$$f(N,x,y,z,l)dl = f(0,x,y,z,l_0)dl_0, \tag{6}$$

which being solved for $f(N,x,y,z,l)$ gives

$$f(N,x,y,z,l)dl = f(0,x,y,z,l_0)\tfrac{dl_0}{dl}, \tag{7}$$

where l_0 and dl_0/d as functions of N and l can be obtain from the solution of (4) in the form

$$\begin{aligned} l_0 &= \{l^{\frac{2-n}{2}} + N(\tfrac{n}{2}-1)g_0[\max_{-\infty<x<\infty} \triangle k_{10}]^n\}^{\frac{2}{2-n}}, \\ \tfrac{dl_0}{dl} &= \{1 + N(\tfrac{n}{2}-1)g_0[\max_{-\infty<x<\infty} \triangle k_{10}]^n l^{\frac{n-2}{2}}\}^{\frac{n}{n-2}}. \end{aligned} \tag{8}$$

Equations (7) and (8) lead to the expression for the crack distribution function f after N loading cycles

$$\begin{aligned} f(N,x,y,z,l) &= f(0,x,y,z,l_0(N,l,y,z))\{1 \\ &+N(\tfrac{n}{2}-1)g_0[\max_{-\infty<x<\infty} \triangle k_{10}]^n l^{\frac{n-2}{2}}\}^{\frac{n}{n-2}}, \end{aligned} \tag{9}$$

where $l_0(N,l,y,z)$ is determined by the first of the equations in (8).

Formula (9) leads to a number of important conclusions. The fatigue crack distribution function $f(N,x,y,z,l)$ depends on the initial crack distribution $f(0,x,y,z,l_0)$ and it changes with the number of applied loading cycles N in such a way that the crack volume density $\rho(N,x,y,z)$ remains constant. Because of crack growth the crack distribution $f(N,x,y,z,l)$ widens with respect to l with number of loading cycles N.

2.5. Local fatigue damage accumulation

Let us design the measure of material fatigue damage. If at a certain point (x,y,z) after N loading cycles radii of all cracks $l < l_c$ then there is no damage at this point and the material local survival probability $p(N,x,y,z) = 1$. On the other hand, if at this point after N loading cycles radii of all cracks $l \geq l_c$, then all cracks reached the critical size and the material at this point is completely damaged and the local survival probability $p(N,x,y,z) = 0$. Obviously, the more fatigue cracks with larger radii l exist at the point the lower is the local survival probability $p(N, x, y, z)$. It is reasonable to assume that the material local survival probability $p(N, x, y, z)$ is a certain monotonic measure of the portion of cracks with radius l below the critical radius l_c. Therefore, $p(N, x, y, z)$ can be represented by the expressions

$$p(N, x, y, z) = \frac{1}{\rho} \int_0^{l_c} f(N, x, y, z, l) dl \; if \; f(0, x, y, z, l_0) \neq 0,$$

$$p(N, x, y, z) = 1 \; otherwise, \tag{10}$$

$$\rho = \rho(N, x, y, z) = \int_0^{\infty} f(N, x, y, z, l) dl = \rho(0, x, y, z).$$

Obviously, the local survival probability $p(N, x, y, z)$ is a monotonically decreasing function of the number of loading cycles N because fatigue crack radii l tend to grow with the number of loading cycles N.

To calculate $p(N, x, y, z)$ from (10) one can use the specific expression for f determined by (9). However, it is more convenient to modify it as follows

$$p(N, x, y, z) = \frac{1}{\rho} \int_0^{l_{0c}} f(0, x, y, z, l_0) dl_0 \; if \; f(0, x, y, z, l_0) \neq 0,$$

$$p(N, x, y, z) = 1 \; otherwise, \tag{11}$$

where l_{0c} is determined by (5) and ρ is the initial volume density of cracks. Thus, to every material point (x, y, z) is assigned a certain local survival probability $p(N, x, y, z)$, $0 \leq p(N, x, y, z) \leq 1$.

Equations (11) demonstrate that the material local survival probability $p(N, x, y, z)$ is mainly controlled by the initial crack distribution $f(0, x, y, z, l_0)$, material fatigue resistance parameters g_0 and n, and external contact and residual stresses. Moreover, the material local survival probability $p(N, x, y, z)$ is a decreasing function of N because l_{0c} from (5) is a decreasing function of N for $n > 2$.

2.6. Global fatigue damage accumulation

The survival probability $P(N)$ of the material as a whole is determined by the local probabilities of all points of the material at which fatigue cracks are present. It is assumed that the material fails as soon as it fails at just one point. It is assumed that the initial crack distribution in the material is discrete. Let $p_i(N) = p(N, x_i, y_i, z_i)$, $i = 1, \ldots, N_c$, where N_c is the total number of points in the material stressed volume V at which fatigue cracks are present. Then based on the above assumption the material survival probability $P(N)$ is equal to

$$P(N) = \prod_{i=1}^{N_c} p_i(N). \tag{12}$$

Obviously, probability $P(N)$ from (12) satisfies the inequalities

$$[p_m(N)]^{N_c} \leq P(N) \leq p_m(N), \; p_m(N) = \min_V p(N,x,y,z). \tag{13}$$

In (13) the right inequality shows that the survival probability $P(N)$ is never greater than the minimum value $p_m(N)$ of the local survival probability $p(N,x,y,z)$ over the material stressed volume V.

An analytical substantiation for the assumption that the first pit is created by the cracks from a small material volume with the smallest survival probability $p_m(N)$ is provided in [1]. Moreover, the indicated analysis also validates one of the main assumptions of the model that new cracks are not being created. Namely, if new cracks do get created in the process of loading, they are very small and have no chance to catch up with already existing and propagating larger cracks. The graphical representation of this fact is given in Fig. 1 [1]. In this figure fatigue cracks are initially randomly distributed over the material volume with respect to their normal stress intensity factor k_1 and are allowed to grow according to Paris' law (see (4)) with sufficiently high value of $n = 6.67 - 9$. In Fig. 1 the values of the normal stress intensity factor k_1 are shown at different time moments (k_0 and L_0 are the characteristic normal stress intensity factor and geometric size of the solid). These graphs clearly show that a crack with the initially larger value of the normal stress intensity factor k_1 propagates much faster than all other cracks, i.e. the value of its k_1 increases much faster than the values of k_1 for all other cracks which are almost dormant. As a result of that, the crack with the initially larger value of k_1 reaches its critical size way ahead of other cracks. This event determines the time and the place where fatigue failure occurs initially. Therefore, in spite of formula (12) which indicates that all fatigue cracks have influence on the survival probability $P(N)$, for high values of n the material survival probability $P(N)$ is a local fatigue characteristic, and it is determined by the material defect with the initially highest value of the stress intensity factor k_1. The higher the power n is the more accurate this approximation is.

Therefore, assuming that it is a very rare occurrence when more than one fatigue initiation/spall happen simultaneously, it can be shown that at the early stages of the fatigue process the material global survival probability $P(N)$ is determined by the minimum of the local survival probability $p_m(N)$

$$P(N) = p_m(N), \; p_m(N) = \min_V p(N,x,y,z), \tag{14}$$

where the maximum is taken over the (stressed) volume V of the solid.

If the initial crack distribution is taken in the log-normal form (1) then

$$P(N) = p_m(N) = \frac{1}{2}\Big\{1 + erf\Big[\min_V \frac{\ln l_{0c}(N,y,z) - \mu_{ln}}{\sqrt{2}\sigma_{ln}}\Big]\Big\}, \tag{15}$$

where $erf(x)$ is the error integral [8]. Obviously, the local survival probability $p_m(N)$ is a complex combined measure of applied stresses, initial crack distribution, material fatigue parameters, and the number of loading cycles.

In cases when the mean μ_{ln} and the standard deviation σ_{ln} are constants throughout the material formula (15) can be significantly simplified

$$P(N) = p_m(N) = \frac{1}{2}\Big\{1 + erf\Big[\frac{\ln \min_V l_{0c}(N,y,z) - \mu_{ln}}{\sqrt{2}\sigma_{ln}}\Big]\Big\}. \tag{16}$$

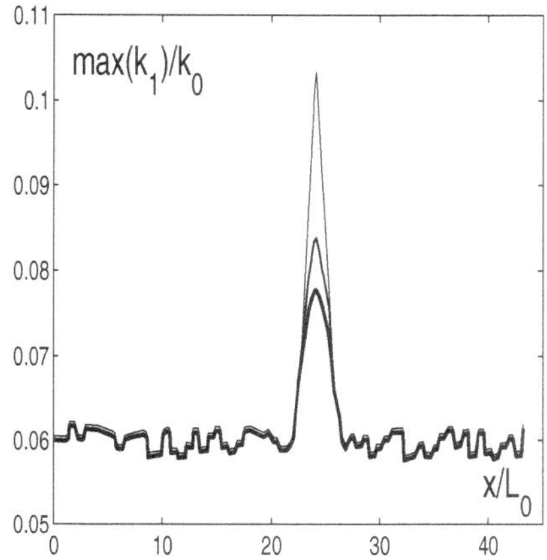

Figure 1. Illustration of the growth of the initially randomly distributed normal stress intensity factor k_1 with time N as initially unit length fatigue cracks grow (after Tallian, Hoeprich, and Kudish [7]). Reprinted with permission from the STLE.

To determine fatigue life N of a contact for the given survival probability $P(N) = P_*$, it is necessary to solve the equation

$$p_m(N) = P_*. \tag{17}$$

2.7. Fatigue life calculation

Suppose the material failure occurs at point (x, y, z) with the probability $1 - P(N)$. That actually determines the point where in (16) the minimum over the material volume V is reached. Therefore, at this point in (16), the operation of minimum over the material volume V can be dropped. By solving (16) and (17), one gets

$$N = \{(\tfrac{n}{2} - 1)g_0[\max_{-\infty<x<\infty} \triangle k_{10}]^n\}^{-1}\{\exp[(1 - \tfrac{n}{2})(\mu_{ln} + \sqrt{2}\sigma_{ln} erf^{-1}(2P_* - 1))] - l_c^{\frac{2-n}{2}}\}, \tag{18}$$

where $erf^{-1}(x)$ is the inverse function to the error integral $erf(x)$. Assuming that the material initially is free of damage, i.e., when $P(0) = 1$, one can simplify the latter equation. Discounting the very tail of the initial crack distribution, one gets $\max\limits_{V} l_0 \leq l_c$. Thus, for well–developed cracks and, in many cases, even for small cracks, the second term in (5) for l_{0c} dominates the first one. It means that the dependence of l_{0c} on l_c and, therefore, on the material fracture toughness K_f can be neglected. Then equation (18) can be approximated by

$$N = \{(\tfrac{n}{2} - 1)g_0[\max_{-\infty<x<\infty} \triangle k_{10}]^n\}^{-1}\{\exp[(1 - \tfrac{n}{2})(\mu_{ln} + \sqrt{2}\sigma_{ln} erf^{-1}(2P_* - 1))]\}. \tag{19}$$

Taking into account that in the case of contact fatigue k_{10} is proportional to the maximum contact pressure q_{max} and that it also depends on the friction coefficient λ and the ratio of residual stress q^0 and $q_{max} = p_H$ as well as taking into account the relationships $\mu_{ln} = \ln \frac{\mu^2}{\sqrt{\mu^2+\sigma^2}}$, $\sigma_{ln} = \sqrt{\ln[1+(\frac{\sigma}{\mu})^2]}$ [1] (where μ and σ are the regular initial mean and standard deviation) one arrives at a simple analytical formula

$$N = \frac{C_0}{(n-2)g_0 p_H^n}\left(\frac{\sqrt{\mu^2+\sigma^2}}{\mu^2}\right)^{\frac{n}{2}-1} \times \exp[(1-\tfrac{n}{2})\sqrt{2\ln[1+(\tfrac{\sigma}{\mu})^2]}erf^{-1}(2P_* - 1)], \quad (20)$$

where C_0 depends only on the friction coefficient λ and the ratio of the residual stress q^0 and the maximum Hertzian pressure p_H. Finally, assuming that $\sigma \ll \mu$ from (20) one can obtain the formula

$$N = \frac{C_0}{(n-2)g_0 p_H^n \mu^{\frac{n}{2}-1}} \exp[(1-\tfrac{n}{2})\tfrac{\sqrt{2}\sigma}{\mu} erf^{-1}(2P_* - 1))]. \quad (21)$$

Also, formulas (20) and (21) can be represented in the form of the Lundberg-Palmgren formula (see [1] and the discussion there).

Formula (21) demonstrates the intuitively obvious fact that the fatigue life N is inversely proportional to the value of the parameter g_0 that characterizes the material crack propagation resistance. Equation (21) exhibits a usual for roller and ball bearings as well as for gears dependence of the fatigue life N on the maximum Hertzian pressure p_H. Thus, from the well–known experimental data for bearings the range of n values is $20/3 \leq n \leq 9$. Keeping in mind that usually $\sigma \ll \mu$, for these values of n contact fatigue life N is practically inverse proportional to a positive power of the mean crack size, i.e. to $\mu^{\frac{n}{2}-1}$. Therefore, fatigue life N is a decreasing function of the initial mean crack (inclusion) size μ. This conclusion is valid for any material survival probability P_* and is supported by the experimental data discussed in [1]. In particular, $\ln N$ is practically a linear function of $\ln \mu$ with a negative slope $1 - \frac{n}{2}$ which is in excellent agreement with the Timken Company test data [9]. Keeping in mind that $n > 2$, at early stages of fatigue failure, i.e. when $erf^{-1}(2P_* - 1) > 0$ for $P_* > 0.5$, one easily determines that fatigue life N is a decreasing function of the initial standard deviation of crack sizes σ. Similarly, at late stages of fatigue failure, i.e. when $P_* < 0.5$, the fatigue life N is an increasing function of the initial standard deviation of crack sizes σ. According to (21), for $P_* = 0.5$ fatigue life N is independent from σ, however, according to (20), for $P_* = 0.5$ fatigue life N is a slowly increasing function of σ. By differentiating $p_m(N)$ obtained from (16) with respect to σ, one can conclude that the dispersion of $P(N)$ increases with σ.

The stress intensity factor k_1 decreases as the magnitude of the compressive residual stress q^0 increases and/or the magnitude of the friction coefficient λ decreases. Therefore, in (20) and 21) the value of C_0 is a monotonically decreasing function of residual stress q^0 and friction coefficient λ.

Being applied to bearings and/or gears the described statistical contact fatigue model can be used as a research and/or engineering tool in pitting modeling. In the latter case, some of the model parameters may be assigned certain fixed values based on the scrupulous analysis of steel quality and quality and stability of gear and bearing manufacturing processes.

In case of structural fatigue the Hertzian stress in formulas (20) and (21) should be replaced the dominant stress acting on the part while constant C_0 would be dependent on the ratios of other external stresses acting on the part at hand to the dominant stress in a certain way (see examples of torsional and bending fatigue below).

2.8. Examples of torsional and bending fatigue

Suppose that in a beam material the defect distribution is space-wise uniform and follows equation (1). Also, let us assume that the residual stress is zero.

First, let us consider torsional fatigue. Suppose a beam is made of an elastic material with elliptical cross section (a and b are the ellipse semi-axes, $b < a$) and directed along the y-axis. The beam is under action of torque M_y about the y-axis applied to its ends. The side surfaces of the beam are free of stresses. Then it can be shown (see Lurye [10], p. 398) that

$$\tau_{xy} = -\frac{2G\gamma a^2}{a^2+b^2}z,\ \tau_{zy} = \frac{2G\gamma b^2}{a^2+b^2}x,\ \sigma_x = \sigma_y = \sigma_z = \tau_{xz} = 0, \tag{22}$$

where G is the material shear elastic modulus, $G = E/[2(1+\nu)]$ (E and ν are Young's modulus and Poisson's ratio of the beam material), and γ is a dimensionless constant. By introducing the principal stresses σ_1, σ_2, and σ_3 that satisfy the equation $\sigma^3 - (\tau_{xy}^2 + \tau_{zy}^2)\sigma = 0$, one obtains that

$$\sigma_1 = -\sqrt{\tau_{xy}^2 + \tau_{zy}^2},\ \sigma_2 = 0,\ \sigma_3 = \sqrt{\tau_{xy}^2 + \tau_{zy}^2}. \tag{23}$$

For the case of $a > b$ the maximum principal tensile stress $\sigma_1 = -\frac{2G\gamma a^2 b}{a^2+b^2}$ is reached at the surface of the beam at points $(0, y, \pm b)$ and depending on the sign of M_y it acts in one of the directions described by the directional cosines

$$\cos(\alpha, x) = \mp\frac{\sqrt{2}}{2},\ \cos(\alpha, y) = \pm\frac{\sqrt{2}}{2},\ \cos(\alpha, z) = 0, \tag{24}$$

where α is the direction along one of the principal stress axes. For the considered case of elliptic beam, the moments of inertia of the beam elliptic cross section about the x- and y-axes, I_x and I_z as well as the moment of torsion M_y applied to the beam are as follows (see Lurye [10], pp. 395, 399) $I_x = \pi ab^3/4,\ I_z = \pi a^3 b/4,\ M_y = G\gamma C,\ C = 4I_x I_z/(I_x + I_z)$. Keeping in mind that according to Hasebe and Inohara [11] and Isida [12], the stress intensity factor k_1 for an edge crack of radius l and inclined to the surface of a half-plane at the angle of $\pi/4$ (see (24)) is $k_1 = 0.705 \mid \sigma_1 \mid \sqrt{\pi l}$, one obtains $k_{10} = \frac{1.41}{\sqrt{\pi}}\frac{|M_y|}{ab^2}$. Then, fatigue life of a beam under torsion follows from substituting the expression for k_{10} into equation (19)

$$N = \frac{2}{(n-2)g_0}\{\frac{1.257ab^2}{|M_y|}\}^n g(\mu, \sigma), \tag{25}$$

$$g(\mu, \sigma) = (\frac{\sqrt{\mu^2+\sigma^2}}{\mu^2})^{\frac{n-2}{2}} \exp[(1 - \frac{n}{2})\sqrt{2\ln[1 + (\frac{\sigma}{\mu})^2]}erf^{-1}(2P_* - 1)\}. \tag{26}$$

Now, let us consider bending fatigue of a beam/console made of an elastic material with elliptical cross section (a and b are the ellipse semi-axes) and length L. The beam is directed along the y-axis and it is under the action of a bending force P_x directed along the x-axis which is applied to its free end. The side surfaces of the beam are free of stresses. The other end $y = 0$

of the beam is fixed. Then it can be shown (see Lurye [10]) that

$$\sigma_x = \sigma_z = 0,\ \sigma_y = -\frac{P_x}{I_z} x(L-y),$$
$$\tau_{xz} = 0,\ \tau_{xy} = \frac{P_x}{2(1+\nu)I_z}\frac{2(1+\nu)a^2+b^2}{3a^2+b^2}\{a^2 - x^2 - \frac{(1-2\nu)a^2z^2}{2(1+\nu)a^2+b^2}\}, \quad (27)$$
$$\tau_{zy} = -\frac{P_x}{(1+\nu)I_z}\frac{(1+\nu)a^2+\nu b^2}{3a^2+b^2}xz,$$

where I_z is the moment of inertia of the beam cross section about the z-axis. By introducing the principal stresses that satisfy the equation $\sigma^3 - \sigma_y\sigma^2 - (\tau_{xy}^2 + \tau_{zy}^2)\sigma = 0$, one can find that

$$\sigma_1 = \frac{1}{2}[\sigma_y - \sqrt{\sigma_y^2 + 4(\tau_{xy}^2 + \tau_{zy}^2)}],\ \sigma_2 = 0,$$
$$\sigma_3 = \frac{1}{2}[\sigma_y + \sqrt{\sigma_y^2 + 4(\tau_{xy}^2 + \tau_{zy}^2)}]. \quad (28)$$

The tensile principal stress σ_1 reaches its maximum $\frac{4|P_x|L}{\pi a^2 b}$ at the surface of the beam at one of the points $(\pm a, 0, 0)$ (depending on the sign of load P_x) and is acting along the y-axis - the axis of the beam. Based on equations (28) and the solution for the surface crack inclined to the surface of the half-space at angle of $\pi/2$ (see Hasebe and Inohara [11] and Isida [12]), one obtains $k_{10} = \frac{4.484}{\sqrt{\pi}}\frac{|P_x|L}{a^2 b}$. Therefore, bending fatigue life of a beam follows from substituting the expression for k_{10} into equation (19)

$$N = \frac{2}{(n-2)g_0}\{\frac{0.395a^2b}{|P_x|L}\}^n g(\mu,\sigma), \quad (29)$$

where function $g(\mu,\sigma)$ is determined by equation (26).

In both cases of torsion and bending, fatigue life is independent of the elastic characteristic of the beam material (see formulas (25), (29), and (26)), and it is dependent on fatigue parameters of the beam material (n and g_0), the initial defect distribution (i.e. on μ and σ), the geometry of the beam cross section (a and b), and its length L and the applied loading (P_x or M_y).

In a similar fashion the model can be applied to contact fatigue if the stress field is known. A more detailed analysis of contact fatigue is presented below for a two-dimensional case.

3. Contact problem for an elastic half-plane weakened by straight cracks

A general theory of a stress state in an elastic plane with multiple cracks was proposed in [13]. In this section this theory is extended to the case of an elastic half-plane loaded by contact and residual stresses [1]. A study of lubricant-surface crack interaction, a discussion of the difference between contact fatigue lives of drivers and followers, the surface and subsurface initiated fatigue as well as fatigue of rough surfaces can be found in [1].

The main purpose of the section is to present formulations for the contact and fracture mechanics problems for an elastic half-plane weakened by subsurface cracks. The problems for surface cracks in an elastic lubricated half-plane are formulated and analyzed in [1]. The problems are reduced to systems of integro-differential equations with nonlinear boundary conditions in the form of alternating equations and inequalities. An asymptotic (perturbation) method for the case of small cracks is applied to solution of the problem and some numerical examples for small cracks are presented.

Let us introduce a global coordinate system with the x^0-axis directed along the half-plane boundary and the y^0-axis perpendicular to the half-plane boundary and pointed in the direction outside the material. The half-plane occupies the area of $y^0 \leq 0$. Let us consider a contact problem for a rigid indenter with the bottom of shape $y^0 = f(x^0)$ pressed into the elastic half-plane (see Fig. 2). The elastic half-plane with effective elastic modulus E' ($E' = E/(1-\nu^2)$, E and ν are the half-plane Young's modulus and Poisson's ratio) is weakened by N straight cracks. The crack faces are frictionless. Besides the global coordinate system we will introduce local orthogonal coordinate systems for each straight crack of half-length l_k in such a way that their origins are located at the crack centers with complex coordinates $z_k^0 = x_k^0 + iy_k^0$, $k = 1, \ldots, N$, the x_k-axes are directed along the crack faces and the y_k-axes are directed perpendicular to them. The cracks are inclined to the positive direction of the x^0-axis at the angles α_k, $k = 1, \ldots, N$. All cracks are considered to be subsurface. The faces of every crack may be in partial or full contact with each other. The indenter is loaded by a normal force P and may be in direct contact with the half-plane or separated from it by a layer of lubricant. The indenter creates a pressure $p(x^0)$ and frictional stress $\tau(x^0)$ distributions. The frictional stress $\tau(x^0)$ between the indenter and the boundary of the half-plane is determined by the contact pressure $p(x^0)$ through a certain relationship. The cases of dry and fluid frictional stress $\tau(x^0)$ are considered in [1]. At infinity the half-plane is loaded by a tensile or compressive (residual) stress $\sigma_{x^0}^{\infty} = q^0$ which is directed along the x^0-axis. In this formulation the problem is considered in [1].

Then the problem is reduced to determining of the cracks behavior. Therefore, in dimensionless variables

$$\begin{gathered}
(x_n^{0\prime}, y_n^{0\prime}) = (x_n^0, y_n^0)/\widetilde{b},\ (p_n^{0\prime}, \tau_n^{0\prime}, p_n') = (p_n^0, \tau_n^0, p_n)/\widetilde{q}, \\
(x_n', t') = (x_n, t)/l_n,\ (v_n', u_n') = (v_n, u_n)/\widetilde{v_n},\ \widetilde{v_n} = \tfrac{4\widetilde{q}l_n}{E'}, \\
(k_{1n}^{\pm\prime}, k_{2n}^{\pm\prime}) = (k_{1n}^{\pm}, k_{2n}^{\pm})/(\widetilde{q}\sqrt{l_n})
\end{gathered} \tag{30}$$

the equations of the latter problem for an elastic half-plane weakened by cracks and loaded by contact and residual stresses have the following form [1]

$$\begin{gathered}
\int\limits_{-1}^{1} \frac{v_k'(t)dt}{t-x_k} + \sum\limits_{m=1}^{N} \delta_m \int\limits_{-1}^{1} [v_m'(t)A_{km}^r(t,x_k) - u_m'(t)B_{km}^r(t,x_k)]dt \\
= \pi p_{nk}(x_k) + \pi p_k^0(x_k),\ v_k(\pm 1) = 0, \\
\int\limits_{-1}^{1} \frac{u_k'(t)dt}{t-x_k} + \sum\limits_{m=1}^{N} \delta_m \int\limits_{-1}^{1} [v_m'(t)A_{km}^i(t,x_k) - u_m'(t)B_{km}^i(t,x_k)]dt \\
= \pi \tau_k^0(x_k),\ u_k(\pm 1) = 0,
\end{gathered} \tag{31}$$

$$\begin{gathered}
p_k^0 - i\tau_k^0 = -\frac{1}{\pi}\int\limits_a^b [p(t)\overline{D}_k(t,x_k) + \tau(t)\overline{G}_k(t,x_k)]dt - \frac{1}{2}q^0(1 - e^{-2i\alpha_k}), \\
p_{nk}(x_k) = 0,\ v_k(x_k) > 0;\ p_{nk}(x_k) \leq 0,\ v_k(x_k) = 0,\ k = 1, \ldots, N,
\end{gathered} \tag{32}$$

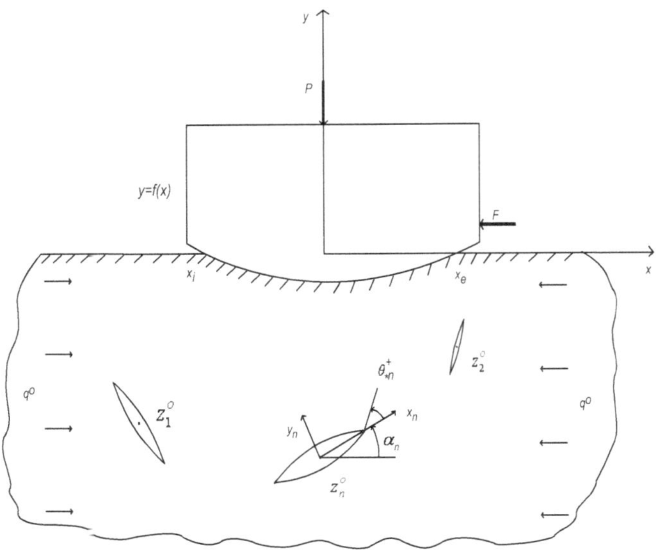

Figure 2. The general view of a rigid indenter in contact with a cracked elastic half-plane.

where the kernels in these equations are described by formulas

$$A_{km} = \overline{R_{km} + S_{km}},\ B_{km} = -i(\overline{R_{km} - S_{km}}),$$

$$(A^r_{km}, B^r_{km}, D^r_k, G^r_k) = Re(A_{km}, B_{km}, \overline{D}_k, \overline{G}_k),$$

$$(A^i_{km}, B^i_{km}, D^i_k, G^i_k) = Im(A_{km}, B_{km}, \overline{D}_k, \overline{G}_k),$$

$$D_k(t, x_k) = \frac{i}{2}\left[-\frac{1}{t-X_k} + \frac{1}{t-\overline{X}_k} - \frac{e^{-2i\alpha_k}(\overline{X}_k - X_k)}{(t-\overline{X}_k)^2} \right],$$

$$G_k(t, x_k) = \frac{1}{2}\left[\frac{1}{t-X_k} + \frac{1-e^{-2i\alpha_k}}{t-\overline{X}_k} - \frac{e^{-2i\alpha_k}(t-X_k)}{(t-\overline{X}_k)^2} \right], \tag{33}$$

$$R_{nk}(t, x_n) = (1-\delta_{nk})K_{nk}(t, x_n) + \frac{e^{i\alpha_k}}{2}\left\{ \frac{1}{X_n - \overline{T}_k} + \frac{e^{-2i\alpha_n}}{\overline{X}_n - T_k} \right.$$

$$\left. +(\overline{T}_k - T_k)\left[\frac{1+e^{-2i\alpha_n}}{(\overline{X}_n - T_k)^2} + \frac{2e^{-2i\alpha_n}(T_k - X_n)}{(\overline{X}_n - T_k)^3} \right] \right\},$$

$$S_{nk}(t, x_n) = (1-\delta_{nk})L_{nk}(t, x_n) + \frac{e^{-i\alpha_k}}{2}\left[\frac{T_k - \overline{T}_k}{(X_n - \overline{T}_k)^2} + \frac{1}{\overline{X}_n - T_k} \right.$$

$$\left. + \frac{e^{-2i\alpha_n}(T_k - X_n)}{(\overline{X}_n - T_k)^2} \right],\ K_{nk}(t_k, x_n) = \frac{e^{i\alpha_k}}{2}\left[\frac{1}{T_k - X_n} + \frac{e^{-2i\alpha_n}}{\overline{T}_k - \overline{X}_n} \right],$$

$$L_{nk}(t_k, x_n) = \frac{e^{-i\alpha_k}}{2}\Big[\frac{1}{\overline{T}_k - \overline{X}_n} - \frac{T_k - X_n}{(\overline{T}_k - \overline{X}_n)^2}e^{-2i\alpha_n}\Big],$$

$$T_k = te^{i\alpha_k} + z_k^0,\ X_n = x_n e^{i\alpha_n} + z_n^0,\ k, n = 1, \ldots, N,$$

where $v_k(x_k)$, $u_k(x_k)$, and $p_{nk}(x_k)$, $k = 1, \ldots, N$, are the jumps of the normal and tangential crack face displacements and the normal stress applied to crack faces, respectively, a and b are the dimensionless contact boundaries, δ_k is the dimensionless crack half-length, $\delta_k = l_k/\tilde{b}$, δ_{nk} is the Kronecker tensor ($\delta_{nk} = 0$ for $n \neq k$, $\delta_{nk} = 1$ for $n = k$), i is the imaginary unit, $i = \sqrt{-1}$.

For simplicity primes at the dimensionless variables are omitted. The characteristic values $\tilde{q}$ and $\tilde{b}$ that are used for scaling are the maximum Hertzian pressure p_H and the Hertzian contact half-width a_H

$$p_H = \sqrt{\frac{E'P}{\pi R}},\ a_H = 2\sqrt{\frac{RP}{\pi E'}}, \tag{34}$$

where R can be taken as the indenter curvature radius at the center of its bottom.

To simplify the problem formulation it is assumed that for small subsurface cracks (i.e. for $\delta_0 \ll 1$, $\delta_0 = \max\limits_{1 \leq k \leq N} \delta_k$) the pressure $p(x^0)$ and frictional stress $\tau(x^0)$ are known and are close to the ones in a contact of this indenter with an elastic half-plane without cracks. It is worth mentioning that cracks affect the contact boundaries a and b and the pressure distribution $p(x^0)$ as well as each other starting with the terms of the order of $\delta_0 \ll 1$.

Therefore, for the given shape of the indenter $f(x^0)$, pressure $p(x^0)$, frictional stress functions $\tau(x^0)$, residual stress q^0, crack orientation angles α_k and sizes δ_k, and the crack positions z_k^0, $k = 1, \ldots, N$, the solution of the problem is represented by crack faces displacement jumps $u_k(x_k)$, $v_k(x_k)$, and the normal contact stress $p_{nk}(x_k)$ applied to the crack faces ($k = 1, \ldots, N$). After the solution of the problem has been obtained, the dimensionless stress intensity factors $k_{1k}^{\pm}$ and $k_{2k}^{\pm}$ are determined according to formulas

$$k_{1n}^{\pm} + ik_{2n}^{\pm} = \mp \lim_{x_n \to \pm 1} \sqrt{1 - x_n^2}[v_n'(x_n) + iu_n'(x_n)],\ 0 \leq n \leq N. \tag{35}$$

3.1. Problem solution

Solution of this problem is associated with formidable difficulties represented by the nonlinearities caused by the presence of the free boundaries of the crack contact intervals and the interaction between different cracks. Under the general conditions solution of this problem can be done only numerically. However, the problem can be effectively solved with the use of just analytical methods in the case when all cracks are small in comparison with the characteristic size $\tilde{b}$ of the contact region, i.e., when $\delta_0 = \max\limits_{1 \leq k \leq N} \delta_k \ll 1$. In this case, it can be shown that the influence of the presence of cracks on the contact pressure is of the order of $O(\delta_0)$ and with the precision of $O(\delta_0)$ the crack system in the half-plane is subjected to the action of the contact pressure $p_0(x^0)$ and frictional stress $\tau_0(x^0)$ that are obtained in the absence of cracks. The further simplification of the problem is achieved under the assumption that cracks are small in comparison to the distances between them, i.e.

$$z_n^0 - z_k^0 \gg \delta_0,\ n \neq k,\ n, k = 1, \ldots, N. \tag{36}$$

The latter assumption with the precision of $O(\delta_0^2)$, $\delta_0 \ll 1$, provides the conditions for considering each crack as a single crack in an elastic half-plane while the crack faces are loaded by certain stresses related to the contact pressure $p_0(x^0)$, contact frictional stress $\tau_0(x^0)$, and the residual stress q^0. The crucial assumption for simple and effective analytical solution of the considered problem is the assumption that all cracks are subsurface and much smaller in size than their distances to the half-plane surface

$$z_k^0 - \overline{z}_k^0 \gg \delta_0,\ k = 1, \ldots, N. \tag{37}$$

Essentially, that assumption permits to consider each crack as a single crack in a plane (not a half-plane) with faces loaded by certain stresses related to $p_0(x^0)$, $\tau_0(x^0)$, and q^0.

Let us assume that the frictional stress $\tau_0(x^0)$ is determined by the Coulomb law of dry friction which in dimensionless variables can be represented by

$$\tau_0(x^0) = -\lambda p_0(x^0), \tag{38}$$

where λ is the coefficient of friction. In (38) we assume that $\lambda \geq 0$, and, therefore, the frictional stress is directed to the left. It is well known that for small friction coefficients λ the distribution of pressure is very close to the one in a Hertzian frictionless contact. Therefore, the expression for the pressure $p_0(x^0)$ in the absence of cracks with high accuracy can be taken in the form

$$p_0(x^0) = \sqrt{1 - (x^0)^2}. \tag{39}$$

Let us consider the process of solution of the pure fracture mechanics problem described by equations (31)-(33), (35), (38), (39). For small cracks, i.e. for $\delta_0 \ll 1$, the kernels from (32) and (33) are regular functions of t, x_n, and x_k and they can be represented by power series in $\delta_k \ll 1$ and $\delta_n \ll 1$ as follows

$$\begin{aligned} &\{A_{km}(t, x_k), B_{km}(t, x_k)\} \\ &= \sum_{j+n=0; j,n\geq 0}^{\infty} (\delta_k x_k)^j (\delta_m t)^n \{A_{kmjn}, B_{kmjn}\}, \end{aligned} \tag{40}$$

$$\{D_k(t, x_k), G_k(t, x_k)\} = \sum_{j=0}^{\infty} (\delta_k x_k)^j \{D_{kj}, G_{kj}\}. \tag{41}$$

In (40) and (41) the values of A_{kmjn} and B_{kmjn} are independent of δ_k, δ_m, x_k, and t while the values of $D_{kj}(t)$ and $G_{kj}(t)$ are independent of δ_k and x_k. The values of A_{kmjn} and B_{kmjn} are certain functions of constants α_k, α_m, x_k^0, y_k^0, x_m^0, and y_m^0 while the values of $D_{kj}(t)$ and $G_{kj}(t)$ are certain functions of α_k, x_k^0, and y_k^0. Therefore, for $\delta_0 \ll 1$ the problem solution can be sought in the form

$$\{v_k, u_k\} = \sum_{j=0}^{\infty} \delta_k^j \{v_{kj}(x_k), u_{kj}(x_k)\},\ p_{nk} = \sum_{j=0}^{\infty} \delta_0^j p_{nkj}(x_k) \tag{42}$$

where functions v_{kj}, u_{kj}, p_{nkj} have to be determined in the process of solution. Expanding the terms of the equations (31)-(33), and (35) and equating the terms with the same powers of δ_0 we get a system of boundary-value problems for integro-differential equations of the first kind which can be easily solved by classical methods [1, 14]. We will limit ourselves to determining only the first two terms of the expansions in (42) in the case of Coulomb's friction law given by (38) and (39). Without getting into the details of the solution process (which can be found in [1]) for the stress intensity factors $k_{1n}^{\pm}$ and $k_{2n}^{\pm}$ we obtain the following analytical formulas [1]

$$k_1^{\pm} = c_0^r \pm \frac{1}{2}\delta_0 c_1^r + \ldots \; if \; c_0^r > 0, \; k_1^{\pm} = 0 \; if \; c_0^r < 0,$$

$$k_1^{\pm} = \frac{\sqrt{3}\delta_0}{9} c_1^r [\pm 7 - 3\theta(c_1^r)] \sqrt{\frac{1 \pm \theta(c_1^r)}{1 \pm 3\theta(c_1^r)}} + \ldots \; if \; c_0^r = 0 \; and \; c_1^r \neq 0,$$

$$k_2^{\pm} = c_0^i \pm \frac{1}{2}\delta_0 c_1^r + \ldots, \tag{43}$$

$$c_j = \frac{1}{\pi} \int\limits_{-1}^{1} [p(x)\overline{D}_j(x) + \tau(x)\overline{G}_j(x)]dx + \frac{\delta_{j0}}{2} q^0 (1 - e^{-2i\alpha}), \; j = 1, 2,$$

$$c_j^r = Re(c_j), \; c_j^i = Im(c_j),$$

where the kernels are determined according to the formulas

$$D_0(x) = \frac{i}{2}\Big[-\frac{1}{x - z^0} + \frac{1}{x - \overline{z}^0} - \frac{e^{-2i\alpha}(\overline{z}^0 - z^0)}{(x - \overline{z}^0)^2} \Big], \; G_0(x) = \frac{1}{2}\Big[\frac{1}{x - z^0}$$
$$+ \frac{1 - e^{-2i\alpha}}{x - \overline{z}^0} - \frac{e^{-2i\alpha}(x - z^0)}{(x - \overline{z}^0)^2} \Big], \; D_1(x) = \frac{ie^{-i\alpha}}{2(x - \overline{z}^0)^2} \Big[1 - e^{-2i\alpha}$$
$$- \frac{2e^{-2i\alpha}(\overline{z}^0 - z^0)}{x - \overline{z}^0} \Big] + \frac{ie^{i\alpha}}{2} \Big[-\frac{1}{(x - z^0)^2} + \frac{e^{-2i\alpha}}{(x - \overline{z}^0)^2} \Big], \; G_1(x) = \frac{e^{-i\alpha}}{2(x - \overline{z}^0)^2} \Big[1 \tag{44}$$
$$- e^{-2i\alpha} - \frac{2e^{-2i\alpha}(x - z^0)}{x - \overline{z}^0} \Big] + \frac{e^{i\alpha}}{2} \Big[\frac{1}{(x - z^0)^2} + \frac{e^{-2i\alpha}}{(x - \overline{z}^0)^2} \Big],$$

and $\theta(x)$ is the step function ($\theta(x) = -1$ for $x < 0$ and $\theta(x) = 1$ for $x \geq 0$).

3.2. Comparison of analytical asymptotic and numerical solutions for small subsurface cracks

Let us compare the asymptotically ($k_{1a}^{\pm}$ and $k_{2a}^{\pm}$) and numerically ($k_{1n}^{\pm}$ and $k_{2n}^{\pm}$) obtained solutions of the problem for the case when $y^0 = -0.4$, $\delta_0 = 0.1$, $\alpha = \pi/2$, $\lambda = 0.1$, and $q^0 = -0.005$. The numerical method used for calculating $k_{1n}^{\pm}$ and $k_{2n}^{\pm}$ is described in detail in [1]. Both the numerical and asymptotic solutions are represented in Fig. 3. It follows from Fig. 3 that the asymptotic and numerical solutions are almost identical except for the region where the numerically obtained $k_1^{+}(x^0)$ is close to zero. The difference is mostly caused by the fact that the used asymptotic solution involve only two terms, i.e., the accuracy of these asymptotic solutions is $O(\delta_0^2)$ for small δ_0. However, according to the two-term asymptotic solutions the maximum values of $k_1^{\pm}$ differ from the numerical ones by no more than 1.4%. One can expect to get much higher precision if $\delta_0 < 0.1$ and $\mid y^0 \mid \gg \delta_0$.

Therefore, formulas (43) and (44) provide sufficient precision for most possible applications and can be used to substitute for numerically obtained values of $k_1^{\pm}$ and $k_2^{\pm}$.

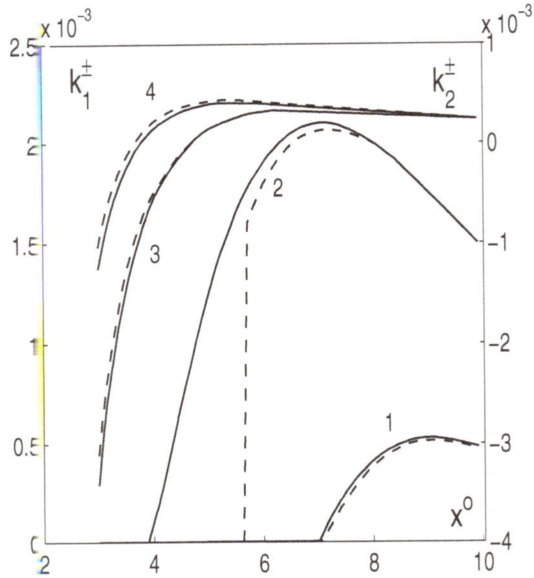

Figure 3. Comparison of the two-term asymptotic expansions $k_{1a}^{\pm}$ and $k_{2a}^{\pm}$ with the numerically calculated stress intensity factors $k_{1n}^{\pm}$ and $k_{2n}^{\pm}$ obtained for $y^0 = -0.4$, $\delta_0 = 0.1$, $\alpha = \pi/2$, $\lambda = 0.1$, and $q^0 = -0.005$. Solid curves are numerical results while dashed curves are asymptotical results. (k_{1n}^{-} group 1, k_{1n}^{+} group 2, k_{2n}^{-} group 3, k_{2n}^{+} group 4) (after Kudish [15]). Reprinted with permission of the STLE.

3.3. Stress intensity factors $k_{1n}^{\pm}$ and $k_{2n}^{\pm}$ behavior for subsurface cracks

Some examples of the behavior of the stress intensity factors $k_{1n}^{\pm}$ and $k_{2n}^{\pm}$ for subsurface cracks are presented below.

It is important to keep in mind that for the cases of no friction ($\lambda = 0$) and compressive or zero residual stress ($q^0 \leq 0$) all subsurface cracks are closed and, therefore, at their tips $k_{1n}^{\pm} = 0$.

Let us consider the case when the residual stress q^0 is different from zero. The residual stress influence on k_{1n}^{+} results in increase of k_{1n}^{+} for a tensile residual stress $q^0 > 0$ or its decrease for a compressive residual stress $q^0 < 0$ of the material region with tensile stresses. From formulas (43), (44), and Fig. 4 (obtained for $y_n^0 = -0.2$, $\alpha_n = \pi/2$, and $\delta_n = 0.1$) follows that for all x_n^0 and for increasing residual stress q^0 (see the curves marked with 3 and 5 that correspond to $\lambda = 0.1$, $q^0 = 0.04$, and $\lambda = 0.2$, $q^0 = 0.02$, respectively) the stress intensity factor k_{1n}^{+} is a non-decreasing function of q^0. Moreover, if at some material point $k_{1n}^{+}(q_1^0) > 0$ for some residual stress q_1^0, then $k_{1n}^{+}(q_2^0) > k_{1n}^{+}(q_1^0)$ for $q_2^0 > q_1^0$ (compare curves marked with 1 and 2 with curves marked with 3 and 4 as well as with curves marked with 5 and 6, respectively). Similarly, for all x_n^0 when the magnitude of the compressive residual stress ($q^0 < 0$) increases (see curves marked with 4 and 6 that correspond to $\lambda = 0.1$, $q^0 = -0.01$ and

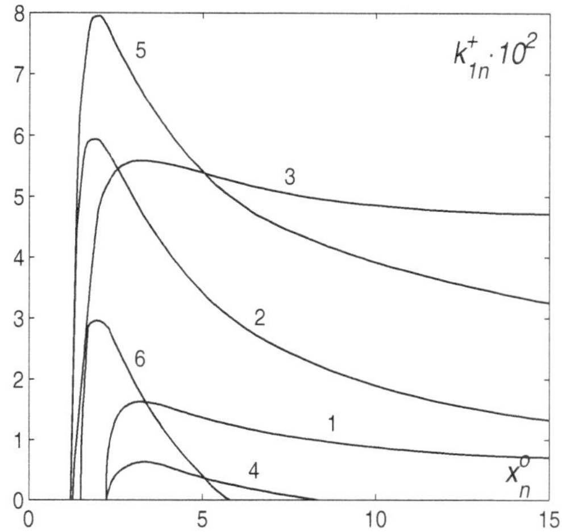

Figure 4. The dependence of the normal stress intensity factor k^+_{1n} on x^0_n for $\delta_n = 0.1$, $\alpha = \pi/2$, $y^0_n = -0.2$ and different levels of the residual stress q^0. Curves 1 and 2 are obtained for $q^0 = 0$ and $\lambda = 0.1$ and $\lambda = 0.2$, respectively. Curves 3 and 4 are obtained for $\lambda = 0.1$, $q^0 = 0.04$ and $q^0 = -0.01$, respectively, while curves 5 and 6 are obtained for $\lambda = 0.2$, $q^0 = 0.01$ and $q^0 = -0.03$, respectively (after Kudish [5]). Reprinted with permission of Springer.

$\lambda = 0.2$, $q^0 = -0.03$, respectively) and at some material point $k^+_{1n}(q^0_1) > 0$ for some residual stress q^0_1 then $k^+_{1n}(q^0_2) < k^+_{1n}(q^0_1)$ for $q^0_2 < q^0_1$. Based on the fact that the normal stress intensity factor k^{0+}_{1n} is positive only in the near surface material layer, we can make a conclusion that this layer increases in size and, starting with a certain value of tensile residual stress $q^0 > 0$, crack propagation becomes possible at any depth beneath the half-plane surface. For increasing compressive residual stresses, the thickness of the material layer where $k^+_{1n} > 0$ decreases.

The analysis of the results for subsurface cracks following from formulas (43) and (44) shows [1] that the values of the stress intensity factors $k^{\pm}_{1n}$ and $k^{\pm}_{2n}$ are insensitive to even relatively large variations in the behavior of the distributions of the pressure $p(x^0)$ and frictional stress $\tau(x^0)$. In particular, the stress intensity factors $k^{\pm}_{1n}$ and $k^{\pm}_{2n}$ for the cases of dry and fluid (lubricant) friction as well as for the cases of constant pressure and frictional stress are very close to each other as long as the normal force (integral of $p(x^0)$ over the contact region) and the friction force (integral of τ over the contact region) applied to the surface of the half-plane are the same [1].

Qualitatively, the behavior of the normal stress intensity factors $k^{\pm}_{1n}$ for different angles of orientation α_n is very similar while quantitatively it is very different. An example of that is presented for a horizontal ($\alpha_n = 0$) subsurface crack in Fig. 5 and for a subsurface crack perpendicular to the half-plane boundary ($\alpha_n = \pi/2$) in Fig 6 for $p(x^0) = \pi/4$, $\tau(x^0) = -\lambda p(x^0)$, $y^0_n = -0.2$, $q^0 = 0$ for $\lambda = 0.1$ and $\lambda = 0.2$.

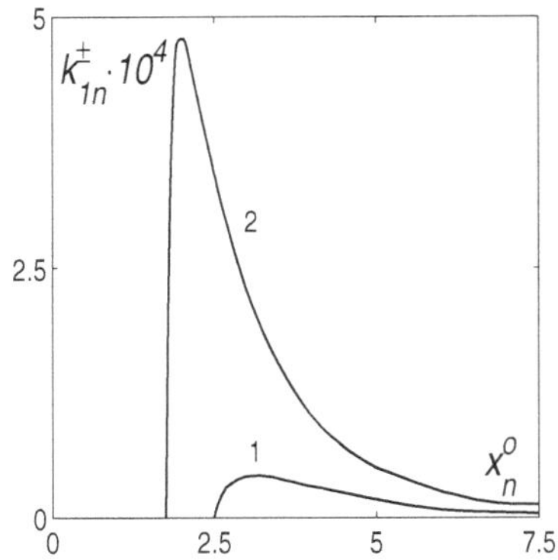

Figure 5. The dependence of the normal stress intensity factor k^{+}_{1n} on the coordinate x^0_n for the case of the boundary of half-plane loaded with normal $p(x^0) = \pi/4$ and frictional $\tau(x^0) = -\lambda p(x^0)$ stresses, $y^0_n = -0.2$, $\alpha_n = 0$, $q^0 = 0$: $\lambda = 0.1$ - curve marked with 1, $\lambda = 0.2$ - curve marked with 2 (after Kudish and Covitch [1]). Reprinted with permission from CRC Press.

For the same loading and crack parameters the behavior of the shear stress intensity factor $k^{\pm}_{2n}$ is represented in Fig. 7 and 8. It is important to observe that the shear stress intensity factors $k^{\pm}_{2n}$ are insensitive to changes of the friction coefficient λ.

Fig. 5 and 6 clearly show that for subsurface cracks with angle $\alpha_n = \pi/2$ for zero or tensile residual stress q^0 the normal stress intensity factors $k^{\pm}_{1n}$ are significantly higher (by two orders of magnitude) than the ones for $\alpha_n = 0$. For $\alpha_n = 0$ and $\alpha_n = \pi/2$ the orders of magnitude of the shear stress intensity factors $k^{\pm}_{2n}$ are the same (see Fig. 7 and 8). Moreover, from these graphs it is clear that the normal stress intensity factors $k^{\pm}_{1n}$ are significantly influenced by the friction coefficient λ while the shear stress intensity factors $k^{\pm}_{2n}$ are insensitive to the value of the friction coefficient λ.

Obviously, in the single-term approximation the behavior of k^{0-}_{1n} and k^{0-}_{2n} is identical to the one of k^{0+}_{1n} and k^{0+}_{2n}, respectively. Generally, the difference between k^{0-}_{1n} and k^{0+}_{1n} as well as between k^{0-}_{2n} and k^{0-}_{2n} is of the order of magnitude of $\delta_0 \ll 1$.

3.4. Lubricant-surface crack interaction. Stress intensity factors $k^{\pm}_{1n}$ and $k^{\pm}_{2n}$ Behavior

The process of lubricant-surface crack interaction is very complex and the details of the problem formulation, the numerical solution approach, and a comprehensive analysis of the results can be found in [1]. Therefore, here we will discuss only the most important features of this phenomenon. It is well known that the presence of lubricant between surfaces in contact

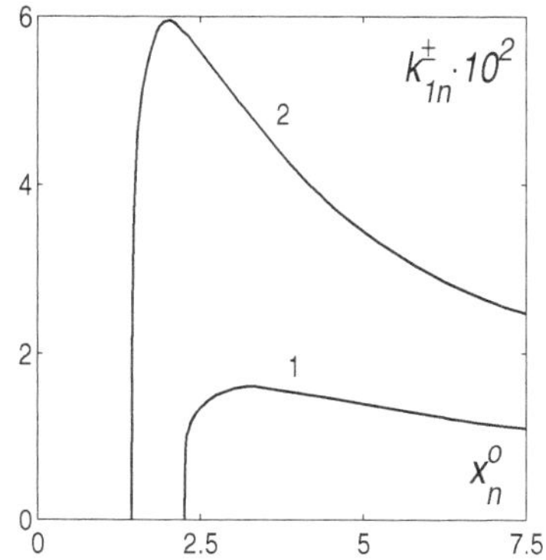

Figure 6. The dependence of the normal stress intensity factor k^{+}_{1n} on the coordinate x^0_n for the case of the boundary of half-plane loaded with normal $p(x^0) = \pi/4$ and frictional $\tau(x^0) = -\lambda p(x^0)$ stresses, $y^0_n = -0.2$, $\alpha_n = \pi/2$, $q^0 = 0$: $\lambda = 0.1$ - curve marked with 1, $\lambda = 0.2$ - curve marked with 2 (after Kudish and Covitch [1]). Reprinted with permission from CRC Press.

is very beneficial as it reduces the contact friction and wear and facilitates better heat transfer from the contact. However, in some cases the lubricant presence may play a detrimental role. In particular, in cases when the elastic solid (half-plane) has a surface crack inclined toward the incoming high contact pressure transmitted through the lubricant. Such a crack may open up and experience high lubricant pressure applied to its faces. This pressure creates the normal stress intensity factor k^{-}_{1n} far exceeding the value of the stress intensity factors $k^{\pm}_{1n}$ for comparable subsurface cracks while the shear stress intensity factors for surface k^{-}_{2n} and and subsurface $k^{\pm}_{2n}$ cracks remain comparable in value. That becomes obvious from the comparison of the graphs from Fig. 4-8 with the graphs from Fig. 10 and 9.

In cases when a surface crack is inclined away from the incoming high lubricant pressure the crack does not open up toward the incoming lubricant with high pressure and it behaves similar to a corresponding subsurface crack, i.e. its normal stress intensity factor k^{-}_{1n} in its value is similar to the one for a corresponding subsurface crack. It is customary to see the normal stress intensity factor for surface cracks which open up toward the incoming high lubricant pressure to exceed the one for comparable subsurface cracks by two orders of magnitude. In such cases the compressive residual stress has very little influence on the crack behavior due to domination of the lubricant pressure.

The possibility of high normal stress intensity factors for surface cracks leads to serious consequences. In particular, it explains why fatigue life of drivers is usually significantly higher than the one for followers [1]. Also, it explains why under normal circumstances fatigue failure is of subsurface origin [1].

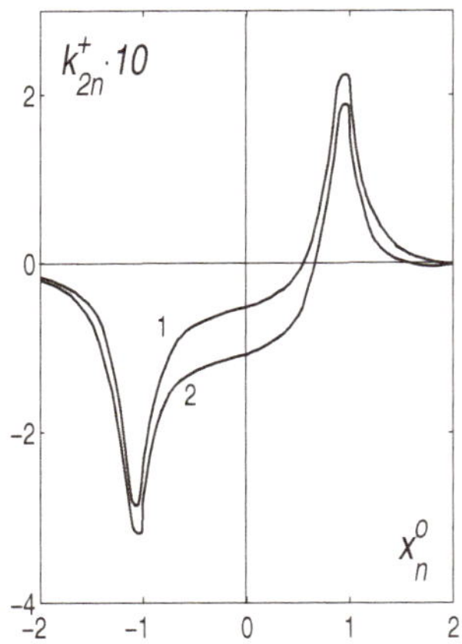

Figure 7. The dependence of the shear stress intensity factor k^+_{2n} on the coordinate x^0_n for the case of the boundary of half-plane loaded with normal $p(x^0) = \pi/4$ and frictional $\tau(x^0) = -\lambda p(x^0)$ stresses, $y^0_n = -0.2$, $\alpha_n = 0$, $q^0 = 0$: $\lambda = 0.1$ - curve marked with 1, $\lambda = 0.2$ - curve marked with 2 (after Kudish and Covitch [1]). Reprinted with permission from CRC Press.

4. Stress intensity factors and directions of fatigue crack propagation

The process of fatigue failure is usually subdivided into three major stages: the nucleation period, the period of slow pre-critical fatigue crack growth, and the short period of fast unstable crack growth ending in material loosing its integrity. The durations of the first two stages of fatigue failure depend on a number of parameters such as material properties, specific environment, stress state, temperature, etc. Usually, the nucleation period is short [1]. We are interested in the main part of the process of fatigue failure which is due to slow pre-critical crack growth. In these cases $\max(k^+_1, k^-_1) < K_f$, where K_f is the material fracture toughness.

During the pre-critical fatigue crack growth cracks remain small. Therefore, they can be modeled by small straight cuts in the material. For such small subsurface cracks it is sufficient to use the one-term asymptotic approximations $k^\pm_1 = k_1$ and $k^\pm_2 = k_2$ for the stress intensity factors from (43) and (44) which in dimensional variables take the form

$$k_1 = \sqrt{l}[Y^r + q^0 \sin^2\alpha]\theta[Y^r + q^0 \sin^2\alpha], \; k_2 = \sqrt{l}[Y^i - \tfrac{q^0}{2}\sin 2\alpha],$$
$$Y = \frac{1}{\pi}\int\limits_{-a_H}^{a_H}[p(t)\overline{D}_0(t) + \tau(t)\overline{G}_0(t)]dt, \; \tau = -\lambda p,$$
$$\{Y^r, Y^i\} = \{Re(Y), Im(Y)\}, \tag{45}$$
$$D_0(t) = \frac{i}{2}\Big[-\frac{1}{t-X} + \frac{1}{t-\overline{X}} - \frac{e^{-2i\alpha}(\overline{X}-X)}{(t-\overline{X})^2}\Big],$$
$$G_0(t) = \frac{1}{2}\Big[\frac{1}{t-X} + \frac{1-e^{-2i\alpha}}{t-\overline{X}} - \frac{e^{-2i\alpha}(t-X)}{(t-\overline{X})^2}\Big], \; X = x + iy,$$

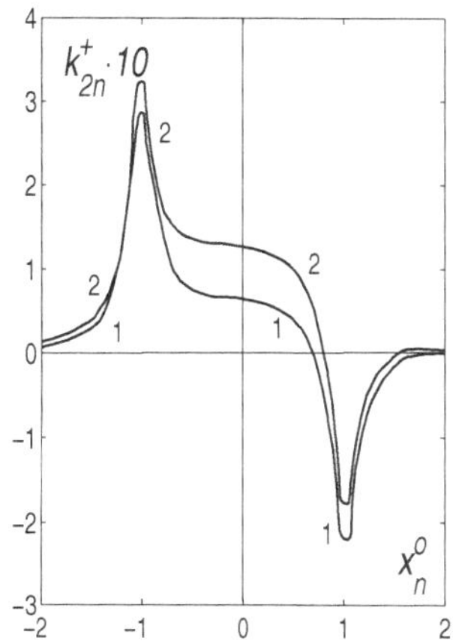

Figure 8. The dependence of the shear stress intensity factor k^+_{2n} on the coordinate x^0_n for the case of the boundary of half-plane loaded with normal $p(x^0) = \pi/4$ and frictional $\tau(x^0) = -\lambda p(x^0)$ stresses, $y^0_n = -0.2, \alpha_n = \pi/2, q^0 = 0$: $\lambda = 0.1$ - curve marked with 1, $\lambda = 0.2$ - curve marked with 2 (after Kudish and Covitch [1]). Reprinted with permission from CRC Press.

where i is the imaginary unit ($i^2 = -1$), $\theta(x)$ is a step function: $\theta(x) = 0,\ x \leq 0$ and $\theta(x) = 1,\ x > 0$. It is important to mention that according to (45) for subsurface cracks the quantities of $k_{10} = k_1 l^{-1/2}$ and $k_{20} = k_2 l^{-1/2}$ are functions of x and y and are independent from l.

Numerous experimental studies have established the fact that at relatively low cyclic loads materials undergo the process of pre-critical failure while the rate of crack growth dl/dN (N is the number of loading cycles) in the predetermined direction is dependent on $k_1^{\pm}$ and K_f. A number of such equations of pre-critical crack growth and their analysis are presented in [6]. However, what remains to be determined is the direction of fatigue crack growth.

Assuming that fatigue cracks growth is driven by the maximum principal tensile stress (see the section on Three-Dimensional Model of Contact and Structural Fatigue) we immediately obtain the equation

$$k_2^{\pm}(N, x, y, l, \alpha^{\pm}) = 0, \quad (46)$$

which determines the orientation angles $\alpha^{\pm}$ of a fatigue crack growth at the crack tips. Due to the fact that a fatigue crack remains small during its pre-critical growth (i.e. practically during its entire life span) and being originally modeled by a straight cut with half-length l at the point with coordinates (x, y) the crack remains straight, i.e. the crack direction is characterized by one angle $\alpha = \alpha^+ = \alpha^-$. This angle is practically independent from crack half-length l because $k_2^{\pm} = k_{20}^{\pm}\sqrt{l}$, where $k_{20}^{\pm}$ is almost independent from l for small l (see (45)). The dependence of $k_2^{\pm}$ on the number of loading cycles N comes only through the dependence of the crack half-length l on N. Therefore, the crack angle α is just a function of the crack location (x, y).

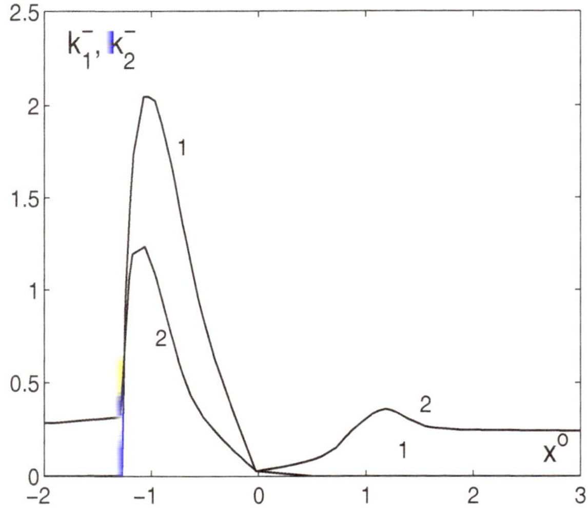

Figure 9. Distributions of the stress intensity factors k_1^- (curve 1) and k_2^- (curve 2) in case of a surface crack: $\alpha = 0.339837$, $\delta_0 = 0.3$, and other parameters as in the previous case of a surface crack (after Kudish [15]). Reprinted with permission of the STLE.

In particular, according to (45) and (46) at any point (x, y) there are two angles α_1 and α_2 along which a crack may propagate which are determined by the equation in dimensional variables

$$\tan 2\alpha = -\frac{2y \int\limits_{-a_H}^{a_H} (t-x)T(t,x,y)dt}{\frac{\pi}{2}q^0 + \int\limits_{-a_H}^{a_H} [(t-x)^2 - y^2]T(t,x,y)dt}, \; T(t,x,y) = \frac{yp(t)+(t-x)\tau(t)}{[(t-x)^2+y^2]^2}. \quad (47)$$

Along these directions k_1 reaches its extremum values. The actual direction of crack propagation α is determined by one of these two angles α_1 and α_2 for which the value of the normal stress intensity factor $k_1(N, x, y, l, \alpha)$ is greater.

A more detailed analysis of the directions of fatigue crack propagation can be found in [1].

5. Two-dimensional contact fatigue model

In a two-dimensional case compared to a three-dimensional case a more accurate description of the contact fatigue process can be obtained due to the fact that in two dimensions it is relatively easy to get very accurate formulas for the stress intensity factors at crack tips [1, 5]. The rest of the fatigue modeling can be done the same way as in the three-dimensional case with few simple changes. In particular, in a two-dimensional case of contact fatigue only subsurface originated fatigue is considered and cracks are modeled by straight cuts with half-length l. That gives the opportunity to use equations (45) and (47) for stress intensity

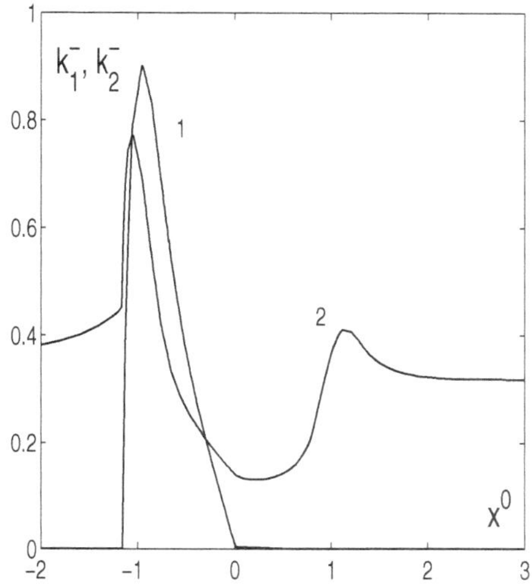

Figure 10. Distributions of the stress intensity factors k_1^- (curve 1) and k_2^- (curve 2) in case of a surface crack: $\alpha = \pi/6$, $\lambda = 0.1$, and $q^0 = -0.5$ (after Kudish and Burris [16]). Reprinted with permission from Kluwer Academic Publishers.

factors k_1, k_2, and crack angle orientations α. The rest of contact fatigue modeling follows exactly the derivation presented in the section on Three-Dimensional Model of Contact and Structural Fatigue. Therefore, the fatigue life N with survival probability P_* of a contact subjected to cyclic loading can be expressed in the form [1, 17]

$$N = \{(\tfrac{n}{2} - 1)g_0[\max_{-\infty<x<\infty} k_{10}]^n\}^{-1}\{\exp[(1 - \tfrac{n}{2})(\mu_{ln} + \sqrt{2}\sigma_{ln} erf^{-1}(2P_* - 1))] - l_c^{\frac{2-n}{2}}\}, \tag{48}$$

where $erf^{-1}(x)$ is the inverse function to the error integral $erf(x)$ [8].

First, let us consider the model behavior in some simple cases. If $f(0, x, y, z, l_0)$ is a uniform crack distribution over the material volume V (except for a thin surface layer where $f = 0$). Then based on (11) it can be shown that $p(N, x, y, z)$ reaches its minimum at the points where k_{10} and the principal tensile stress reach their maximum values. This leads to the conclusion that the material local failure probability $(1 - p)$ reaches its maximum at the points with maximal tensile stress. Therefore, for a uniform initial crack distribution $f(0, x, y, z, l_0)$ the survival probability $P(N)$ from (16) is determined by the material local survival probability at the points at which the maximal tensile stress is attained.

However, the latter conclusion is not necessarily correct if the initial crack distribution $f(0, x, y, z, l_0)$ is not uniform over the material volume. Suppose, $k_{10}(x, y, z)$ is maximal at the point (x_m, y_m, z_m) and at the initial time moment $N = 0$ at some point (x_*, y_*, z_*) there exist cracks larger than the ones at the point (x_m, y_m, z_m), namely,

$$\int_0^{l_c} f dl_0 \mid_{(x_*,y_*,z_*)} < \int_0^{l_c} f dl_0 \mid_{(x_m,y_m,z_m)} .$$

Then after a certain number of loading cycles $N > 0$ the material damage at point (x_*, y_*, z_*) may be greater than at point (x_m, y_m, z_m), where l_{0c} reaches its maximum value. Therefore, fatigue failure may occur at the point (x_*, y_*, z_*) instead of the point (x_m, y_m, z_m), and the material weakest point is not necessarily is the material most stressed point.

If μ_{ln} and σ_{ln} depend on the coordinates of the material point (x, y, z), then there may be a series of points where in formula (16) for the given number of loading cycles N the minimum over the material volume V is reached. The coordinates of such points may change with N. This situation represents different potentially competing fatigue mechanisms such as pitting, flaking, etc. The occurrence of fatigue damage at different points in the material depends on the initial defect distribution, applied stresses, residual stress, etc.

In the above model of contact fatigue the stressed volume V plays no explicit role. However, implicitly it does. In fact, the initial crack distribution $f(0, x, y, z, l_0)$ depends on the material volume. In general, in a larger volume of material, there is a greater chance to find inclusions/cracks of greater size than in a smaller one. These larger inclusions represent a potential source of pitting and may cause a decrease in the material fatigue life of a larger material volume.

Assuming that μ_{ln} and σ_{ln} are constants, and assuming that the material failure occurs at the point (x, y, z) with the failure probability $1 - P(N)$ following the considerations of the section on Three-Dimensional Model of Contact and Structural Fatigue from (48) we obtain formulas (19)-(21). Actually, fatigue life formulas (20) and (21) can be represented in the form of the Lundberg-Palmgren formula, i.e.

$$N = \frac{C_*}{p_H^n}, \tag{49}$$

where parameter n can be compared with constant c/e in the Lundberg-Palmgren formula [1]. The major difference between the Lundberg-Palmgren formula and formula (49) derived from this model of contact fatigue is the fact that in (49) constant C_* depends on material defect parameters μ, σ, coefficient of friction λ, residual stresses q^0, and probability of survival P_* in a certain way while in the Lundberg-Palmgren formula the constant C_* depends only on the depth z_0 of the maximum orthogonal stress, stressed volume V, and probability of survival P_*.

Let us analyze formula (21). It demonstrates the intuitively obvious fact that the fatigue life N is inverse proportional to the value of the parameter g_0 that characterizes the material crack propagation resistance. So, for materials with lower crack propagation rate, the fatigue life is higher and vice versa. Equation (21) exhibits a usual for gears and roller and ball bearings dependence of fatigue life N on the maximum Hertzian pressure p_H. Thus, from the well-known experimental data for bearings, the range of n values is $20/3 \leq n \leq 9$. Keeping in mind that usually $\sigma \lll \mu$, for these values of n contact fatigue life N is practically inverse proportional to a positive power of the crack initial mean size, i.e., to $\mu^{n/2-1}$. Therefore, fatigue life N is a decreasing function of the initial mean crack (inclusion) size μ and $\ln N = -(n/2 - 1) \ln \mu + constant$. This conclusion is valid for any value of the material survival

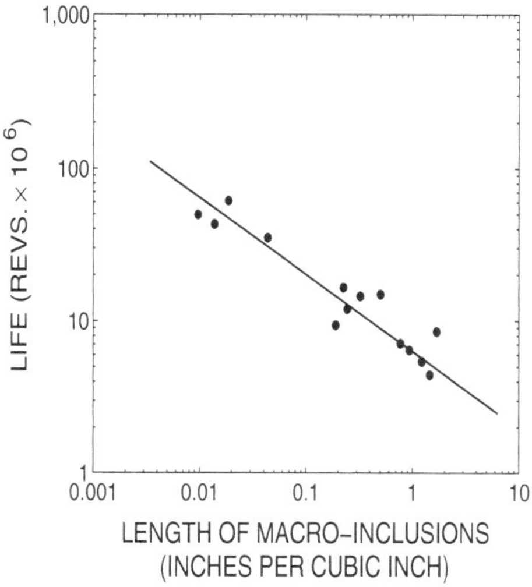

Figure 11. Bearing life-inclusion length correlation (after Stover and Kolarik II[18], COPYRIGHT The Timken Company 2012).

probability P_* and is supported by the experimental data obtained at the Timken Company by Stover and Kolarik II [18] and represented in Fig. 11 in a log-log scale. If $P_* > 0.5$ then $erf^{-1}(2P_* - 1) > 0$ and (keeping in mind that $n > 2$) fatigue life N is a decreasing function of the initial standard deviation of crack sizes σ. Similarly, if $P_* < 0.5$, then fatigue life N is an increasing function of the initial standard deviation of crack sizes σ. According to (20), for $P_* = 0.5$ fatigue life N is independent from σ, and, according to (21), for $P_* = 0.5$ fatigue life N is a slowly increasing function of σ. By differentiating $p_m(N)$ obtained from (16) with respect to σ, we can conclude that the dispersion of $P(N)$ increases with σ.

From (45) (also see Kudish [1]) follows that the stress intensity factor k_1 decreases as the magnitude of the compressive residual stress q^0 increases and/or the magnitude of the friction coefficient λ decreases. Therefore, it follows from formulas (45) that C_0 is a monotonically decreasing function of the residual stress q^0 and friction coefficient λ. Numerical simulations of fatigue life show that the value of C_0 is very sensitive to the details of the residual stress distribution q^0 versus depth.

Let us choose a basic set of model parameters typical for bearing testing: maximum Hertzian pressure $p_H = 2\ GPa$, contact region half-width in the direction of motion $a_H = 0.249\ mm$, friction coefficient $\lambda = 0.002$, residual stress varying from $q^0 = -237.9\ MPa$ on the surface to $q^0 = 0.035\ MPa$ at the depth of 400 μm below it, fracture toughness K_f varying between 15 and 95 $MPa \cdot m^{1/2}$, $g_0 = 8.863\ MPa^{-n} \cdot m^{1-n/2} \cdot cycle^{-1}$, $n = 6.67$, mean of crack initial half-lengths $\mu = 49.41\ \mu m$ ($\mu_{ln} = 3.888 + ln(\mu m)$), crack initial standard deviation $\sigma = 7.61\ \mu m$ ($\sigma_{ln} = 0.1531$). Numerical results show that the fatigue life is practically independent from the material fracture toughness K_f, which supports the assumption used for the derivation of formulas (19)-(21). To illustrate the dependence of contact fatigue life on some of the model

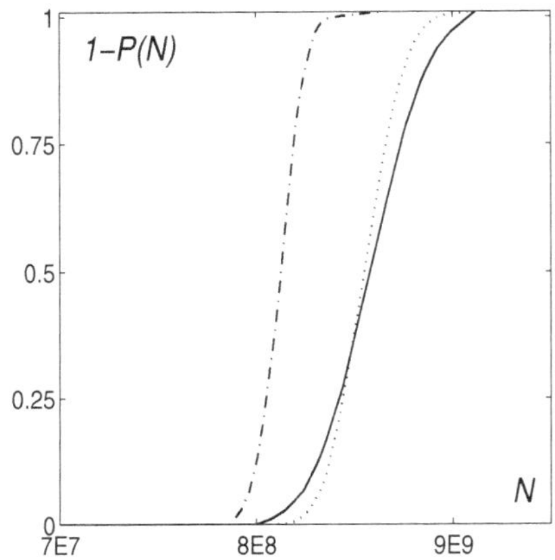

Figure 12. Pitting probability $1 - P(N)$ calculated for the basic set of parameters (solid curve) with $\mu = 49.41\ \mu m$, $\sigma = 7.61\ \mu m$ ($\mu_{ln} = 3.888 + \ln(\mu m)$, $\sigma_{ln} = 0.1531$), for the same set of parameters and the increased initial value of crack mean half-lengths (dash-dotted curve) $\mu = 74.12\ \mu m$ ($\mu_{ln} = 4.300 + \ln(\mu m)$, $\sigma_{ln} = 0.1024$), and for the same set of parameters and the increased initial value of crack standard deviation (dotted curve) $\sigma = 11.423\ \mu m$ ($\mu_{ln} = 3.874 + \ln(\mu m)$, $\sigma_{ln} = 0.2282$) (after Kudish [17]). Reprinted with permission from the STLE.

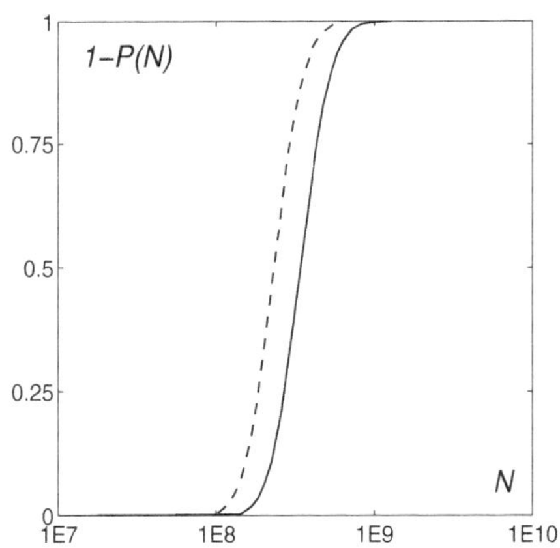

Figure 13. Pitting probability $1 - P(N)$ calculated for the basic set of parameters including $\lambda = 0.002$ (solid curve) and for the same set of parameters and the increased friction coefficient (dashed curve) $\lambda = 0.004$ (after Kudish [17]). Reprinted with permission from the STLE.

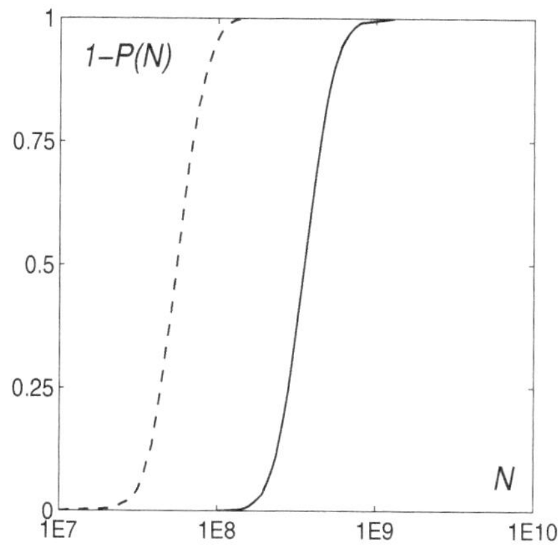

Figure 14. Pitting probability $1 - P(N)$ calculated for the basic set of parameters (solid curve) and for the same set of parameters and changed profile of residual stress q^0 (dashed curve) in such a way that at points where q^0 is compressive its magnitude is unchanged and at points where q^0 is tensile its magnitude is doubled (after Kudish [17]). Reprinted with permission from the STLE.

parameters, just one parameter from the basic set of parameters will be varied at a time and graphs of the pitting probability $1 - P(N)$ for these sets of parameters (basic and modified) will be compared. Figure 12 shows that as the initial values of the mean μ of crack half-lengths and crack standard deviation σ increase contact fatigue life N decreases. Similarly, contact fatigue life decreases as the magnitude of the tensile residual stress and/or friction coefficient increase (see Fig. 4 and 13). The results show that the fatigue life does not change when the magnitude of the compressive residual stress is increased/decreased by 20% of its base value while the tensile portion of the residual stress distribution remains the same. Obviously, that is in agreement with the fact that tensile stresses control fatigue. Moreover, the fatigue damage occurs in the region with the resultant tensile stresses close to the boundary between tensile and compressive residual stresses. However, when the compressive residual stress becomes small enough the acting frictional stress may supersede it and create new regions with tensile stresses that potentially may cause acceleration of fatigue failure.

μ $[\mu m]$	σ $[\mu m]$	$N_{15.9}$ $[cycles]$
49.41	7.61	$2.5 \cdot 10^8$
73.13	11.26	$1.0 \cdot 10^8$
98.42	15.16	$5.0 \cdot 10^7$
147.11	22.66	$2.0 \cdot 10^7$
244.25	37.62	$6.0 \cdot 10^6$

Table 1. Relationship between the tapered bearing fatigue life $N_{15.9}$ and the initial inclusion size mean and standard deviation (after Kudish [17]). Reprinted with permission from the STLE.

Let us consider an example of the further validation of the new contact fatigue model for tapered roller bearings based on a series of approximate calculations of fatigue life. The main simplifying assumption made is that bearing fatigue life can be closely approximated by taking into account only the most loaded contact. The following parameters have been used for calculations: $p_H = 2.12\ GPa$, $a_H = 0.265\ mm$, $\lambda = 0.002$, $g_0 = 6.009\ MPa^{-n} \cdot m^{1-n/2} \cdot cycle^{-1}$, $n = 6.67$, the residual stress varied from $q^0 = -237.9\ MPa$ on the surface to $q^0 = 0.035\ MPa$ at the depth of 400 μm below the surface, fracture toughness K_f varied between 15 and $95 MPa \cdot m^{1/2}$. The crack/inclusion initial mean half-length μ varied between 49.41 and 244.25 μm ($\mu_{ln} = 3.888 - 5.498 + ln(\mu m)$), the crack initial standard deviation varied between $\sigma = 7.61$ and 37.61 μm ($\sigma_{ln} = 0.1531$). The results for fatigue life $N_{15.9}$ (for $P(N_{15.9}) = P_* = 0.159$) calculations are given in the Table 1 and practically coincide with the experimental data obtained by The Timken Company and presented in Fig. 19 by Stover, Kolarik II, and Keener [?] (in the present text this graph is given as Fig. 11). One must keep in mind that there are certain differences in the numerically obtained data and the data presented in the above mentioned Fig. 11 due to the fact that in Fig. 11 fatigue life is given as a function of the cumulative inclusion length (sum of all inclusion lengths over a cubic inch of steel) while in the model fatigue life is calculated as a function of the mean inclusion length.

It is also interesting to point out that based on the results following from the new model, bearing fatigue life can be significantly improved for steels with the same cumulative inclusion length but smaller mean half-length μ (see Fig. 12). In other words, fatigue life of a bearing made from steel with large number of small inclusions is higher than of the one made of steel with small number of larger inclusions given that the cumulative inclusion length is the same in both cases. Moreover, bearing and gear fatigue lives with small percentage of failures (survival probability $P > 0.5$) for steels with the same cumulative inclusion length can also be improved several times if the width of the initial inclusion distribution is reduced, i.e., when the standard deviation σ of the initial inclusion distribution is made smaller (see Figure 12). Figures 13 and 14 show that the elevated values of the tensile residual stress are much more detrimental to fatigue life than greater values of the friction coefficient.

Finally, the described model is flexible enough to allow for replacement of the density of the initial crack distribution (see (1)) by a different function and of Paris's equation for fatigue crack propagation (see (4)) by another equation. Such modifications would lead to results on fatigue life varying from the presented above. However, the methodology, i.e., the way the formulas for fatigue life are obtain and the most important conclusions will remain the same.

This methodology has been extended on the cases of non-steady cyclic loading as well as on the case of contact fatigue of rough surfaces [1]. Also, this kind of modeling approach has been applied to the analysis of wear and contact fatigue in cases of lubricant contaminated by rigid abrasive particles and contact surfaces charged with abrasive particles [19] as well as to calculation of bearing wear and contact fatigue life [20].

In conclusion we can state that the presented statistical contact and structural fatigue models take into account the most important parameters of the contact fatigue phenomenon (such as normal and frictional contact and residual stresses, initial statistical defect distribution, orientation of fatigue crack propagation, material fatigue resistance, etc.). The models allows for examination of the effect of variables such as steel cleanliness, applied stresses, residual stress, etc. on contact fatigue life as single or composite entities. Some analytical results illustrating these models and their validation by the experimentally obtained fatigue life data for tapered bearings are presented.

6. Closure

The chapter presents a detailed analysis of a number of plane crack mechanics problems for loaded elastic half-plane weakened by a system of cracks. Surface and subsurface cracks are considered. All cracks are considered to be straight cuts. The problems are analyzed by the regular asymptotic method and numerical methods. Solutions of the problems include the stress intensity factors. The regular asymptotic method is applied under the assumption that cracks are far from each other and from the boundary of the elastic solid. It is shown that the results obtained for subsurface cracks based on asymptotic expansions and numerical solutions are in very good agreement. The influence of the normal and tangential contact stresses applied to the boundary of a half-plane as well as the residual stress on the stress intensity factors for subsurface cracks is analyzed. It is determined that the frictional and residual stresses provide a significant if not the predominant contribution to the problem solution. Based on the numerical solution of the problem for surface cracks in the presence of lubricant the physical nature of the "wedge effect" (when lubricant under a sufficiently large pressure penetrates a surface crack and ruptures it) is considered. Solution of this problem also provides the basis for the understanding of fatigue crack origination site (surface versus subsurface) and the difference of fatigue lives of drivers and followers. New two- and three-dimensional models of contact and structural fatigue are developed. These models take into account the initial crack distribution, fatigue properties of the solids, and growth of fatigue cracks under the properly determined combination of normal and tangential contact and residual stresses. The formula for fatigue life based on these models can be reduced to a simple formula which takes into account most of the significant parameters affecting contact and structural fatigue. The properties of these contact fatigue models are analyzed and the results based on them are compared to the experimentally obtained results on contact fatigue for tapered bearings.

7. References

[1] Kudish, I.I. and Covitch, M.J., 2010. *Modeling and Analytical Methods in Tribology*. Boca Raton, London, New York: CRC Press, Taylor & Francis Group.

[2] Kudish, I.I. 2007. Fatigue Modeling for Elastic Materials with Statistically Distributed Defects. *ASME J. Appl. Mech.* 74:1125-1133.

[3] Bokman, M.A., Pshenichnov, Yu.P., and Pershtein, E.M. 1984. The Microcrack and Non-metallic Inclusion Distribution in Alloy D16 after a Plastic Strain. *Plant Laboratory*, Moscow, 11:71-74.

[4] Cherepanov, G.P. 1979. *Mechanics of Brittle Fracture*. New York: McGraw–Hill.

[5] Kudish, I.I. 1987. Contact Problem of the Theory of Elasticity for Pre-stressed Bodies with Cracks. *J. Appl. Mech. and Techn. Phys.* 28, 2:295-303.

[6] Yarema, S.Ya. 1981. Methodology of Determining the Characteristics of the Resistance to Crack Development (Crack Resistance) of Materials in Cyclic Loading. *J. Soviet Material Sci.* 17, No. 4:371-380.

[7] Tallian, T., Hoeprich, M., and Kudish, I. 2001. Author's Closure. *STLE Tribology Trans.* 44, No. 2:153-155.

[8] *Handbook of Mathematical Functions with Formulas, Graphs and Mathematical Tables*, Eds. M. Abramowitz and I.A. Stegun, National Bureau of Standards, 55, 1964.

[9] Stover, J.D., Kolarik II, R.V., and Keener, D.M. 1989. The Detection of Aluminum Oxide Stringers in Steel Using an Ultrasonic Measuring Method. *Mechanical Working and Steel*

Processing XXVII: Proc. 31st Mechanical Working and Steel Processing Conf., Chicago, Illinois, October 22-25, 1989, Iron and Steel Soc., Inc., 431-440.

[10] Lurye, A.I. 1970. *Theory of Elasticity*. Moscow: Nauka.

[11] Hasebe, N. and Inohara, S. 1980. Stress Analysis of a Semi-infinite Plate with an Oblique Edge Crack. *Ing. Arch.* 49:51-62.

[12] Isida, M. 1979. Tension of a Half Plane Containing Array Cracks, Branched Cracks and Cracks Emanating from Sharp Notches. *Trans. Jpn. Soc. Mech. Engrs.* 45, No. 392:306-317.

[13] Savruk M.P. 1981. *Two – –Dimensional Problems of Elasticity for Bodies with Cracks.* Kiev: Naukova Dumka.

[14] Vorovich, I.I., Alexandrov, V.M., and Babesko, V.A. 1974. *Non–Classic Mixed Problems in the Theory of Elasticity*. Moscow: Nauka.

[15] Kudish I.I. 2002. Lubricant-Crack Interaction, Origin of Pitting, and Fatigue of Drivers and Followers. *STLE Tribology Trans.* 45:583-594.

[16] Kudish I.I. and Burris, K.W. 2004. Modeling of Surface and Subsurface Crack Behavior under Contact Load in the Presence of Lubricant. *Intern. J. Fract.* 125:125-147.

[17] Kudish, I.I. 2000. A New Statistical Model of Contact Fatigue. *STLE Tribology Trans.* 43:711-721.

[18] Stover, J.D. and Kolarik II, R.V. 1987. The Evaluation of Improvements in Bearing Steel Quality Using an Ultrasonic Macro-Inclusion Detection Method. *The Timken Company Technical Note*, (January 1987), 1-12.

[19] Kudish, I.I. 1991. Wear and Fatigue Pitting Taking into Account Contaminated Lubricants and Abrasive Particles Indented into Working Surfaces, *Journal of Friction and Wear, Vol.* 12, No. 3:713-725.

[20] Kudish, I.I. 1990. Statistical Calculation of Rolling Bearing Wear and Pitting. *Soviet J. Frict. and Wear* 11:71-86.

Chapter 7

Fracture Mechanics Analysis of Fretting Fatigue Considering Small Crack Effects, Mixed Mode, and Mean Stress Effect

Kunio Asai

Additional information is available at the end of the chapter

1. Introduction

Like sharp-notch fatigue, fretting fatigue strength is mostly determined by whether small cracks propagate, which originate at the local high stress area. Hence, applying fracture mechanics is expected to be effective in evaluating fretting strength (Asai, 2010; Attia, 2005; Edward, 1984; Kondo et al., 2004; Makino et al., 2000; Nicholas et al., 2003). In these methods, the fretting fatigue limit is predicted by evaluating whether the stress intensity factor range ΔK is greater than its threshold value ΔK_{th}. Kondo (Kondo et al., 2004) developed a model for evaluating micro-crack propagation, which is shown in Fig. 1. In this model, when ΔK is lower than ΔK_{th} at a certain crack depth, the crack is thought to stop propagation and to remain as a non-propagating crack (O). On the other hand, when ΔK is larger than ΔK_{th} along the entire crack length, it is thought to propagate to failure. The objective of this study is to evaluate fretting fatigue strength quantitatively using this model under various test conditions including different material strengths, contact pressure, and mean stress by overcoming the following difficulties.

The following two major difficulties need to be addressed when quantitatively applying the micro-crack propagation model.

1. Small crack and mean stress effects on ΔK_{th},
2. Mixed modes of tensile and shear ΔK.

Regarding the small crack effects on ΔK_{th}, El Haddad (El Haddad et al., 1979) proposed the correlation factor, a_0, for the crack length, a, and the threshold of a long crack, $\Delta K_{th,\,l}$, as expressed in Eq. (1),

$$\Delta K_{th} = \Delta K_{th,\,l}\sqrt{a/\left(a+a_0\right)}. \quad (1)$$

The empirical rule proposed by Murakami (Murakami & Endo, 1986) is also well known, where ΔK_{th} is proportional to one-third power of the square root of the micro-crack surface area, equivalent to crack length. Although these approaches are effective in estimating ΔK_{th} for micro cracks, there are few data available for the mean stress effects on micro-crack ΔK_{th} (Usami & Shida, 1979), especially under a high negative stress ratio (R), which is indispensable in evaluating the fretting fatigue strength.

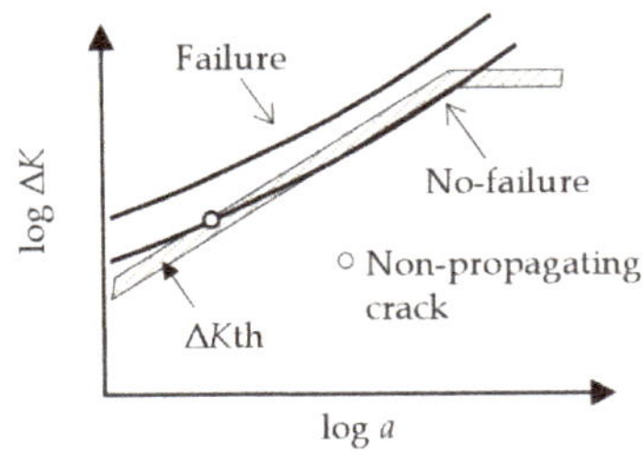

Figure 1. Schematic of small-crack propagation model at fretting fatigue

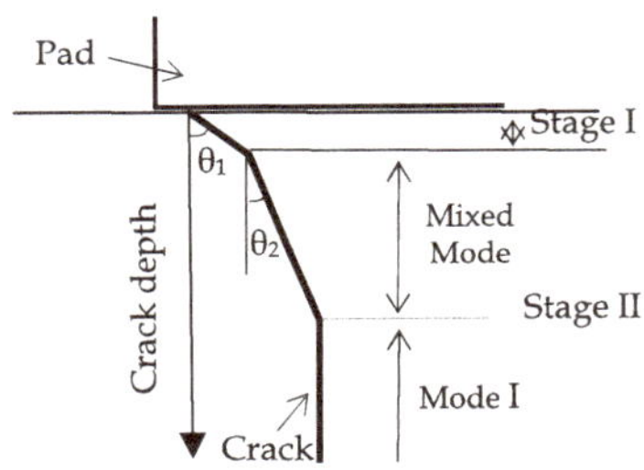

Figure 2. Schematic view of fretting crack propagation

Mixed modes of tensile and shear ΔK should be considered because most fretting fatigue cracks incline under multi-axial stress fields caused by the contact pressure and tangential force (Lamacq et al., 1996; Qian & Fatemi, 1996; Zhang & Fatemi, 2010). According to Mutoh (Mutoh, 1997), the crack path of fretting fatigue is classified into two stages, as shown in Fig. 2. Stage I is an initial crack stage where a crack inclines greatly against the normal direction, and stage II is where a crack is thought to propagate in the direction perpendicular to the maximum principal stress amplitude. As many researchers state (Dubourg & Lamacq, 2000; Faanes, 1995; Mutoh & Xu, 2003), maximum tangential stress theory is considered to be effective for expressing the crack propagation in stage II; hence, one problem is how to model its propagation in stage I. To solve this problem, Pook's failure mechanism map (Pook, 1985) in the $\Delta K_{I}-\Delta K_{II}$ plane is informative for separating crack propagation patterns into shear and tensile modes. Although it was proposed to define equivalent stress intensity factors, such as $\sqrt{\Delta K_I^2 + \Delta K_{II}^2}$ based on the strain energy release rate, and $(\Delta K_I^4+8\Delta K_{II}^4)^{1/4}$ from Tanaka (Tanaka, 1974), there seems to be no unified model applicable to various test results (Hannes & Alfredsson, 2012). Summarizing the studies on mixed modes, what makes it difficult to explain the crack propagation in stage I are the difficulties in experimentally obtaining the mode II thresholds (Murakami et al., 2002) and quantitatively estimating the actual ΔK_{II} considering the crack surface friction effects (Bold et al., 1992). However, from the standpoint of practical use, it is thought useful to apply the maximum tangential stress theory in stage I if its estimation is satisfactorily accurate.

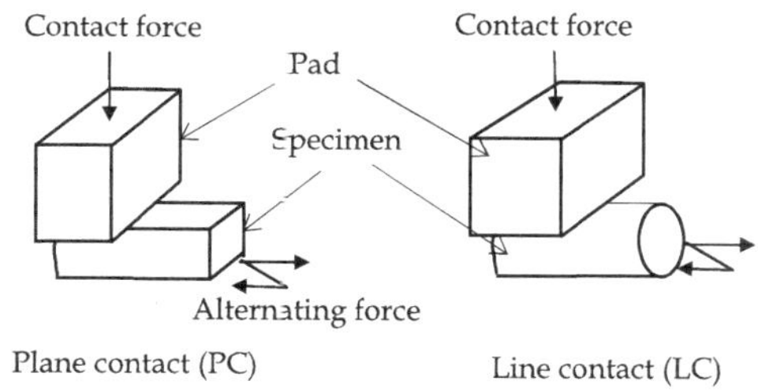

Figure 3. Schematic view of fretting fatigue tests

In this study, fretting fatigue tests under different contact conditions; plane-contact (PC) and line-contact (LC), shown in Fig. 3, were carried out using two 12% Cr steel samples with different static strengths and the effects of the material strength and mean stress were investigated. Fretting fatigue strength was evaluated quantitatively by applying the micro-crack propagation model under various test conditions while considering the above-mentioned difficulties. The practical effectiveness is discussed in applying the maximum tangential stress theory in stage I by obtaining non-propagating crack lengths of run-out specimens and ΔK_{th} from fretting pre-cracks under various R, including negative mean stress.

2. Fretting test method

The test materials were two 12% Cr steel samples (A and B) that had different static strengths, as shown in Table 1. Tensile and 0.2% yield strengths of sample B were approximately 40% higher than those of sample A. Figure 4 shows the shapes of the test specimens and a contact pad. Two kinds of tests were undertaken using rectangular-cross-section specimens (5 mm × 5 mm) for PC conditions and circular-cross-section specimens (8 mm in diameter) for LC conditions. Sample A was used for the contact pad. Heat treatments were applied to the specimens at 600°C × 4 h to relieve the residual stress caused by machining.

After first applying an axial mean load, the contact force was applied using cramping bolts, and axial alternative loads were then applied. The contact force was measured and adjusted by the cylindrical load cell with an uncertainty of 5% to the target value during the tests. Contact pressure was 80 MPa for PC conditions, and LC loads were 60, 150, 300, and 450 N/mm, which respectively corresponded to 584, 923, 1306, and 1569 MPa of the average elastic contact pressure calculated from Hertz's formula. Mean stresses were 0 and 400 MPa for all test cases and –100 MPa for PC conditions of sample A. Tests were carried out using an electro-magnetic-resonance machine in air at ambient temperature. The frequencies were about 125 Hz for LC and about 110 Hz for PC, which were determined by the stiffness of the specimens and the machine.

	0.2% proof stress	Tensile strength	Elongation (%)	Reduction of area	Vickers hardness
Sample A	610	745	26.3	65.5	238
Sample B	842	1037	15.4	51.0	329

Table 1. Mechanical properties of materials

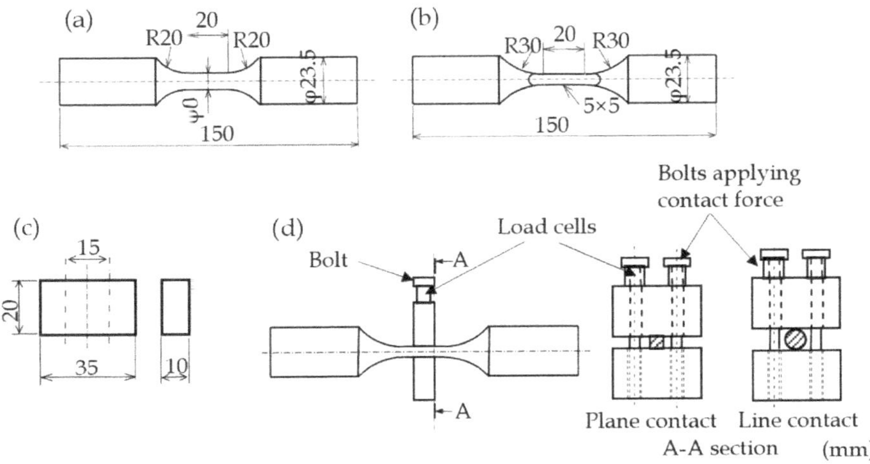

Figure 4. Shapes of specimens and test apparatus: (a) specimen for LC tests, (b) specimen for PC tests, (c) contact pad, and (d) test apparatus

Plain fatigue tests were also carried out without contact pads using run-out fretting specimens at 2×10^7 cycles, and the size of fretting non-propagating cracks, which were the cause of plain fatigue fracture in most cases, were investigated. In addition to the crack length, the plain fatigue tests were aimed at obtaining ΔK_{th} from the fretting pre-crack under constant R, −3, −1, 0, 0.5, by increasing the applied stress step by step until the specimens broke. In these tests, the number of run-out cycles was defined as 10^7 and maximum and minimum applied nominal stresses were not to exceed 0.2% yield strength. The crack-profile path from the initial point was also measured at the fracture surface by using a laser microscope to analyze the behavior of the crack propagation. When the specimen did not break from the fretting non-propagating crack, its depth was measured by polishing the crack surface until it disappeared.

3. Fretting fatigue test results

3.1. Fretting fatigue strength

Figure 5 shows the stress amplitude σ_a against the number of cycles (S–N) diagrams for PC and LC conditions. Figure 6 shows the effect of the contact pressure on fretting fatigue strength when mean stresses σ_m were 0 and 400 MPa, indicating failure- and non-failure-stress amplitudes at 2×10^7 cycles, respectively. The fatigue limits for LC conditions decreased as the contact pressure increased and minimized at a certain contact pressure. The minimum strength pressure, *MSP*, when fretting fatigue strength minimized, depended on the material strength, i.e., *MSP* of sample B (higher static strength) was higher than that of sample A. The average Hertz's contact pressure at *MSP* almost corresponded to about 1.5 times 0.2% proof stress $\sigma_{0.2}$ for both samples A and B. Under PC, sample B exhibited 10-25% higher fretting-fatigue strength than sample A. On the other hand, the minimum strengths of samples A and B differed little (about 5%) under LC conditions; this tells us that a high-static-strength material does not necessarily improve the fretting fatigue strength when local high contact pressure arises. Fretting fatigue strength depended on the mean stress in a high contact pressure region; the strength over *MSP* increased more drastically at σ_m=0 MPa than that at σ_m =400 MPa.

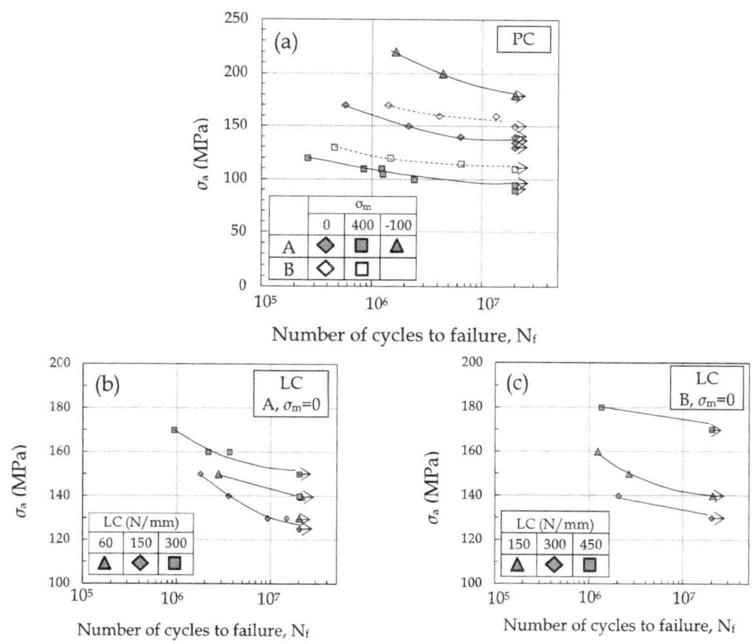

Figure 5. S-N diagram of fretting tests: (a) PC, (b) LC for sample A, and (c) LC for sample B

Figure 7 shows an observed contact surface near the contact edge under LC conditions at 150 N/mm-pressure. The crack edge was located about 0.12 mm inside the contact edge. The width of the wear region was about 0.5 mm, greater than the elastic contact width calculated from Hertz's formula (about 0.16 mm). This was caused by plastic deformation at the contact edge under high local pressure.

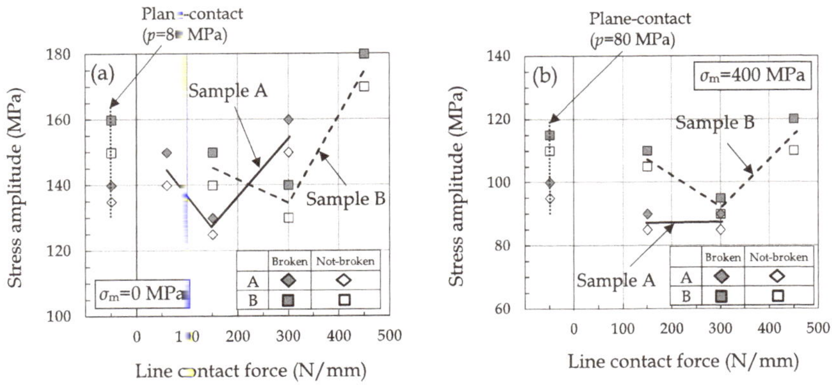

Figure 6. Effect of contact pressure on fretting fatigue strength at mean stress, (a) 0 MPa, and (b) 400 MPa. (Open: not broken; closed: broken in less than 2×10^{7}cycles

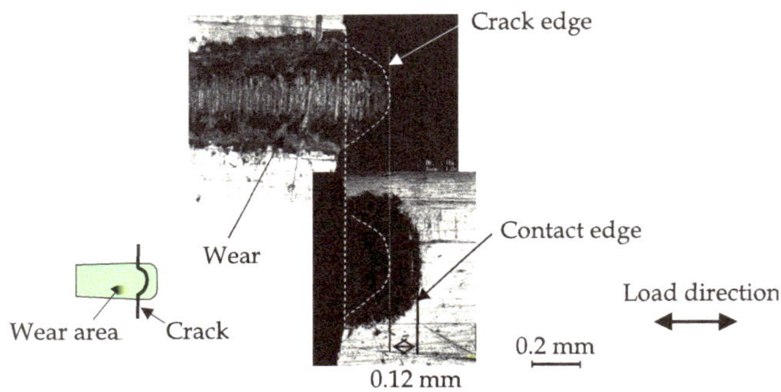

Figure 7. Side view around contact edge of failure specimen. (Sample A: LC, 150 N/mm, σ_m=0 MPa, σ_a=130 MPa, N_f=9.23×10^6)

3.2. Dimensions of non-propagating cracks

Figure 8(a) shows an example of a non-propagating crack at the fracture surface. Its depth a and surface length l, projected in the plane perpendicular to the axial direction, were obtained from the fracture surfaces. The value of a/l was almost 0.15, as shown in Fig. 8(b). The relationship between non-propagating crack length a_{eq} and stress amplitude is summarized in Fig. 9, where equivalent crack length a_{eq} was calculated from Eq. (2).

$$a_{eq}=a/Q,\ Q=1+4.593(a/l)^{1.65}\ . \qquad (2)$$

The value of a_{eq} corresponds to half the center crack length of the infinite plate under uniform stress field. Figure 9 suggests the following three characteristics:

- The value of a_{eq} for sample B (higher static strength) is smaller than that of sample A on the same σ_a under FC.

- Higher mean stress leads to smaller a_{eq} at the fatigue limit under PC
- Under LC, the relations of $\sigma_a - a_{eq}$ are almost the same under various contact pressures and mean stresses

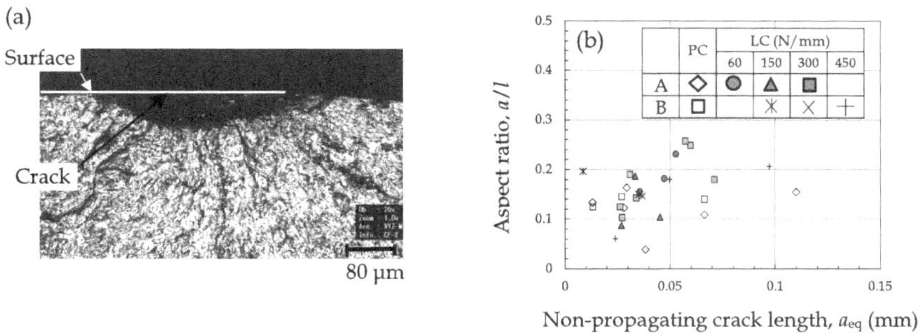

Figure 8. (a) Example of non-propagating crack and (b) aspect ratio of non-propagating cracks

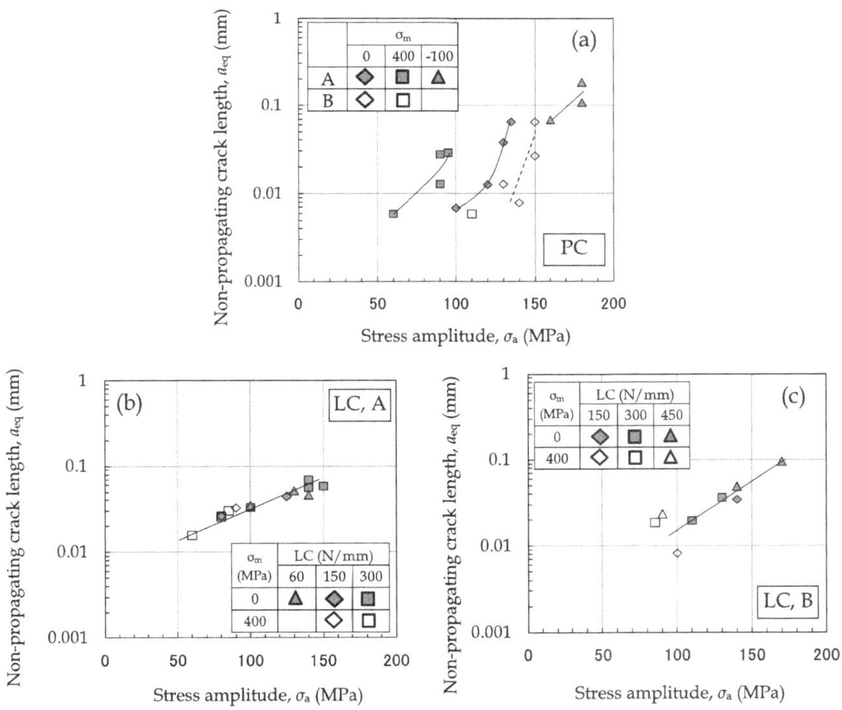

Figure 9. Non-propagating crack depth obtained from fretting fatigue tests: (a) PC, (b) LC (sample A), and (c) LC (sample B)

3.3. Profile path of crack propagation

Figure 10 shows profile paths of crack propagation from the initial crack measured using a laser-microscope, where the stress amplitude is shown in parentheses for each case. The angle of crack inclination against the normal direction was about 50-70° in stage I and about 20° at the mixed mode region in stage II, as also shown in Fig. 10.

Non-propagating cracks under PC conditions were almost all located in stage II except one case (sample B at 400 MPa-mean stress) when no profile data were obtained because the run-out specimen was not fractured from the fretting pre-crack. On the other hand, under LC conditions, all non-propagating cracks were located near the boundary between stages I and II. The boundary crack depth between stages I and II, d_1, depended on the mean stress, contact pressure, and material strength. The following explains why this occurred.

- The values of d_1 of sample B (higher static strength) were smaller than those of sample A under the same test conditions.
- The values of d_1 under PC condition were smaller than those under LC conditions at the same mean stress.
- 400 MPa-mean stress led to lower d_1 than 0 MPa-mean stress under LC conditions.
- When mean stress was –100 MPa in the PC condition, d_1 was extremely small (less than 5 μm).

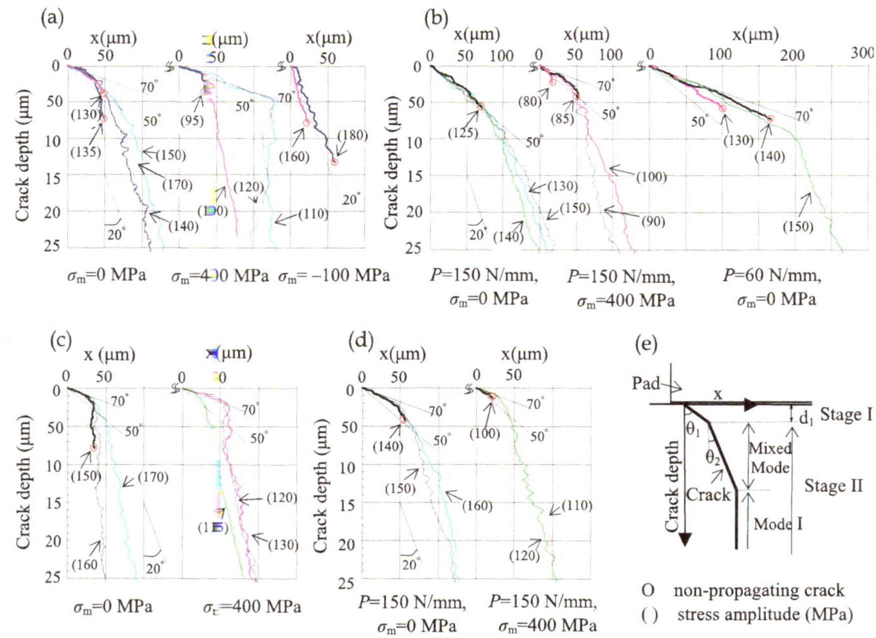

Figure 10. Non-propagating and propagating crack profiles obtained from fretting fatigue tests: (a) PC (sample A), (b) LC (sample A), (c) PC (sample B), (d) LC (sample B), and (e) schematic view of fretting crack

These results regarding d_I and its inclined angles can be described by considering crack propagation under mixed modes in stage I, but the details are to be discussed in future work. The objective with this study was to quantitatively evaluate the micro-crack propagation in stages I and II by applying the maximum tangential stress theory.

4. Discussions

4.1. Analysis condition for calculating stress intensity factor

The relationship between crack depth and the stress intensity factor was calculated by carrying out three-dimensional elastic Finite Element (FE) analysis. Figure 11 shows the analysis models under PC and LC conditions where an inclined elliptical surface crack was introduced. The aspect ratio (the ratio of crack depth to surface length) was 0.15 determined from the results shown in Fig. 8(b). The crack depths were 0.03, 0.06, 0.1, and 0.2 mm (a_{eq}=0.025, 0.05, 0.083, and 0.17 mm) and the oblique angle against the normal direction, α, was 20° on the basis of the test results in the mixed mode region of stage II. Furthermore, to investigate the effect of α for a small crack, analysis was done when α=0, 20, 50, 70° at a_{eq}= 0.025, and 0.05 mm. A crack was introduced 0.1 mm inside the contact edge to prevent the edge effect in contact analysis and to be consistent with the test results shown in Fig. 7. The friction coefficient was 0.8, determined from gross slip tests. Calculated accuracy was compared with the analytical solution through analysis without contact.

The stress intensity factor ranges ΔK_I and ΔK_{II} were calculated using the extrapolation method of stress distribution from the deepest point of the crack. By substituting ΔK_I and ΔK_{II} into Eq. (3), tensile ΔK_θ and shear ΔK_τ were obtained in the local coordinate system at any evaluation angle θ.

$$\begin{aligned}\Delta K_\theta &= \Delta K_I\left(\frac{3}{4}\cos\frac{\theta}{2}+\frac{1}{4}\cos\frac{3\theta}{2}\right)+\Delta K_{II}\left(-\frac{3}{4}\sin\frac{\theta}{2}-\frac{3}{4}\sin\frac{3\theta}{2}\right)\\ \Delta K_\tau &= \Delta K_I\left(\frac{1}{4}\sin\frac{\theta}{2}+\frac{1}{4}\sin\frac{3\theta}{2}\right)+\Delta K_{II}\left(\frac{1}{4}\cos\frac{\theta}{2}+\frac{3}{4}\cos\frac{3\theta}{2}\right)\end{aligned} \tag{3}$$

Alternating axial loads over one cycle were applied after applying the mean axial load and the contact force. The stress intensity factor range ΔK_θ is the difference of K_θ at the maximum and minimum loads, and the mean value $K_{\theta,\,mean}$ is the average of K_θ at these loads.

Elasto-plastic analysis was also done under LC conditions to investigate the effects of local plastic deformation using the non-crack model whose minimum mesh size was 10 μm at the contact edge. Cyclic stress–strain test data were used in the calculation to consider the cyclic softening effects of test materials, where the cyclic 0.2% yield strengths were about 82% of the static ones for both samples A and B. The alternating force was applied after applying the mean stress and contact pressure in three cycles to obtain the convergence stress distribution.

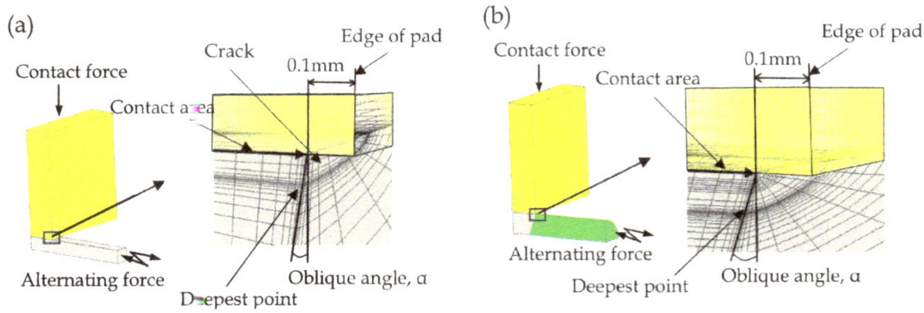

Figure 11. FE analysis model with crack of (a) PC and (b) LC (1/8 symmetry)

4.2. Analysis results on stress intensity factor

Figure 12 shows the variation of ΔK_θ and ΔK_τ against β, defined as the angle of the evaluation direction against the normal direction, when a crack was 0.03 mm deep under PC conditions (p=80 MPa, σ_a=100 MPa, and σ_m=0 MPa). The angle β is 20° when ΔK_θ maximizes, which slightly depends on the crack α. This angle of β corresponds well to the inclined crack angle confirmed by tests at the mixed mode in stage II, as shown in Fig. 10. This supports the maximum tangential stress theory that the fretting fatigue crack propagates in a direction perpendicular to the maximum principal stress amplitude in stage II. When β is 50-70°, corresponding to the inclined angle of the initial crack in stage I, ΔK_τ is not zero. This suggests that both ΔK_τ and ΔK_θ affect crack propagation in stage I, unlike in stage II. As shown in Fig. 12(b), $K_{\theta,\text{mean}}$ is negative, and its absolute value decreases with β when β is less than about 60°.

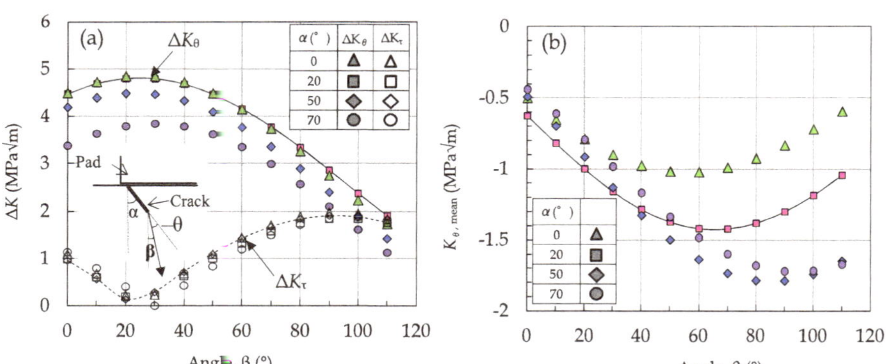

Figure 12. Calculated stress intensity factors for oblique cracks as function of β: (a) ΔK_θ and ΔK_τ , and (b) mean value of K_θ (PC, p=80 MPa, σ_m=0 MPa, σ_a=100 MPa, a_{eq}=0.025 mm)

Next, Figs. 13(a) and (b) show the relationships between a_{eq} and $\Delta K_{\theta max}$ and mean value $K_{\theta max,\text{ mean}}$ when σ_m=0 MPa and σ_a=100 MPa. The value of $\Delta K_{\theta max}$ is the maximum value of

ΔK_θ with the variation of β and $K_{\theta max, mean}$ is the mean K_θ when ΔK_θ is $\Delta K_{\theta max}$. When the crack is short, $\Delta K_{\theta max}$ is strongly affected by the contact and, as the crack grows, it asymptotically reaches the value calculated under the uniform stress distribution without fretting effects. $\Delta K_{\theta max}$s at α=0° and 20° without contact were confirmed to coincide within 3% of error with the solution of the Raju-Newman equation (Raju & Newman, 1981). The values of ΔK at α=70° were found to be about 30% smaller than those at α=0° under both contact and non-contact conditions. The absolute value of $K_{\theta max, mean}$ decreases as a crack grows, as shown in Fig. 13(b). This is because the compression stress caused by the contact force decreases as the distance from the surface increases.

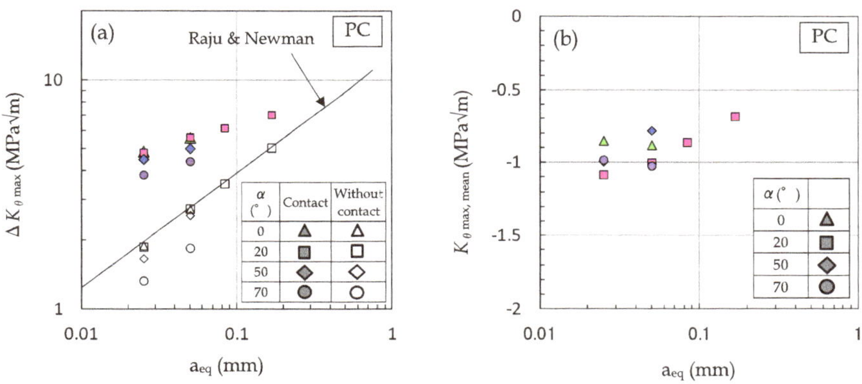

Figure 13. Relationship between a_{eq} and (a) $\Delta K_{\theta max}$ and (b) $K_{\theta max, mean}$ calculated using FE analysis (PC, p=80 MPa, σ_m=0 MPa, σ_a=100 MPa)

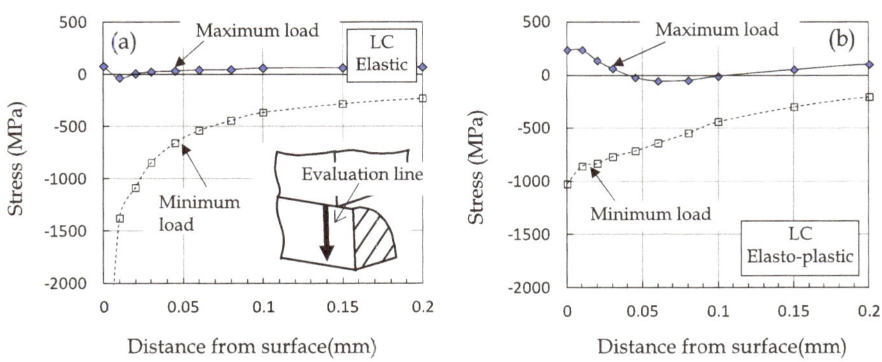

Figure 14. Axial stress distribution from the surface with non-crack model calculated by (a) elastic and (b) elasto-plastic analyses (LC, P=150 N/mm, σ_m=0 MPa, σ_a=100 MPa)

The stress distributions calculated using elastic and elasto-plastic analyses are compared in Fig. 14 for non-crack models under LC conditions for sample A when P=150 N/mm, σ_m=0

MPa, and σ_a=100 MPa. This figure shows the maximum and minimum axial stress distributions along the line from the surface of the contact edge. While high compressive stress arises when using the elastic analysis, the minimum stress almost saturates when calculated using the elasto-plastic analysis. Using the polynomial approximation of the elasto-plastic stress distributions without a crack, ΔK and K_{mean} were calculated using the American Society of Mechanical Engineers (ASME) section XI method (ASME, 2001). The values of ΔK and K_{mean} against a_{eq} are shown in Fig. 15, where the solid diamond were calculated by the elastic analysis with a crack at α=20° using the stress extrapolation method and the dashed lines were calculated from the elasto-plastic analysis using the above-mentioned ASME method. As shown in Fig. 15(a), ΔKs are almost the same in two calculations except a_{eq}=0.025 mm, where local stress is higher from elastic analysis than that from elasto-plastic analysis because the former does not take into account the yield effects. On the other hand, K_{mean}s are different from the two analyses, especially when a_{eq} is small. Since K_{mean} affects ΔK_{th}, it is necessary to evaluate stress redistribution using elasto-plastic analysis considering the actual yield behavior under LC conditions.

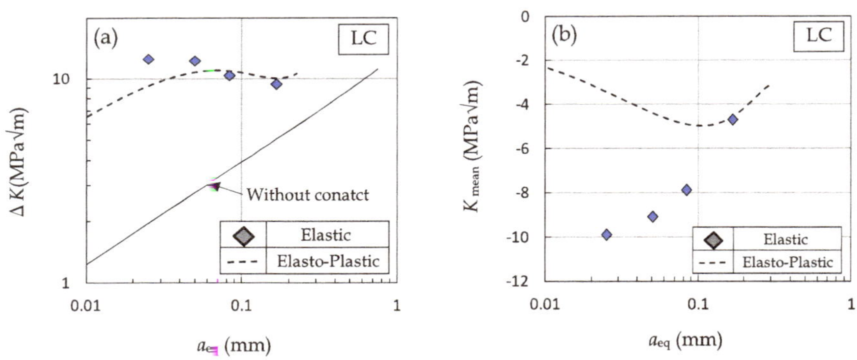

Figure 15. Comparison of (a) ΔK and (b) K_{mean} calculated using elastic analysis with crack model and elasto-plastic analysis with non-crack model (LC, P=150 N/mm, σ_m=0 MPa, σ_a=100 MPa)

4.3. Qualitative evaluation of small crack propagation

Figure 16(a) shows a schematic view of the material strength's effect on ΔK under PC. The value of ΔK_{th} in the small crack region increases as the material strengthens: ΔK_{th} of sample B was higher than that of sample A. On the other hand, ΔK from the applied stress does not depend on the material strength when the local plastic deformation is ignorable. From this evaluation, the fatigue limit of sample B was higher than that of sample A, which correlates well with the experimental results. Supposing that a crack stops propagating when ΔK is smaller than ΔK_{th}, the non-propagating crack depth of sample B is estimated to be smaller than that of sample A under the same stress amplitude. This also agrees well with the experimental results shown in Fig. 9(a).

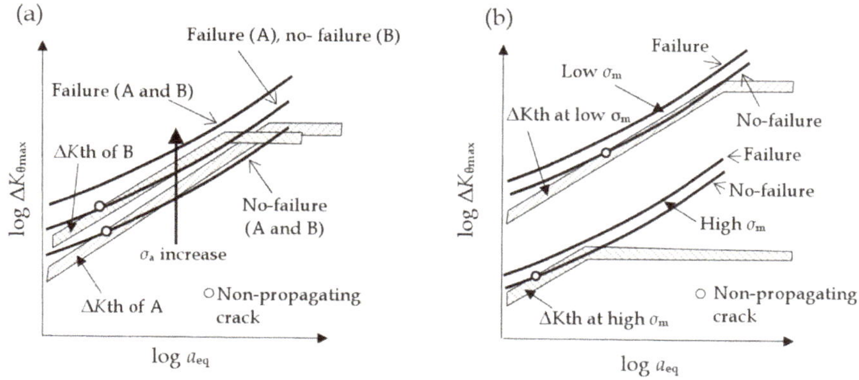

Figure 16. Schematics of small-crack propagation model under PC condition on effects of (a) material strength, and (b) mean stress.

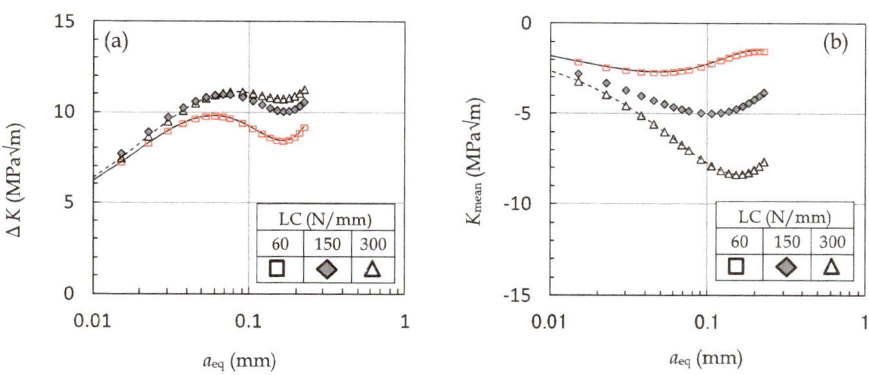

Figure 17. Effects of LC pressure on (a) ΔK, and (b) K_{mean} cacalculated using elasto-plastic analysis with non-crack model (LC, σ_m=0 MPa, σ_a=100 MPa, sample A)

Figure 16(b) schematically shows the effect of mean stress under PC. Because higher mean stress results in smaller ΔK_{th}, larger σ_m leads to smaller fatigue strength and a smaller non-propagating crack at the fatigue limit from this model. This was also confirmed to correspond well with the results shown in Fig. 9(a).

The fretting fatigue strength minimizes at a certain contact pressure under LC conditions, at 150 N/mm for sample A, as shown in Fig. 6. The relations of $a_{eq}-\Delta K$ and $a_{eq}-K_{mean}$ are shown in Fig. 17 for sample A at various contact pressures when σ_m =0 and σ_a= 100 MPa. As Fig. 17(a) shows, ΔK increases with the contact pressure, but ΔKs differ little between 150 and 300 N/mm. On the other hand, K_{mean} decreases monotonically with the increase in the LC pressure. This is due to two conflicting effects, the increase in ΔK accelerates crack propagation and negative large K_{mean} delays its propagation as the contact pressure increases, that is, 150 N/mm-pressure results in the minimum fretting fatigue strength for sample A.

4.4. ΔK_{th} from small fretting pre-cracks

Figure 18 shows ΔKs obtained from plain fatigue tests by using fretting pre-crack specimens, where open marks mean non-fracture and closed marks mean fracture. In this figure, ΔK_{th}s for long crack are also shown. Estimated ΔK_{th}s for small cracks, as boundaries between open and closed marks, were confirmed to depend on the crack length as a slope of 1/3 in the double logarithmic plots under various R. This slope of 1/3 agrees well with Murakami's empirical rule (Murakami & Endo, 1986). Some data slightly deviated from the approximate line, which was probably caused by inclined pre-crack effects and residual compressive stress due to previous fretting tests.

Threshold values for small cracks are modelled as Eq. (4) using $\Delta K_{th,\,0.1}$, ΔK_{th} at a_{eq}=0.1mm on the 1/3 slope line, and threshold for long cracks, $\Delta K_{th,l}$. The variation of $\Delta K_{th,\,0.1}$ is shown in Fig. 19 as a function of R for samples A and B obtained from plain fatigue tests using fretting pre-crack specimens. It was confirmed that the values of $\Delta K_{th,\,0.1}$ for sample B (higher static strength) are higher than those of sample A under all R.

$$\begin{aligned} \Delta K_{th,s} &= \Delta K_{th,\,0.1}\left(a_{eq}/0.1\right)^{1/3}, \\ \Delta K_{th} &= \Delta K_{th,s}, \ \text{when}\ \Delta K_{th,s} < \Delta K_{th,l}, \\ \Delta K_{th} &= \Delta K_{th,l}, \ \text{when}\ \Delta K_{th,s} > \Delta K_{th,l}. \end{aligned} \tag{4}$$

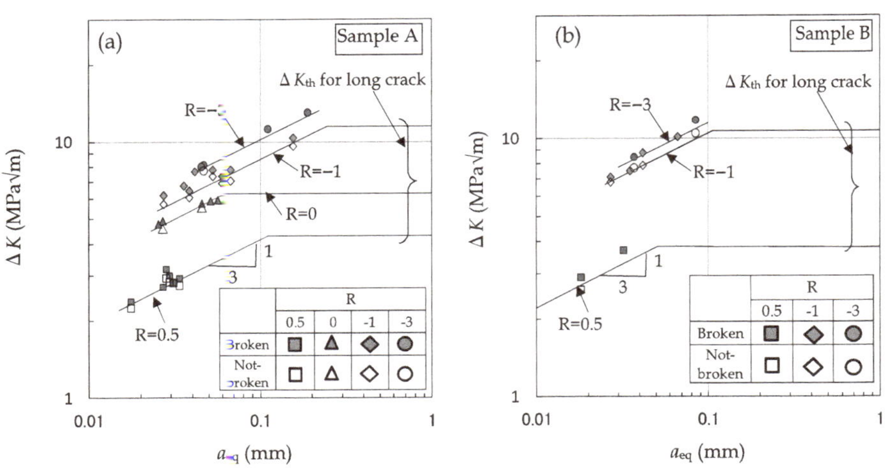

Figure 18. The value of ΔK_{th} obtained by plain fatigue tests using fretting pre-crack specimens for (a) Sample A and (b) Sample B under various stress ratios

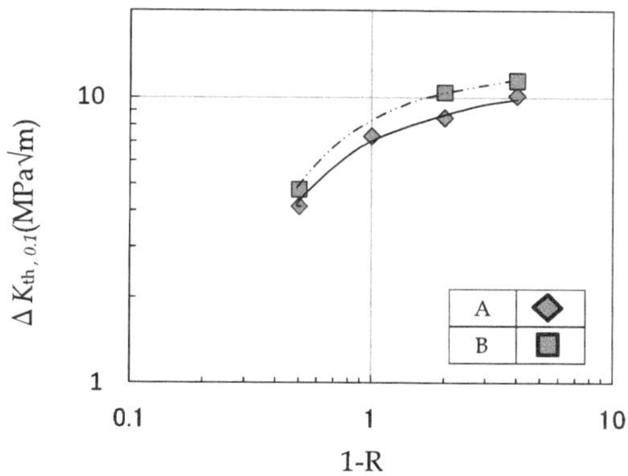

Figure 19. The value of $\Delta K_{th,0.1}$ obtained by tests for steels A and B under various stress ratios as a function of 1-R

4.5. Quantitative evaluation of small crack propagation

Micro-crack propagation behavior using the experimental and analytical results under various contact pressures and mean stresses is discussed. Figure 20 shows the evaluation results of fretting fatigue crack propagation under PC conditions when mean stresses are 0, 400, and –100 MPa for sample A and 0, and 400 MPa for sample B. In this figure, K_{max}, K_{min}, and ΔK were calculated based on the maximum tangential stress theory at α=20° using the minimum stress, leading tc fracture in the experiments. The value of ΔK_{th} was evaluated using Eq. (4) and test results shown in Fig. 19 corresponding to the calculated R. The analysis results are in good agreement with the test results in all cases since the calculated ΔK under the fracture condition is greater than ΔK_{th} through almost the entire crack length.

The results are shown in Fig. 21 under LC conditions for sample A evaluated using the micro-crack propagation model when mean stress was 0 and 400 MPa and P=60, 150, 300 N/mm. In this figure, K_{max}, K_{min}, and ΔK were calculated using the ASME method with elasto-plastic stress distribution withcut crack, as discussed in Section 4.2, under minimum stress leading to fracture. The analysis results also quantitatively agree well with the tests under LC conditions.

Finally, the length of non-propagating cracks is quantitatively discussed. The ratios of ΔK to ΔK_{th} calculated using the above-mentioned model are summarized in Fig. 22 using non-propagating crack length observed in the experiments. Under PC conditions, the ratios of ΔK to ΔK_{th} were almost 1, as shown in Fig. 22(a). This indicates that the micro crack propagation model estimates non-propagating crack length with considerable accuracy. Under LC conditions, the ratios were 0.8-1.4, as shown in Fig. 22(b), which is also satisfactorily accurate. The test results can be successfully explained using the micro-crack propagation model, as shown in Figs. 20-22. Therefore, the maximum tangential stress theory is effective for satisfactorily evaluating fretting fatigue strength in stage I as well as in stage II for practical use.

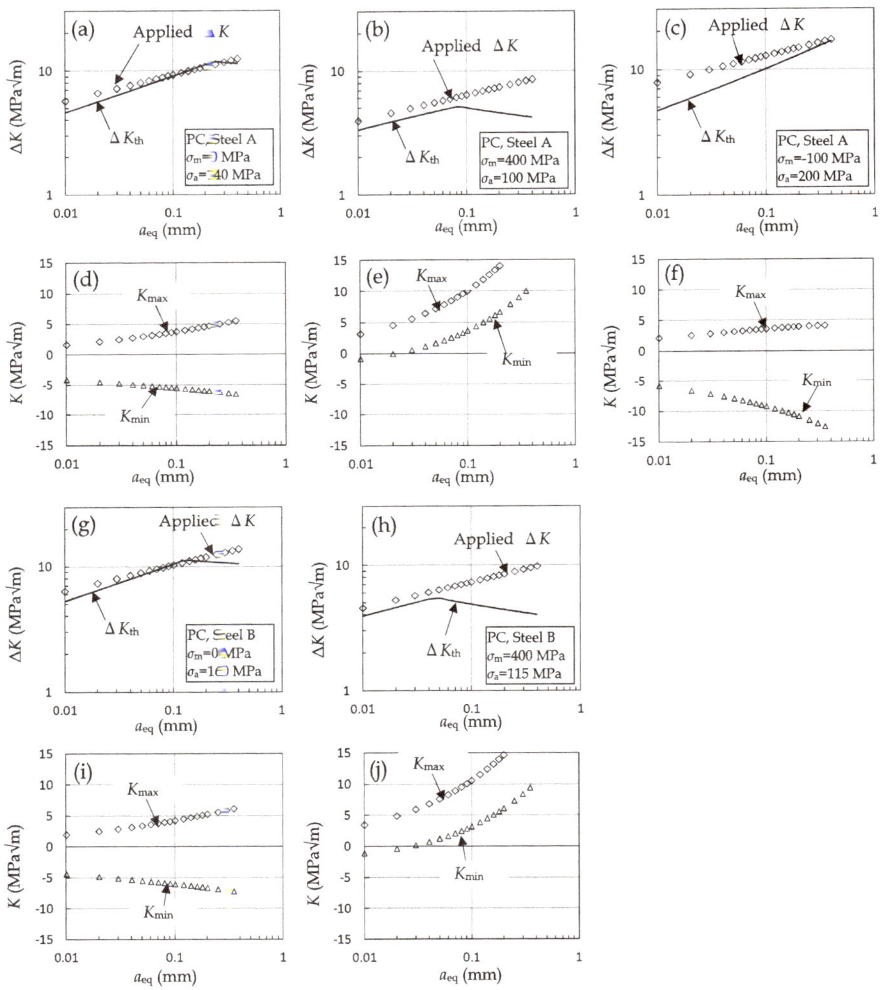

Figure 20. Evaluation results on ΔK, K_{max} and K_{min} using minimum stress leading to fracture in experiments under PC at p=80 MPa when σ_m= (a)(d) 0 MPa, (b)(e) 400 MPa, (c)(f) -100 MPa for sample A, and σ_m= (g)(i) 0 MPa, (h)(j) 400 MPa for sample B

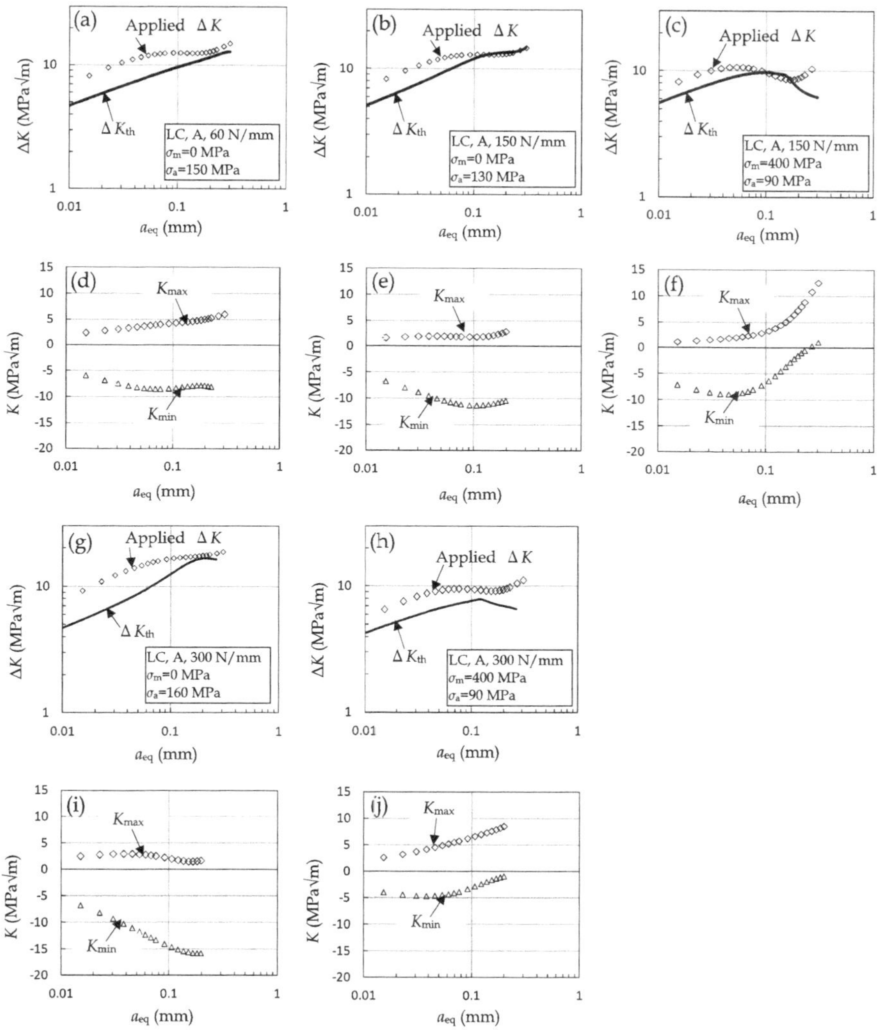

Figure 21. Evaluation results on ΔK, K_{max} and K_{min} using minimum stress leading to fracture in experiments under LC for sample A when (a)(d) P=60 N/mm, σ_m= 0 MPa, (b)(e) P=150 N/mm, σ_m= 0 MPa, (c)(f) P=150 N/mm, σ_m= 400 MPa, (g)(i) P=300 N/mm, σ_m= 0 MPa, and (h)(j) P=300 N/mm, σ_m= 400 MPa

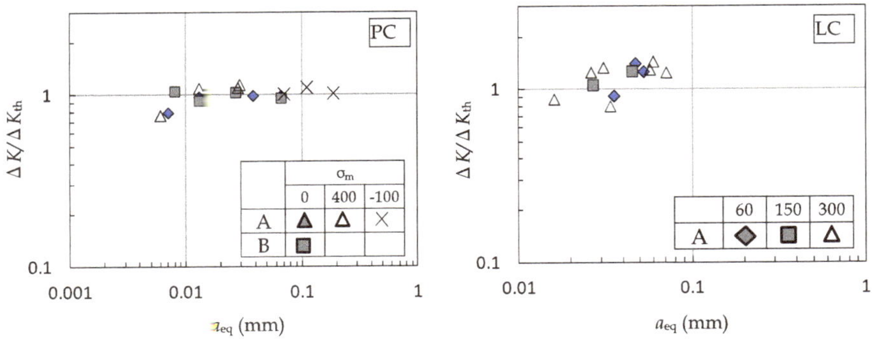

Figure 22. Stress intensity factor ratios of ΔK to ΔK_{th} of non-propagating cracks for (a) PC and (b) LC

5. Conclusion

Fretting fatigue tests were undertaken in LC conditions as well as PC conditions using 12% Cr steel samples with parameters of mean stress, contact pressure, and material strength. The strengths were evaluated quantitatively by applying the micro-crack propagation model under various test conditions considering small crack and mean stress effects on ΔK_{th} and mixed modes of tensile and shear ΔK. Crack propagation behavior was also examined quantitatively by obtaining non-propagating crack lengths of run-out specimens and ΔK_{th} from fretting pre-cracks under several R, including negative mean stress. The results obtained are as follows.

1. Test results concerning the fretting fatigue strength could be successfully explained using the micro-crack propagation model by applying the maximum tangential stress theory in both stages I and II under PC conditions at different mean stresses for samples A and B. Under LC conditions, where high contact pressure arises, it was found that elasto-plastic analysis is necessary for calculating ΔK and K_{mean} considering the actual yield behavior, and the proposed method is effective for expressing test results when σ_m=0 and 400 MPa under 60, 150, and 300 N/mm contact pressure for sample A.
2. Cracks were confirmed to propagate in stage II at the angle where the maximum stress intensity factor range $\Delta K_{\theta max}$ occurred by observing the propagation profile. This model also confirmed the experimental results that the depth of non-propagating cracks decreases as the mean stress and the material strength increase.
3. Under PC at 80-MPa pressure, sample B (the static strength of which was about 40% higher than that of sample A) exhibited 10-25% higher fretting-fatigue strength than sample A. Under LC conditions, the fretting fatigue strength decreased as contact pressure increased and minimized when Hertz's average contact pressure was about 1.5 times 0.2% proof stress. This behavior is explained as two conflicting effects; the increase in ΔK accelerates crack propagation and negative large K_{mean} delays its propagation as the contact pressure increases.

4. Crack propagation data in stage I, such as inclined angles and the boundary depth from stages I to II, were obtained under various test conditions, which is expected to be clarified quantitatively by analyzing mixed mode effects .

Author details

Kunio Asai
Hitachi Research Laboratory/Hitachi, Ltd., Japan

6. References

Asai, K. (2010). Fretting fatigue strength of 12% Cr steel under high local contact pressure and its fracture mechanics analysis. *Schience Direct, Procedia Engineering*, Vol. 2, pp. 475-484

ASME section XI appendix A-3000. (2001). Method for K_I determination, pp. 372-373

Attia, M. H. (2005). Prediction of fretting fatigue behavior of metals using a fracture mechanics approach with special consideration to the contact problem. *Journal of Tribology*, Vol. 127, pp. 685-6930

Bold, P. E., Brown, M. W., & Allen, R. J. (1992). A review of fatigue crack growth in steels under mixed mode I and II loading. *Fatigue Fracture Engineering Materials Structure*, Vol. 15, No. 10, pp. 965-977

Dubourg, M. C., & Lamacq, V. (2000), Stage II crack propagation direction determination under fretting fatigue loading: A new approach in accordance with experimental observations. *ASTM STP* 1367, pp. 436-450

Edwards, P. R. (1984), Fracture mechanics application to fretting in joints. *Advances in Fracture research*, Pergamon Press, pp. 3813-3836

El Haddad, M. H., Smith, K. N., & Topper, T. H. (1979). Fatigue crack propagation of short cracks. *Transactions of the ASME*, Vol. 101, pp. 42-46

Faanes, S. (1995). Inclined cracks in fretting fatigue. *Engineering Fracture Mechanics*, Vol. 52, No. 1, pp. 71-82

Hannes, D., & Alfredsson, B. (2012). A fracture mechanical life prediction method for rolling contact fatigue based on the asperity point load mechanism. *Engineering Fracture Mechanics*, Vol. 83, pp. 62-74

Kondo, Y., Sakae, C., Kubota, M., Nagasue, T., & Sato, S. (2004). Fretting fatigue limit as a short crack problem at the edge of contact. *Fatigue Fracture Engineering Materials Structure*, Vol. 27, pp. 361-368

Lamacq, V., Dubourg, M. C., & Vincent, L. (1996). Crack path prediction under fretting fatigue- A theoretical and experimental approach. *Journal of Tribology*, Vol. 118, pp. 711-720

Makino, T., Yamamoto, M., & Hirakawa, K. (2000), Fracture mechanics approach to the fretting fatigue strength of axle assemblies. *ASTM STP* 1367, pp. 509-522

Murakami, Y., & Endo, M. (1986). Effects of hardness and crack geometry on ΔK_{th} of small cracks. *Journal of the Society of Materials Science, Japan*, Vol. 35, No. 395, pp. 911-917

Murakami, Y., Fukuhara, T., & Hamada, S. (2002). Measurement of Mode II threshold stress intensity factor range ΔK_{IIth}, *Journal of the Society of Materials Science, Japan*, Vo. 51, No. 8, pp. 918-925

Mutoh, Y. (1997). Fracture mechanics of fretting fatigue. *Journal of the Society of Materials Science, Japan*, Vol. 46, No. 11, pp. 1233-1241

Mutoh, Y., & Xu, J.-Q. (2003). Fracture mechanics approach to fretting fatigue and problems to be solved. *Tribology International*, Vol. 36, pp. 99-107

Nicholas, T., Hutson, A., John, R., & Olson, S. (2003). A fracture mechanics methodology assessment for fretting fatigue. *International Journal of Fatigue*, Vol. 25, pp. 1069-1077

Pook, L. P. (1985). A failure mechanism map for mixed mode I and II fatigue crack growth thresholds. *International Journal of Fracture*, Vol. 28, pp. 21-23

Qian, J., & Fatemi, A. (1996). Mixed mode fatigue crack growth: A literature survey. *Engineering Fracture Mechanics*, Vol. 55, No. 6, pp. 969-990.

Raju, IS. & Newman, JC. (1981). An empirical stress intensity factor equation for the surface crack. *Engineering Fracture Mechanics*, Vol. 15, pp. 185-192

Tanaka, K. (1974). Fatigue crack propagation from a crack inclined to the cyclic tensile axis. *Engineering Fracture Mechanics*, Vol. 6, No. 3, pp. 493-507

Usami, S., & Shida, S. (1979). Elastic-plastic analysis of the fatigue limit for a material with small flaws. *Fatigue of engineering materials and structures*, Vol. 1, No. 4, pp. 471-481.

Zhang, H., & Fatemi, A. (2010). Short fatigue crack growth behavior under mixed-mode loading. *International Journal of Fracture*, Vol. 165, pp. 1-19

Chapter 8

Early Corrosion Fatigue Damage on Stainless Steels Exposed to Tropical Seawater: A Contribution from Sensitive Electrochemical Techniques

Narciso Acuña-González, Jorge A. González-Sánchez,
Luis R. Dzib-Pérez and Aarón Rivas-Menchi

Additional information is available at the end of the chapter

1. Introduction

A major concern in the design of engineering structures is the ability of components to maintain their integrity during their entire life service even when subjected to a combination of fluctuating loads and aggressive environments. The requirement for reliable structural integrity is particularly important for structures, involved in fundamental fields such as transportation, oil and gas production and energy generation. Fatigue and corrosion fatigue failures take place in components and structures as result of complex loading histories. There are different stages of fatigue damage in engineering components where defects may nucleate on initially smooth or undamaged sections, followed by microstructural crack formation, stable propagation and finally unstable crack propagation where catastrophic failure occurs.

Total fatigue life is often considered as a process of fourth major stages [1,2]:

a. Initiation or nucleation of fatigue cracks: substructural and microstructural which cause nucleation or permanent damage,
b. Small-Crack growth: the creation and growth of microscopic cracks,
c. Macro-Crack propagation: the growth and coalescence of microscopic flaws to form "dominant cracks" and stable propagation of dominant macrocracks
d. Final fracture: structural instability or complete fracture.

Then, fatigue life of engineering components and structures can be defined as the period during which cracks initiate from defects and propagate. The largest fraction of fatigue life is spent in the crack propagation stage. However, when engineering structures operate under severe conditions, the problem of fatigue failure is raised, especially in presence of aggressive environments. Under such severe conditions, the crack initiation stage is dramatically reduced as the case of structures in marine environments.

On the other hand, the stresses operating on structures can be mechanically applied (constant or cyclic), residual, and in some cases, thermal which depend on the operation conditions.

Stainless steels are an important group of materials for diverse engineering applications. This kind of corrosion resistant materials was introduced with the aim of reducing maintenance costs involved with materials such as carbon steel. However, there are several cases around the world, [3-8] in which the failure frequency of stainless steel piping systems on offshore platforms is very high, resulting in unacceptable maintenance costs.

Unfortunately the degradation process is more complex than just localized electrochemical reactions at the metal surface. It is documented [5,9,10] that the corrosion of stainless steels (SS) is more severe in natural seawater than in sterile or synthetic seawater due to microorganisms activity. There is agreement among researchers that the increased corrosion is due to biofilm formation [4,5,7,8]. Electrochemical reactions are influenced by the chemical micro-environments generated by formation of biofilms at the metal surface. In seawater applications, the most widely used stainless steel grade, namely, 316L, suffers from localized corrosion. Even the substantial amount of data published about studies of localized corrosion of 316 and 316L stainless steels, frequently affected by pitting and crevice corrosion, the problem of structural failure in marine environments remains.

An outstanding problem in corrosion science and engineering involves the prediction, of damage caused by corrosion fatigue during the early stages of damage. This problem is particularly important because localized attack, such as pitting corrosion fatigue (PCF), normally occurs on passive alloys with minimal metal weight loss, and is commonly detected when the failure of the component is complete.

The most important aspect of corrosion fatigue assessment on components and structures in real service is the prediction of their residual life when there is a risk of undergoing localized attack as precursor of fatigue cracks. The conventional fracture mechanics tools are limited to provide information concerning the type and rate of damage occurring during the early stages of damage and consequently provide limited insight into the mechanism of pitting corrosion fatigue failure process. It is important to understand the role of the environment on the early stages of corrosion fatigue damage. In this sense, very sensitive electrochemical techniques have been used in corrosion fatigue studies by the authors of this chapter, where the main aim has been to determine the early stages of corrosion fatigue damage. This chapter presents a review of information obtained from applied sensitive electrochemical techniques to stainless steels exposed to seawater and chloride ions media to observe Early Corrosion Fatigue Damage.

2. Corrosion fatigue

The *S-N* approach to fatigue studies is usually appropriate for situations where a component or structure can be considered a continuum (i.e., cracks-free assumption). However, the fatigue process leading to failure is generally represented in terms of crack initiation stage followed by crack propagation to critical size.

It must be recognized, however, that there is no generally accepted definition as to what constitutes initiation or when the initiation phase is complete and propagation commences. The net cycles to failure represents the sum of both, crack initiation and propagation.

$$Total\ fatigue\ life\ = Ni + Np \quad (1)$$

where **Ni** represents the number of cycles for crack initiation and **Np** the number of cycles for crack propagation to failure. A major difficulty with regard to fatigue research involves the large number of variables which influence the phenomenon. These may be divided into three general variables categories [11,12]:

a. Mechanical
b. Material
c. Environmental

The complex task associated with fatigue resistance evaluation is the fact that individual factors may be synergically interactive, so a change in one must be assumed to affect the role of the others. Thus, the accumulation of damage during the fatigue lifetime of a component depends not only on its load history, but also of the synergistic effects of stress and surrounding environment [13,14]. Then, corrosion fatigue is defined as a synergistic effect in which corrosion and fatigue occur simultaneously. The combined effect of an aggressive environment, such as seawater, with a cyclic stress or strain is, invariably, more severe than the sum of the two effects of corrosion and fatigue acting separately. The effect of the environment on fatigue resistance is well documented in the form of stress vs number of cycles to failure (*S-N*) curves which show that, for the majority of the cases, the environment removes the fatigue limit (Figure 1). It has been shown that many corrosion fatigue failures of stainless steels in seawater are induced in the early stages of damage by pitting corrosion [15-18].

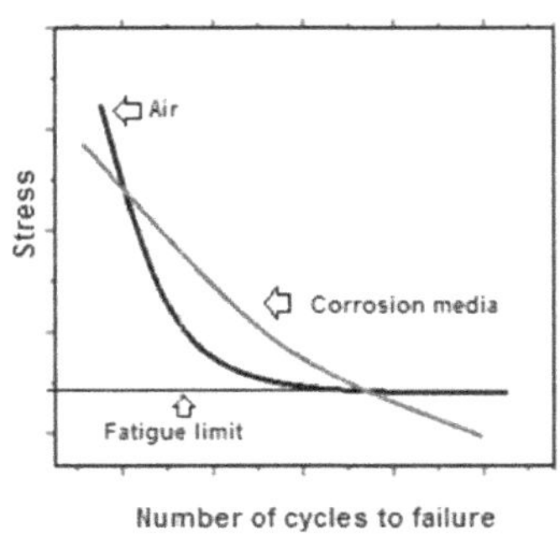

Figure 1. Schematic of *S-N* curve presenting air fatigue and corrosion fatigue behavior.

Fatigue and CF studies address the following regimes:

a. LEFM (long crack) regime, were cracks are of the order of millimeters (i.e. propagation)
b. EPFM (physical short cracks) regime, were cracks of the order hundreds of micrometers
c. MFM (short crack) regime, were cracks of the order of tends of micrometers develop early in the lifetime and exhibit growth rate variations through the microstructure (i.e. nucleation).

The most important aspect of corrosion fatigue assessment on components and structures in real service is the prediction of their residual life when there is a risk of undergoing localized attack as precursor of fatigue cracks.

In passive alloys, it is recognized that pitting corrosion plays a very important role in CF crack initiation [16, 17, 19]. CF lifetime prediction turns complex because of the need to account main processes leading to fatigue failure from a 'defect-free' surface [14, 20], where the early corrosion fatigue damage includes:

i. breakdown of the surface passive film
ii. pit initiation and growth,
iii. transition from pit to short crack nucleation,
iv. short crack growth

The selection of the techniques used to evaluate and study corrosion depends on the kind of corrosion taking place. The principal aim of the corrosion scientist is to determine the mechanism of the corrosion process including intermediate reactions and the kinetics of these processes. Sometimes it is necessary to use intrusive methods (electrochemical methods) in which the system is stimulated externally and the response of the system to that perturbation is measured. Other techniques are able to measure electrochemical activity without the necessity of any kind of external perturbation to the system, the electrochemical noise technique (ENT), the scanning reference electrode technique (SRET) and scanning vibrating electrode technique (SVET) are examples of non-intrusive techniques. The most important aspect of corrosion assessment on components and structures in real service is the prediction of their residual life when there is a risk of undergoing localized attack. In that respect the use of electrochemical methods for studying localized corrosion is an important part of the evaluation since the kinetics of localized corrosion is different with respect to general (uniform) corrosion.

From a structural integrity point of view general corrosion does not represent a serious problem for the corrosion engineer due to the fact that stress concentrators are not generated. Therefore general corrosion can be assessed using basic electrochemistry concepts and techniques [21,22]. In the case of localized corrosion, the measurement of the corrosion rate becomes more complicated due to unequal anodic and cathodic area ratios and therefore the effects of this kind of damage on the structural integrity of structures are much more severe.

2.1. Localized corrosion concepts

Due to their recurrent nature, localized corrosion processes often cause major practical problems affecting the performance of technologically important metallic materials, like stainless steels, nickel, aluminium, and many others metals and their alloys in different environments, especially those containing chlorides.

It can be considered that the corrosion process results from changes in the nature and composition of metals exposed to environments of homogeneous composition. The point of interest is "what happens when metals with approximately uniform composition are exposed to electrolytic environments in which composition changes makes them heterogeneous, e.g. differential aeration?" In this case we are dealing with concentration cells, i.e. localised corrosion.

One of the established essential principles of corrosion is that: The sum of the rates of the cathodic reactions must be equal to the sum of the rates of the anodic reactions, irrespective of whether the attack is uniform or localized [21], so the following equation must be satisfied:

$$\sum I_a = \sum I_c \tag{2}$$

where I_c and I_a denote the cathodic and anodic currents respectively which has a direct relation with the reaction rate of the electrochemical process.

If it is considered that the attack is uniform and assuming that there is only a single predominant anodic and cathodic reaction, then:

$$I_a / S_a = I_c / S_c \ or \ i_a = i_c \tag{3}$$

where i_a and i_c are the anodic and cathodic current densities assuming the area of the cathode S_c equals the area of the anode S_a.

When the corrosion attack is localized the anode area is very small compared with the cathode area, $S_a < S_c$ and as a consequence $i_a > i_c$, and the larger the ratio i_a:i_c, the more intensive the attack. Thus localized attack usually involves a corrosion cell consisting of a large cathodic area and a small anodic area.

It is now known that many aqueous corrosion processes of great technical interest occur under conditions in which the access of electrolyte is restricted. This can be due to the special geometry of the corroding material e.g. structures with riveted plates, flange joints, gaskets; and also due to the existence of some deposits such as corrosion products or scales on the corroding surface. In the cases where corrosion occurs under conditions of restricted diffusion, the chemical composition of the corroding environment inside the occluded cavity may be very different from that of the composition of the bulk solution. One of the most important effects is that changes in oxygen or ionic concentration in the electrolyte give rise to changes in corrosion potential. Under normal circumstances where the aggressive environment is aerated, the surface of the corroding alloy outside the occluded cavities is often passive as a result of the formation of an oxide film of corrosion products. Passivation takes place at these sites because they are in direct contact with the oxygen dissolved within the electrolyte. These surfaces act as aerated cathodes where oxygen takes part in the reduction reaction of the corrosion process. In this reaction the oxygen is reduced with the consequent increase of pH according to the following reaction for the case of basic or neutral solutions [23]:

$$O_2 + 2H_2O + 4e^- \rightarrow 4OH^- \tag{4}$$

or in a solution of low pH:

$$O_2 + 4H^+ + 4e^- \rightarrow 2H_2O \tag{5}$$

The surface, which is inside the occluded cavity, is active and acts as anode as a consequence of the very low concentration of oxygen in the solution inside the cavity. At these surfaces the metal undergoes dissolution. Additionally, the hydrolysis reaction takes place with a decrease in the pH of the solution according to the following reaction sequence:

for iron,

$$Fe \rightarrow Fe^{++} + 2e^- and\ 3Fe^{++} + 4H_2O \rightarrow Fe_3O_4 + 8H^+ + 2e^- \tag{6}$$

Much effort has been made in the last 40 years for a better understanding of the processes involved in localized corrosion, the majority of these studies being focused principally on aspects of the growth and stability of the sites undergoing localized attack. On the other hand, the mechanisms of nucleation and early stages of damage are not yet completely understood.

An important improvement on the present state of knowledge is achieved when the global process of localized corrosion can be partitioned into a nucleation stage (initiation), and a successive metastable growth stage, which eventually is followed either by rapid repassivation or stable pit growth. If the active condition inside a crevice or pit is stable over a longer period of time, rapid metal dissolution usually takes place. The rate controlling reaction during the localized corrosion processes such as activation, diffusion, or ohmic control, has a significant influence on the shape and geometry of the crevices or pits produced.

The theoretical models, that have been proposed to describe the initiation of the localized corrosion process, can be grouped into [24,25]:

- Adsorption mechanisms
- Ion migration and penetration models
- Mechanical film breakdown theories.

Substantial research has been carried out to study the structural parameters involved in localized corrosion processes. From these, it has been demonstrated that defects in the metal structure, such as non-metallic inclusions or dislocations may generally act as sites for pit initiation on passive metal surfaces [26]. Stainless Steels (SS's) with higher concentrations of alloying elements were found to have an increased pitting resistance due to the formation of passive films containing fewer defects and greater stability.

Despite the efforts made to understand this phenomenon, localized corrosion is a non-predictable degradation process of metals and alloys in contact with aggressive environments. Given that most of the information related to the stable stage of localized corrosion has been obtained throughout studies of pitting and crevice corrosion, this aspect of the localized attack will be presented in the next section dedicated to pitting corrosion.

3. Pitting corrosion fatigue

3.1. Breakdown of the surface passive film

Corrosion is an electrochemical reaction process between a metal or metal alloy and its environment [27], which involves complex mass and charge transfer taking place at the metal-electrolyte interface. These charge transfer reactions at the interface are the origin of the instability of metals. The high corrosion resistance of Austenitic SS's is primarily attributed to the adherent metal-oxide film formed on its surface called passive film. This natural coating acts as a barrier that avoids the contact between fresh metal surface and the electrolyte limiting the corrosion reaction and such alloy is said to be passive. When stainless steels are exposed to oxygen containing electrolytes, a chemically stable, non soluble film is formed from a mixture of iron and chromium oxides, with hydroxide and water-containing compounds located in the outermost region of the film, and chromium oxide enrichment at the metal-film interface [28]. However, the resistance of this passive film is determined by the environmental conditions to which the SS is exposed to, as well as by the alloy composition. Many metals and alloys become covered with oxide films on exposure to aqueous environments. Copper, aluminum or tin and their alloys develop thick films, while metals which exhibit passive behaviour like stainless steels have a very thin film of the order of 9 to 20 nm. The passive films are cathodic to the metal matrix. However, the susceptibility of passive films to suffer local breakdown depends upon the nature of the film, the quantity and nature of non metallic inclusions, the chemical composition of the electrolyte or solution, and the electrochemical state at the metal/solution interface, but also to a possible stress state.

For CF test, when the passive metal or alloy is subjected to cyclic loading, the passive film is damaged mechanically due to the creation of slip steps from beneath and following a repeated process of microdeformation-activation-dissolution-repassivation.

During the initial cyclic loading, fine slip lines appear on the specimen surface because dislocations, at or near the surface, the stress in localized sites can exceed the nominal stress. By using microscopic techniques has been possible to observe localized plasticity at microscopic scale in the grains, where fatigue damage is caused by microplasticity produced by large shear stresses.

Processes of nucleation and formation of fatigue slips bands often precedes at grain boundaries as is shown in figure 2. Mughrabi et al. [29] showed that nucleation at grain boundaries occur at sites where persistent slip bands (PSBs) impinge.

In smooth surfaces of SS's, during the early corrosion fatigue damage, the permanent slip bands are associated with breakdown of the passive film. The breakdown of the passive film causes dissolution of the non-covered active metal in contact with the electrolyte, which acts as a small anode against the large cathodic. This small anode is a vulnerable area on the metal surface, where a high local metal dissolution takes place, and leading cavities such as pits and/or crevice. Extensive development of techniques to study electrochemical reactions have simulated the analysis of potential and current fluctuations taking place during electrochemical processes which are translated in to electrochemical noise (EN) signals.

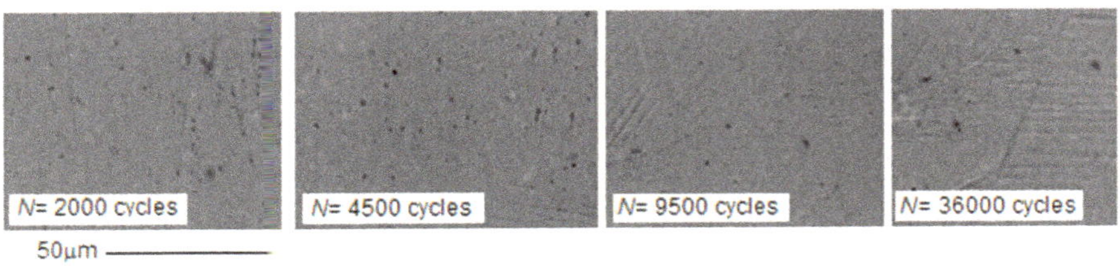

Figure 2. Evolution of fatigue slip bands in a 316L SS during cycling stress in air, under σ_{max}=217 MPa (90%$\sigma_{0.2}$), load frequency ω=1.0 Hz and a stress ratio R=0 conditions.

EN is defined as spontaneous fluctuations in current and potential generated by corrosion reactions [30-34]. In this context, the study of spontaneous current or potential fluctuations to characterize corrosion processes have received considerable attention, such as the study of corrosion potential (E_{corr}) and current fluctuations applied to monitor the onset of events characterizing pitting [35-36] or stress corrosion cracking (SCC) [37-39] The noise is typically measured potentiostatically, galvanostatically, or in a Zero Resistance Ammeter (ZRA) mode [30,31,34]. With ZRA mode, both the potential and current fluctuations can be measured using two nominally identical electrodes, connected through a ZRA; where a net current and changes in the potential are observed at free corrosion conditions according to the corrosion process.

During the Corrosion Fatigue tests conducted in natural seawater [16], the potential time series obtained from electrochemical noise measurements had a common characteristic pattern of quick drop and slow recovery. Common explanations for these patterns of quick drop and slow recovery have been associated to anodic dissolution, passive film breakdown, metastable and stable pitting corrosion, or with the crack initiation and crack growth due to microstrain and local mechanical stress conditions [16].

The figure 3 shows patterns of quick drop and slow recovery, indicating that passive film eventually is broken, due to applied mechanical stress and the interaction with the electrolyte, turning it in an active dissolution of metal. Also, on figure 3, the surface damage after N=43200 cycles of corrosion fatigue tests can be observed. In 316L SS-natural seawater system, current transients for unstressed conditions reaches average values in current density around 0.20 μA.cm^{-2}, while under cyclic stress conditions it ranges between 0.25 to 0.85 μA.cm^{-2} [16].

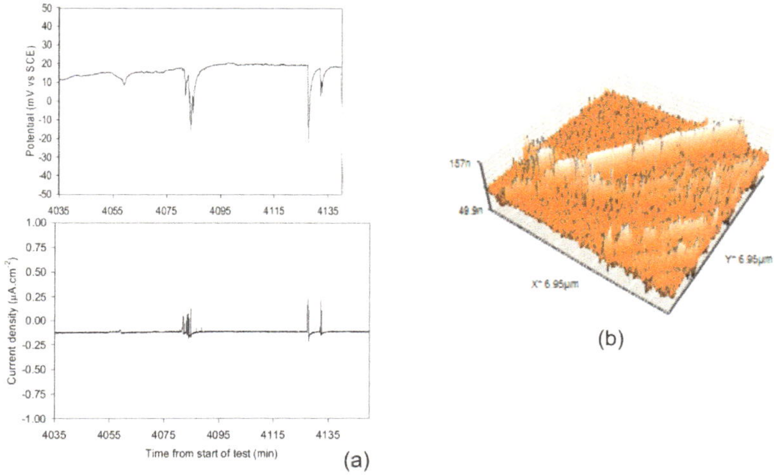

Figure 3. a) Potential and current noise profiles [16] and b) AFM surface image, for a 316L SS exposed to natural seawater under an induced cyclic stress of $\Delta\sigma$ = 140 MPa (58%$\sigma_{0.2}$), after N=43200 cycles, at a load frequency ω=0.17 Hz and stress ratio R=0.

3.2. Pit nucleation and growth

Pit nucleation

According to Szklarska-Smialowska [34], the susceptibility of a metal or alloy to pitting can be estimated by determination of one of the following criteria:

- Characteristic pitting potential
- Critical pitting temperature,
- Number of pits per unit area, or weight loss and
- The lowest concentration of chloride ions that may cause pitting

One of the most important parameter to determine an alloy's susceptibility to pitting corrosion is the pitting potential, i. e., the potential at which the passive film starts to break down locally. The potential above which pits nucleate is denoted by E_p and the potential below which pitting does not occur and above which the nucleated pits can grow is often indicated by E_{pp} (Figure 4 and Table 1).

As mentioned above, pitting corrosion involves localised attack of very small areas of metal surface undergo localised attack whilst the rest of the surface is largely unaffected and remains passive. Pitting is a particularly insidious form of localized corrosion because

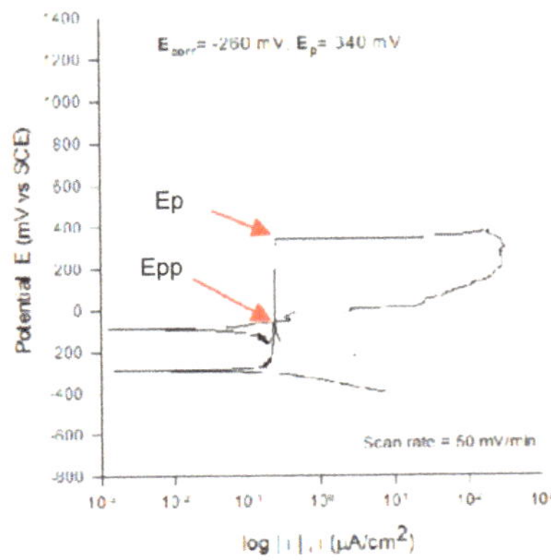

Figure 4. Arrows indicate the E_P and E_{PP} on the polarization curve [40].

SS Alloy	(E_{corr})* (mV vs SCE)	(E_P)* (mV vs SCE)	$(i_{passive})$* (μA/cm^2)
304 SS	- 260 to – 270	290 to 350	0.3 to 0.4
316L SS	-220 to -260	270 to 320	0.3 to 0.5
Duplex SS	-280 to -310	1060 to 1110	0.3 to 0.5

* Range of values for Ecorr, EP and *i*passive were obtained from five tests under identical experimental conditions.

Table 1. Values of electrochemical parameters for Stainless Steels in artificial seawater, at 26°C and pH 8.2 [40]

the extent of metal dissolution reaction is small but the attack is rapid and penetrates into the metal. The location of pits on metals that develop passive films is often unpredictable, and the pits tend to be randomly dispersed on the metal surface [41-43]. However, Sedriks [44,45] indicates the importance of different types of heterogeneities at the material surface during pit initiation, because the location of pits is to some extent defined by the microstructure of the alloy and the geometry of the system. In chloride containing aqueous solutions, the uniform corrosion rate of stainless steels in the passive state is insignificant, i.e. 0.15 to 15.0 μm per year [46]. However, pitting corrosion of this type of alloys is very common in these environments [45], where the local dissolution rate of metal can be up to 12 mm per year [41].

Pitting corrosion can occur when a local breakdown of the passive film takes place in an aqueous solution on a microscopic scale (nucleation process). This process involves small and sudden increments of anodic current, which are characterized by current spikes, leading to oxidation and dissolution of less than 0.01 μm^3 of metal. The nucleation process is unstable and in most cases it will stop by the regeneration of the passive film. Another possibility is that a nucleation event develops into the second stage of initiation, i.e. the metastable pit growth. During this step, a gradual and bigger increase of the anodic current takes place. This stage stops if repassivation of the micro-pit occurs. A volume of up to several μm^3 of metal can be dissolved in stainless steel during the metastable growth. In some cases the metastable pit growth precedes stable pitting. On the other hand, if repassivation of metastable micro-pit does not occur, the growth of stable pits can take place (figure 5).

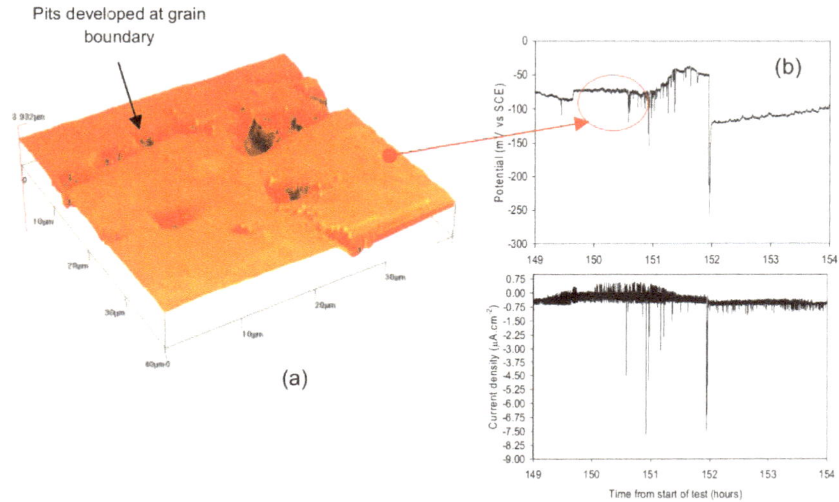

Figure 5. (a) AFM surface image and (b) Potential and current noise profiles for a 316L SS exposed to natural seawater under an induced cyclic stress of $\Delta\sigma$= 140 MPa (58% $\sigma_{0.2}$), after N=86400 cycles, at a load frequency ω=0.17 Hz and stress ratio R=0 [16].

Measurements of the physical separation of anodic and cathodic areas, the currents flowing between them as well as the mapping of potentials in electrolytic solutions have been successfully used for the study of the processes of localized corrosion of different systems [47-53]. A schematic drawing of a local corrosion cell is shown in figure 6.

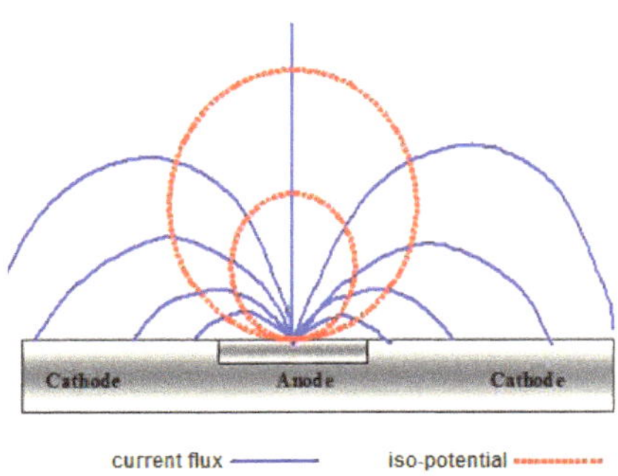

Figure 6. Schematic of current and potential distribution in solution during localized corrosion [40].

With the aim of determining the velocity of metal dissolution directly in active pits during pitting corrosion, Rosenfeld and Danilov [54], designed an apparatus to measure the field strength in the electrolyte directly above an active pit. They employed a twin probe method by using two reference electrodes, which makes it possible to measure the potential difference ΔE in any direction between two points in the electrolyte with the aid of two non-polarizable electrodes, for example calomel electrodes.

The equipment used for the measurements of the potential difference ΔE (ohmic potential gradients) is called the Scanning Reference Electrode (SRET). With the measurement of the electric field strength in the electrolyte over the pits it was possible to determine the current flowing from the anode points, based on the fact that, the vector of the normal component of the current density at a pre-determined point (i') in a uniform field is equal to the product of the electric field strength E and the specific conductivity of the medium κ. The resolution of the SRET depends upon the proximity of the scanning probes to the corroding sites and the magnitude of the corrosion currents from each site. As shall be shown later, the distance between the probe and specimen surface and the conductivity of the solution governs the sensitivity of the technique. It has been reported the capability of the SRET to identify the position of localized activity in the metal surface however did not report any assessment of pit size or shape from the performed SRET measurements [48-51]. The growth of corrosion pits is the next stage of the localized attack process.

Pit growth law

Pit growth studies have received less attention than its initiation. The kinetics of pit growth is generally assessed by electrochemical and metallographic methods. However, a combination of the two methods is usually preferred. A wide variety of pit growth laws have been reported for different systems metal - electrolyte, but the theory of pits growth remains unclear. Although stainless steel 304 is a very common and extensively used alloy, the determination of pit growth in NaCl solutions at open circuit potential has not been possible in terms of reproducible and reliable measurements. This is essentially due to the tendency for spontaneous repassivation of the pits when the steel is at open circuit potential [5]. An approach has been the determination of the pit growth kinetics on potentiostatically generated pits in specific metal – electrolyte systems. Dzib-Pérez [55] generated corrosion pits potentiostatically on specimens type 304 SS in stationary conditions in two electrolytes: 3.5% wtNaCl solution (κ=55.8 mS/cm) and natural seawater (κ=50.6 mS/cm). Specimens were initially polarized potentiostatically in natural seawater for 105 s at a potential of 365 mV vs SCE (to induce the formation of pits) and immediately after, the electrode potential was stepped down to a value of 300 mV vs SCE (to induce the growth of the initially formed pits avoiding the formation of new pits). The chronoamperometries obtained for 304 steel in natural seawater are shown at Figure 7.

All chronoamperometries obtained showed a rapid increase in current during the time the stainless steel specimen was polarized potentiostatically above the pit potential, which is associated to induced nucleation and fast pit growth.

Most of the generated pits quickly repassivated when the potential was stepped down to a potential level below the pitting potential. Under this condition 1 to 3 pits grew steadily until the end of polarisation. The increase of current as a function of time can be associated with two processes taking place under potentiostatic polarization: mainly the active dissolution of nucleated pits (one to three pits) and the simultaneous dissolution of the metal surface in the passive state. From the integration of the area beneath the curve current vs time the electric charge was obtained, from which the quantity of dissolved metal per pit was calculated using Faraday's law according to the chemical composition of the 304 stainless steel [55].

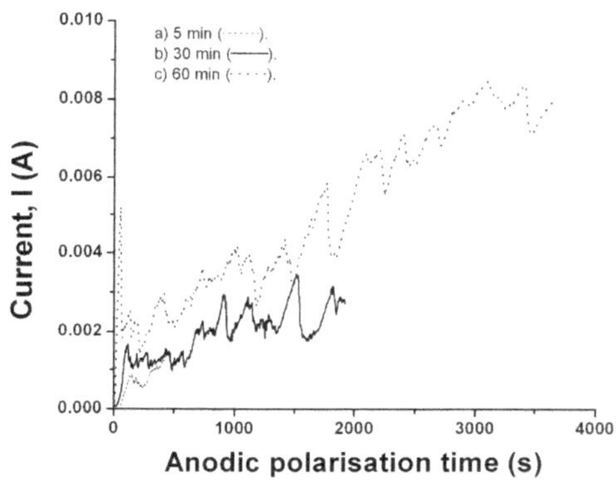

Figure 7. Chronoamperometries obtained during potentiostatic generation of pits in a 304 stainless steel in natural seawater in stationary conditions. Polarisation at 300 mV vs SCE for: a) 5, b) 30 and c) 60 minutes.

And the other hand, Figure 8 shows the amount of material dissolved as function of time of anodic polarization, also includes the fit (solid and dotted lines) performed through the following power equation:

$$DM_{AC} = ct^{d} \tag{7}$$

where DM_{AC} is the amount of dissolved material per pit in μg obtained from chronoamperometries, t is the pit growth time (or anodic polarization time) in minutes and, c and d are constants.

Once determined the amount of dissolved material per pit as a function of anodic polarization time and knowing the density of stainless steel (ρ= 8.03 g/cm³), the volume of dissolved metal was determined. Using a material removal procedure described in detail by Gonzalez Sanchez and Dzib-Pérez [40-55] to determine the pit depth as a function of time of anodic polarization the empirical pit growth law was established for corrosion pits in 304 stainless steel. Figure 9 shows the Maximum Pit Depth (*MPD*), generated in 304 steel samples, as a function of time under anodic polarisation.

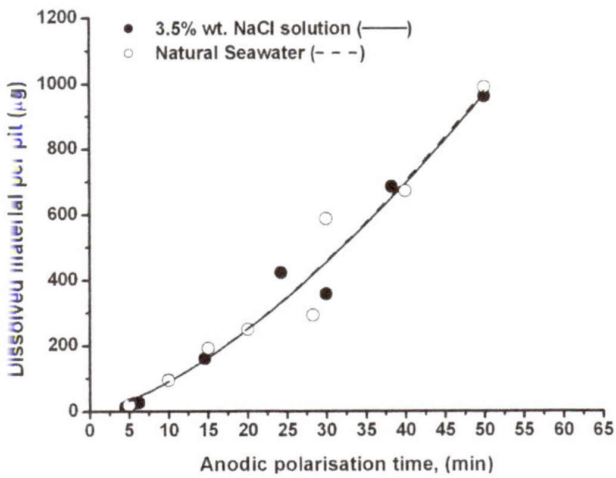

Figure 8. Amount of dissolved material per pit as a function of polarization time for a 304 stainless steel in 3.5% wt NaCl solution (κ=55.8 mS /cm) and in natural seawater (κ=50.6 mS /cm).

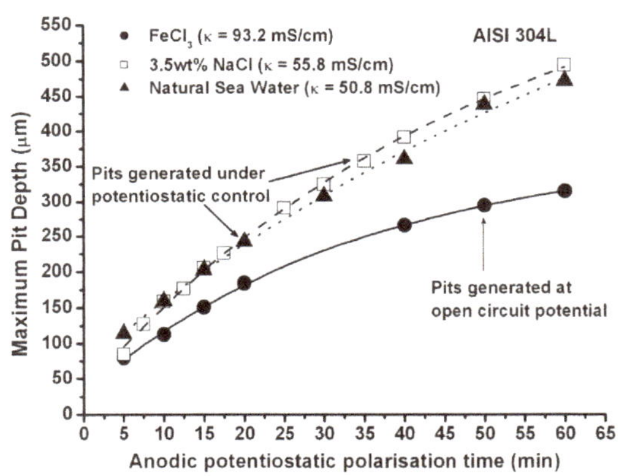

Figure 9. Maximum Pit Depth, as a function of pit growth time, for a 304L stainless steel in electrolytes of different conductivity.

From fitting of these relationships, represented by solid and dotted line, was obtained the empirical equation for the pits growth law for each experimental condition, which has the shape of an exponential equation [55]:

$$P_p = a - bc^t \tag{8}$$

Where *P*p represents the pit depth in μm, and t is the pit growth time in minutes. The values of the constants for the equation 8 are presented in Table 2.

Material	Electrolyte	Conductivity (mS/cm)	Parameter		
			a	b	c
304 SS	Natural seawater	50.6	852.99687	782.72174	0.98796
	3.5% wtNaCl	55.8	692.82813	659.34217	0.98054
	$FeCl_3$	93.2	378.94312	347.10403	0.97224

Table 2. Values of the constants a, b y c from equation 8.

Has been reported that the growth rate of a pit is characterized by a gradual increase in the current at constant potential, which is proportional to t_2 or t_3 [34]. Gonzalez-Sanchez [40] generated pitting on stainless steel 304 and 316L using the same electrochemical process used in this study. This author reported an empirical equation for the pit-growth law of the form $P_d=Kt^{\beta}$ generated under potentiostatic control where P_d is pit depth, t, time and K and β are constants depending on the material and the composition of the environment:

- 316L in artificial seawater,

$$P_d = 26t^{0.6} \tag{9}$$

- and for 304 stainless steel, also in artificial seawater, the following equation:

$$P_d = 27.9t^{0.67} \tag{10}$$

where P_d is similar to *MPD* in μm and t is the polarization time in minutes.

The MPD generated on the 304 stainless steel in artificial sea water, reported by Gonzalez-Sanchez [40] were slightly lower, but with a trend similar to those obtained by Dzib-Pérez [55]. These forms contribute to assess the first stage of damage (pitting).

Newman [43] reported the pit growth under potentiostatic polarization on a 304 stainless steel in a 1M NaCl solution containing 0.04 M $Na_2S_2O_3$. The pits generated had a hemispherical morphology. These authors found good agreement between results obtained via Faraday's law and those obtained from microscopic analysis. Researchers had reported the pits growth under potentiostatic polarization on 304 and 316L stainless steels in artificial seawater (Gonzalez-Sanchez, 2002) and natural seawater [56]. The shape of pits generated at both stainless steels was semi-elliptical [40, 56]. Gonzalez-Sanchez [40] reported a good agreement between the results obtained from Scanning Reference Electrode Technique (SRET) measurements with those obtained of a method of material removal.

In the present document, pits were generated under potentiostatic polarization on 304 stainless steels in 3.5% wt NaCl solution and in natural seawater as well as at open circuit potential in a solution of 1M $FeCl_3$. The pits generated had hemispherical morphology as seen in Figure 10.

In order to determine the correct morphology (semi-elliptical or hemispherical) of pits generated in this investigation the volume of each pit was calculated according to the geometry of an ellipse or an sphere and compared with results obtained from the chronoamperometries. The results indicate that the volumes calculated based on a spherical geometry are closer to the volume calculated from the chronoamperometry curves, as seen in Figure 11. Therefore, the volume of a pit, obtained from removal of material, calculated according to a spherical geometry was used for comparison and analysis with the results obtained using electrochemical techniques. The values of the volume per pit as a function of growth time were used to calculate the corresponding amount of dissolved material for a pit grown under potentiostatic control on 304 SS in 3.5 wt% NaCl solution and in seawater natural. In Figure 12 shows the amount of dissolved material, determined chronoamperometries and by a method of material removal, as a function of pits growth time on the 304 stainless steel in NaCl solution and in water natural sea. Also, the figure 13 illustrates the pit nucleation and growth process obtained from SRET Maps.

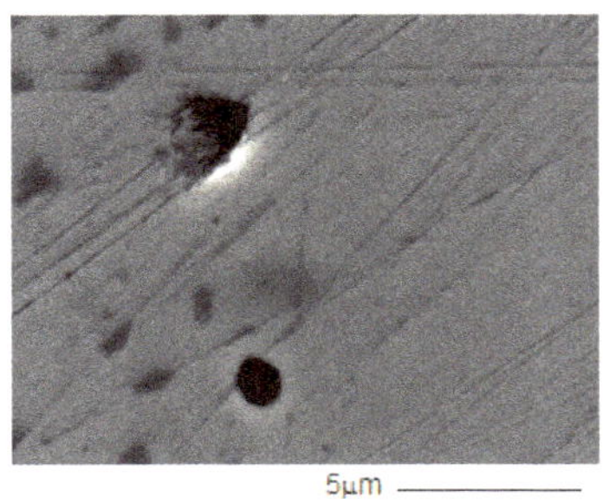

Figure 10. SEM image shows the pits potentiostatically generated in artificial seawater had hemispherical morphology on 316L steel surface under [40].

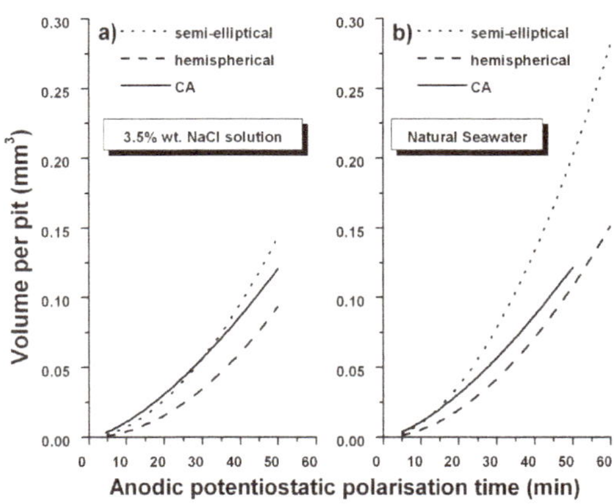

Figure 11. Comparison of the pits volume determined from the material removal method and from the integration of chronoamperometries for 304 SS in a) 3.5% NaCl and b) natural seawater [55].

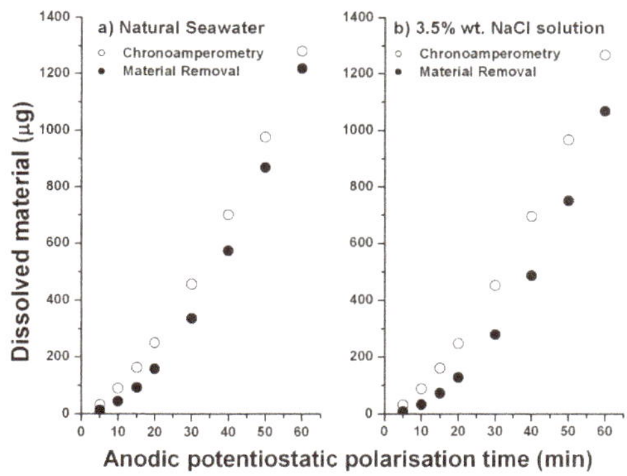

Figure 12. Amount of dissolved material per pit, determined from the chronoamperometries and by a method of material removal as a function of anodic polarization time for the 304 SS in a) natural seawater and b) 3.5% wt NaCl solution [55].

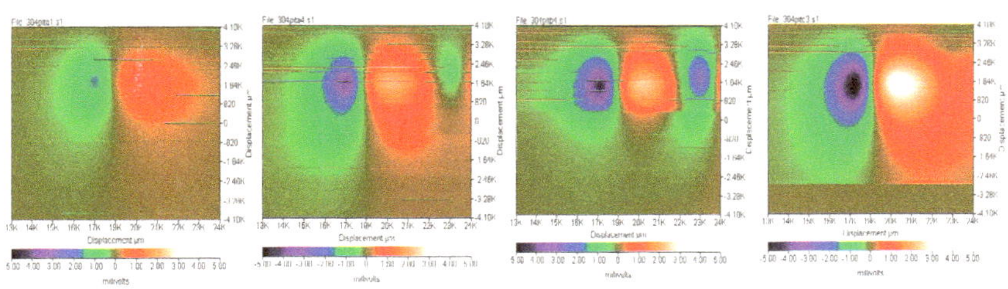

Figure 13. SRET map scans during pitting initiation and growth on 304 SS specimens polarized at 290 mV vs SCE for 2.3, 17.5, 37.5 and 56.5 minutes under fatigue loading conditions [40].

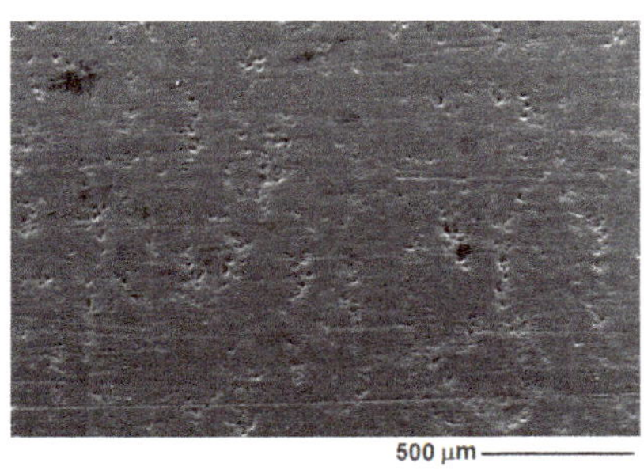

Figure 14. Pits formed at grain boundaries of a 316L SS undergoing cyclic loading in natural seawater. $\Delta\sigma$ = 140 MPa (58%σ0.2), after N=86400 cycles, at a load frequency ω=0.17 Hz, stress ratio R=0, T_{Ave}= 29°C and pH_{Ave} = 7.9. Patterns of grain boundaries are reveled practically in the image [56].

Acuña [56] has observed that on the maximum tensile stress zone of bent-beam specimens, after *N* loading cycles in natural seawater, the pits nucleation sites generally were located at grain boundaries on the specimen surface (Figure 14).

3.3. Pit to small crack transition

Pit to short crack transition involves the determination of the critical pit size for the transition from pit to fatigue crack. In this stage, the microstructure and the state of stresses around the corrosion pit play a very important role. The conditions for the pit to crack transition are generally defined phenomenologically by using a Linear Elastic Fracture Mechanics (LEFM) approach. A major concern is that the threshold stress intensity factor range defined for corrosion fatigue cracking is determined considering long cracks and can not be applied to microscopically small cracks (MSC) initiated from pits.

Two criteria are used to describe the transition processes:

- the stress intensity factor associated with the equivalent surface crack growth at which the corrosion pit reaches the threshold stress intensity factor for the fatigue crack growth,

$$(\Delta K)_{pit} = (\Delta K)_{crack}, \tag{11}$$

- Corrosion fatigue crack growth rate also exceeds the pit growth rate,

$$\left(\frac{da}{dt}\right)_{crack} \geq \left(\frac{da}{dt}\right)_{pit} \tag{12}$$

Akid et. al. [57] have shown that in some cases pit development overcomes the nucleation stage of the corrosion fatigue cracking process. However, it appears that transition from pit to short fatigue crack occurs at some critical size associated with microstructure, applied stress level, mode of load and stress state around the pit. In order to describe the stress state and the stress intensity conditions around the pit, several models, based in pitting corrosion fatigue mechanisms have been proposed [58-61]. They have been assumed that corrosion pits behave as sites for fatigue crack initiation through a purely mechanical micro-notch effect in smooth surfaces. Real components are normally undergoing complex stress conditions which induce mixed loading modes; where it has been reported that the threshold stress intensity factor for short fatigue cracks in structures subjected to mixed loading modes is lower than for structures subject to simple loading mode conditions [62].

The general aim of this section is dedicated to determine the influence of mixed stress conditions on the initiation of fatigue cracks from corrosion pits. Here we present the results obtained from the analysis of the stress state and the threshold stress intensity around hemispherical corrosion pits developed on austenitic stainless steel samples immersed in natural seawater. The concept of pits as surface semi-elliptic cracks is introduced throughout a linear elastic fracture mechanics analysis. Pit depth to major mouth axes ratio (a/c) and the effect of stress orientation were analyzed in terms of variations of the stress state and mixed load modes. Also the pit to short crack transition and short crack growth is reported through sensitive electrochemical techniques.

From the SEM analysis conducted on the metallic specimens after 15 days of immersion in natural seawater, semi-elliptic corrosion pits with different depth /major mouth axes (a/c aspect ratio) values: 0.33, 0.47, 0.56, 0.62 and 0.74 were found. Figure 15 presents corrosion pits formed on the surface of a specimen after N= 130200 cycles of test in flowing natural seawater and the potential and current noise profiles.

Pits developed on grain boundaries, mainly at triple point grain boundary. Small cracks (~ 5 μm long) had grown from nucleation sites (Figure 15c). From the earliest stage, the cracks grew by interconnecting several smaller cracks following the grain boundary pattern. Crack nucleation events and the early crack growth could be associated with the patterns of quick drop and slow recovery, of amplitudes of 20 and 70 mV and current density between 0.10 and 0.60 μA/cm^2 where these signals had higher intensity and higher frequency. Small cracks could grow from a multitude of nucleation sites and coalesce to form large cracks in a relatively short time. Large intergranular crack growth events can be associated with those patterns of 200 mV in amplitude and a cathodic current density of 8.0 μA/cm^2 (in the figure 15 (a) indicated with ellipses on the potential and current EN patrons). The potential transients could be the result of crack wall dissolution and a possible long active dissolution out of the crack, associated to current transients related to cracking by the reduction processes, where the cathodic reaction consumes electrons that are removed from the surface and the electrochemical double layer present on metal surface, both external to the crack and on the crack walls.

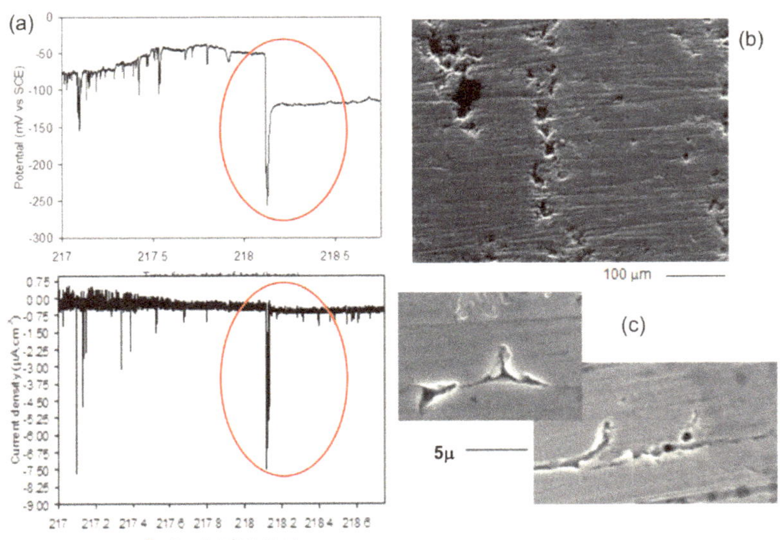

Figure 15. Corrosion pits developed on the stainless steel surface after N=130200 cycles of test in flowing natural seawater. $\Delta\sigma$= 140 MPa (58%$\sigma_{0.2}$), after N=130200 cycles, at a load frequency ω=0.17 Hz, stress ratio R=0, T_{Ave}= 29°C and pH_{Ave}= 7.9. (a) Potential and current noise (b) corrosion pits developed on grain boundaries [16] and (c) SEM photomicrograph showing CF small cracks formed at triple point and along grain boundaries.

In order understand the role of cyclic stress on pit development and growth, local electrochemical measurements were made using SRET on tubular samples of 316L steel immersed in 0.05% $FeCl_3$ solution pH~1.9, conductivity 11.6 mS/cm; under fatigue conditions with an applied stress range of 185 MPa (75%$\sigma_{0.2}$), ω=0.27 Hz, R=0. The crack appeared after ~700 load cycles. On figure 16 it can be seen that pit electrochemical activity increases with increasing number of applied cycles until the pit to crack transition occurs, which is associated with a drop in local pit current density [40].

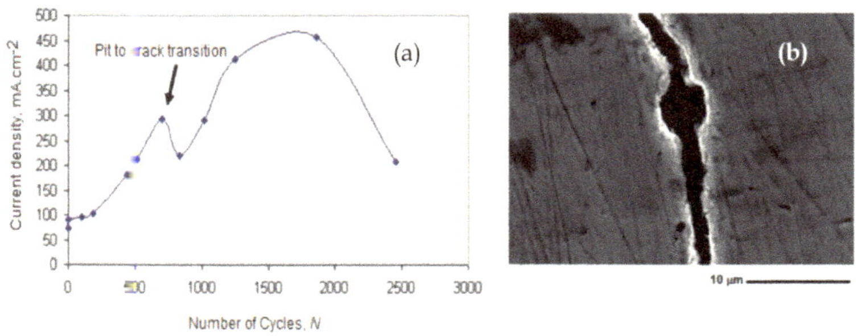

Figure 16. (a) Change in pit current density as function of number of fatigue cycles, indicated by the arrow, and (b) Crack growth from pit observed during corrosion fatigue test [40].

The current density at which the pit–crack transition occurs can then be used to determine the threshold stress intensity for the onset of cracking for damage from smooth surfaces. Figures 15(a) and 16(a) indicate that the first period was dominated by pitting corrosion and the CF cracks incubation took place; the second period corresponds to the nucleation and initial growth of short CF cracks from grain bonding triple points which is associated with a drastic decrease in potential (15 (a)) and current (16(a)). And finally, the third period where the electrochemical current noise signal recovered a synchronized pattern and also the current density from SRET measurements presented a change to higher values that may be related to the arrest of the CF cracks as their Faradaic contribution disappeared.

As above mentioned, the pit to a small crack transition is not governed solely by the stress distribution around the defect (pit), but is also determined by the local electrochemistry which controls pit growth, all together in a synergic phenomenon. We suggest that despite the solution chemistry is not be the same in large cracks as that found in inside a pit, and the microplasticity is generated by the stress concentration generated by the pit geometry for the case of austenitic stainless steels (susceptible to work hardening), the ΔK_{th} is a parameter that could be used to determine the stress intensity around of pit and pit contour. There are basically two reasons why the cyclic plastic deformation is higher just at the surface: concentration of plastic deformation due to higher stresses near the surface, and the low constraint on near-surface volumes of cyclically loaded material. In complex engineering components, higher surface stresses result from either micronotches (i.e. pits) or bending and twisting, both of which lead to stress gradients with the highest stress on the surface.

Usually, under conditions of uniaxial load, the geometry of a pit-like crack tends to keep a semicircular shape, whereas for the case of bending load, the surface cracks adopt a semi-elliptic configuration during their propagation [56]. It is for this reason that the direction of the applied stress plays an important role in the nucleation and propagation of fatigue cracks from corrosion pits.

In this section, the threshold stress intensity factor range (ΔK_{th}) was evaluated as a function of the aspect ratio (a/c) and the orientation of the corrosion pit with respect to the direction of the applied stress assuming tilted pits with hemispherical geometry, where in the case of fatigue limits based on crack propagation, corrosion pits act as cracks under conditions of linear elastic behavior [63-65]. The ΔK_I, ΔK_{II}, ΔK_{III} and ΔG were evaluated as a function of the applied stress direction (θ) assuming hemispherical pits, the position along the pit contour and also of the aspect ratio (a/c). Figure 17 presents the schematic of the stress element from which the transformed stresses as a function of the pit orientation angle are obtained.

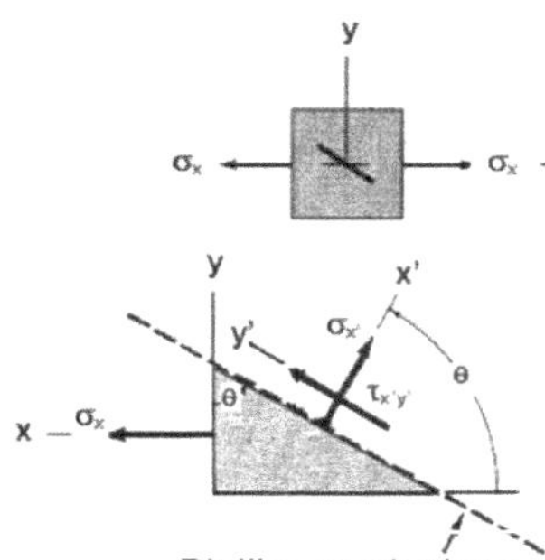

Figure 17. Stress state acting on a tilted hemispherical pit-like crack plane

Redefining the coordinate system to make the new axes coincide with the pit-like crack orientation, it is possible to resolve the applied stress on normal and tangential components. The relationship between the transformed stresses and the normal applied stress are given for equations (13) and (14).

$$\sigma_{x'} = \sigma_x \cos^2\theta \tag{13}$$

$$\tau_{x'y'} = \sigma_x \cos\theta \sin\theta \tag{14}$$

were $\sigma_{x'}$ and $\tau_{x'y'}$ are the normal and tangential stresses respectively as indicated in figure 18.

The stress components associated to the plane x'y' where the crack is located are $\sigma_{x'}$ y $\tau_{x'y'}$. The component $\sigma_{x'}$ induces load mode I whereas $\tau_{x'y}$ induces load mode II in the surface and load mode III at the bottom of the pit-like crack. Raju and Newman [66-67] proposed a solution for superficial hemispherical cracks; such a solution was based on finite element calculations.

$$\Delta K_I = \sigma_{x'} \sqrt{\frac{\pi a}{Q}} f(\phi) \tag{15}$$

$$f(\phi) = \left[\sin^2 \phi + \left(\frac{a}{c} \right)^2 \cos^2 \phi \right]^{\frac{1}{4}}$$

where a/c is ≤ 1, and φ specifies a particular position the stress level at around the tip of a hemispherical crack. Q is the shape factor of the defect or pit. The maximum value of ΔK_I occurs for $\varphi = 0°$, where $f(\varphi) = \sqrt{a/c}$, this is, near the surface over the major axis of the ellipse.

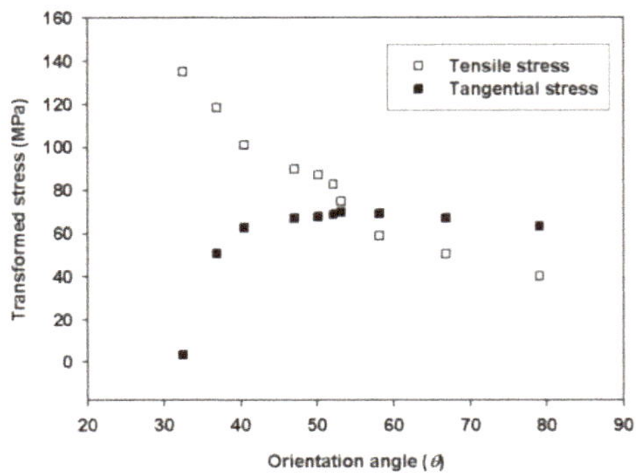

Figure 18. Relationship between the transformed stress acting around a pit-like crack and the orientation angle

The value of the threshold stress intensity factor for load mode II (ΔK_{II}) and III (ΔK_{III}) were calculated using the solution proposed for long elliptical surface cracks under mixed load mode [68]:

$$\Delta K_{II(\phi=0)} = \frac{\tau_{x'y'} \sqrt{\pi a} k^2 (a/c)^{1/2}}{B} \tag{16}$$

$$\Delta K_{III(\phi>0)} = \frac{\tau_{x'y'} \sqrt{\pi a} k^2 (1-\nu) \sin \phi}{B \left[\sin^2 \phi + (a/c)^2 \cos^2 \phi \right]^{1/4}} \tag{17}$$

where $k^2 = 1-(a/c)^2$ $B = (k^2 - \nu)E(k) + \nu(a/c)2K(k)$ and $E(k)$ is the second order elliptic integral defined by:

$$E(k) = \sqrt{Q} = \int_0^{\pi/2} \sqrt{1 - k^2 \sin^2 \phi}\, d\phi \,,\; K(k) = \int_0^{\pi/2} \frac{d\varphi}{\sqrt{1 - k^2 \sin^2 \varphi}}$$

Equations 15, 16 and 17 were used to determine the stress intensity factor for load mode I, II and III respectively developed at the pit contour defined by the angle ϕ: ($\phi = 0°$, $\phi = 45°$ and

$\phi = 90^\circ$). Considering two different normal applied stress values $\Delta\sigma_x = 217$ MPa (90%$\sigma_{0.2}$) and $\Delta\sigma x = 140$ MPa (58%$\sigma_{0.2}$), with constant amplitude and a stress ratio $R = 0$. Normal and tangential stresses were simulated taking into account a system used by Acuña et al. [69], which allows changing sample orientation with respect to the applied stress direction for orientation angles $\theta > 0°$(Figure 17). Five different orientation angles (θ) were used: 0°, 30°, 45°, 60° and 75°. Figure 19 presents a schematic of the corrosion pit profile developed on the specimen surface using the electrochemical pit generation procedure.

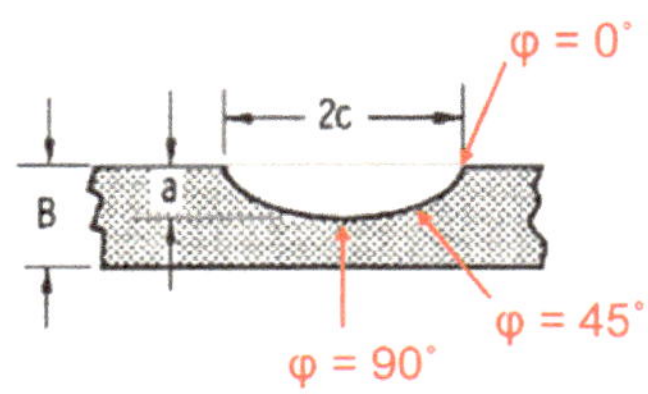

Figure 19. Schematic Corrosion pit contour developed on the stainless steel surface after electrochemical pre-pitting procedure in natural seawater.

Table 3, presents the values of ΔK_I, ΔK_{II} and ΔK_{III} as a function of the orientation angle θ, the aspect ratio a/c and the position around the pit profile ($\varphi = 0^\circ$, 45° and 90°).

In order to define the conditions for the initiation of fatigue cracks from pits on components subject to mixed load modes, a direct equivalence of stress intensity factor to the energy approach was considered. The approach where the elastic strain energy release rate (ΔG) related to the three load modes was considered.

$$\Delta G = \frac{(1-\nu^2)}{E}(\Delta K_I^2 + \Delta K_{II}^2) + \frac{(1-\nu)}{E}\Delta K_{III}^2 \tag{18}$$

Equation 18 indicates that ΔG depends upon the values of ΔK_I, ΔK_{II} and ΔK_{III} which are function of the orientation angle θ. It is worth mentioning that ΔK_{II} and ΔK_{III} do not have any contribution on ΔG for the case where θ is equal to zero.

Figure 20 shows the values of the ΔG as a function of the orientation angle for different locations along the pit contour: at the surface $\phi = 0^\circ$, at $\phi = 45^\circ$ and at the bottom of the pit where $\phi = 90^\circ$ for three different aspect ratio a/c values.

The fracture resistance value R in mode I, was also calculated and indicated in the graphics of figure 18. For the three considered a/c values, the maximum ΔG value was reached at $\theta = 45^\circ$in which the tangential stress predominates over the normal stress, independently of the position along the pit contour, which is more clearly showed in figure 21.

The orientation of the applied stress plays an important role in the nucleation and propagation of fatigue cracks from this kind of surface flaws (Figure 21). The asymmetry of the applied load induces normal and tangential stresses which along with the aspect ratio have a strong influence on the stress intensity factors which is observed in the variation of

ΔG vs orientation angle θ. Tangential stress has a coupled influence with tension stress for angles from 30° to 50° (load modes I, II and III).

Pit Contour		ϕ=0°			ϕ=90°		
		Maximum Stress Intensity Factor (MPa $m^{1/2}$)			Maximum Stress Intensity Factor (MPa $m^{1/2}$)		
Aspect ratio a/c	Mode Mixity (θ)	ΔK_I	ΔK_{II}	ΔK_{III}	ΔK_I	ΔK_{II}	ΔK_{III}
0.33	0°	100.87	0.00	0.00	176.94	0.00	0.00
	30°	75.66	36.91	52.99	132.71	36.91	52.99
	45°	50.44	42.62	61.19	88.47	42.62	61.19
	60°	25.22	36.91	52.99	44.24	36.91	52.99
	75°	6.76	21.31	30.59	11.85	21.31	30.59
	90°	0.00	0.00	0.00	0.00	0.00	0.00
0.47	0°	123.58	0.00	0.00	180.26	0.00	0.00
	30°	92.69	40.68	48.57	135.20	40.68	48.57
	45°	61.79	46.98	56.09	90.13	46.98	56.09
	60°	30.90	40.68	48.57	45.07	40.68	48.57
	75°	8.28	23.49	28.04	12.08	23.49	28.04
	90°	0.00	0.00	0.00	0.00	0.00	0.00
0.56	0°	134.36	0.00	0.00	179.55	0.00	0.00
	30°	100.77	41.42	45.30	134.66	41.42	45.30
	45°	67.18	47.83	52.31	89.77	47.83	52.31
	60°	33.59	41.42	45.30	44.89	41.42	45.30
	75°	9.00	23.91	26.16	12.03	23.91	26.16
	90°	0.00	0.00	0.00	0.00	0.00	0.00
0.62	0°	139.97	0.00	0.00	178.49	0.00	0.00
	30°	104.98	41.22	43.03	133.87	41.22	43.03
	45°	69.99	47.60	49.69	89.24	47.60	49.69
	60°	34.99	41.22	43.03	44.62	41.22	43.03
	75°	9.38	23.80	24.84	11.96	23.80	24.84
	90°	0.00	0.00	0.00	0.00	0.00	0.00
0.74	0°	150.23	0.00	0.00	175.24	0.00	0.00
	30°	112.68	38.65	36.90	131.43	38.65	36.90
	45°	75.12	44.62	42.61	87.62	44.62	42.61
	60°	37.56	38.65	36.90	43.81	38.65	36.90
	75°	10.06	22.31	21.30	11.74	22.31	21.30
	90°	0.00	0.00	0.00	0.00	0.00	0.00

Table 3. Stress Intensity Factor (MPa√m), ΔK_I, ΔK_{II} and ΔK_{III} as a function of θ, φ and a/c.

For angles lower than 30, tension stress are favoured (load mode I) and become the dominant stress. The orientation of the stress in an angle relative to the pit-like crack plane generates a normal tension stress at the microscopic plane of the crack that can be even higher than the externally applied stress. This induces a stress singularity that makes the perpendicular tension stress component sufficiently high to activate the crack initiation from the pit.

The analysis indicates that ΔG is higher in the surface where $\phi = 0^{\circ}$ (pit mouth) than at $\phi = 45^{\circ}$ and than at $\phi = 90^{\circ}$, (at the middle and at the bottom of the pit respectively) for inclination angles from 20° to 65° for a/c = 0.30. For the case of a/c = 0.60 and a/c = 0.75, the highest values of ΔG were observed at orientation angles from 5° to 75°. This result would indicate that the transition from pit to small crack takes place predominantly at the surface ($\phi = 0^{\circ}$), when the aspect ratio a/c→1 and the orientation angle θ increases, compared with those for ($\phi = 45^{\circ}$ and $\phi = 90^{\circ}$).

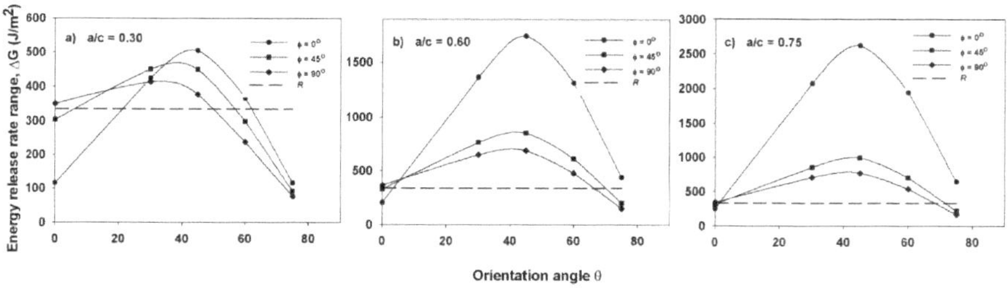

Figure 20. Energy release rate on semi-elliptic pits as a function of the orientation angle for three different values of a/c.: a) 0.30, b) 0.60 and c) 0.75.

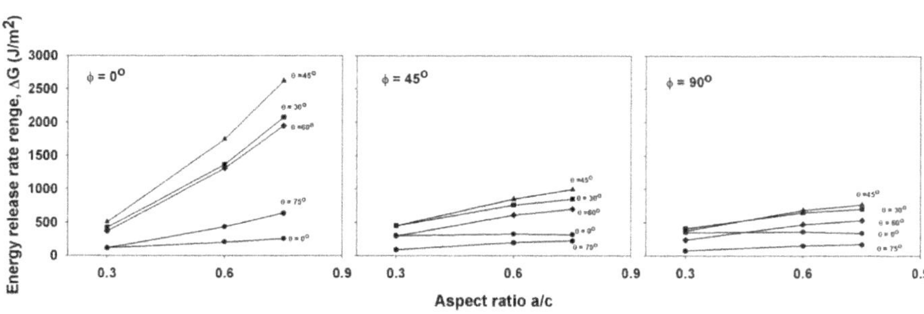

Figure 21. Energy release rate as a function of the aspect ratio for the 5 orientation angles at three different sites along the pit contour ($\varphi = 0^{\circ}$, $\varphi = 45^{\circ}$ and $\varphi = 90^{\circ}$).

Stress and strain distributions are viewed perpendicular to loading axis. Equivalent Von-Mises Stress and (b) total deformation of a pit with pit depth P_d=168μm. The plane stress had been assumed in the present analysis and the stress distribution around of semi-elliptic pit-like crack was done by considering biaxial loading conditions. Figures 22 and 23 show the Finite Element Analysis performed using ANSYS 11.0 [70]. A 3-D version of a thin-walled 350 μm tube specimen was modeled and pits were created half way along the length of the specimen. A refined mesh was developed in the region around the pit to enhance the accuracy of stress and strain predictions. The von Mises criterion model was used in the analysis to characterize the elastic-plastic behaviour of the material. The FE model was applied to a hemispherical shaped pit of 168 μm, 274 μm and 291 μm deep. The aspect ratio is found to generally increase with increasing pit depth. Table 4 reports the results of ΔK and ΔG obtained for 3 pits depth configuration. Pit shape was obtained by a method of material removal and after modelled using Finite Element Method.

Specimen under cyclic loading at a stress level around 90% $\sigma_{0.2}$ exhibited a number of developed surface pits of 168μm deep. High stress and strain is observed on the figure 22. In the figure 22 (a), for biaxial stress conditions, the maximum stress is observed ϕ~45° at pit contour, near to the wall of the pit mouth, whereas the higher plastic strain is observed at the bottom of pit than just near to wall pit surface.

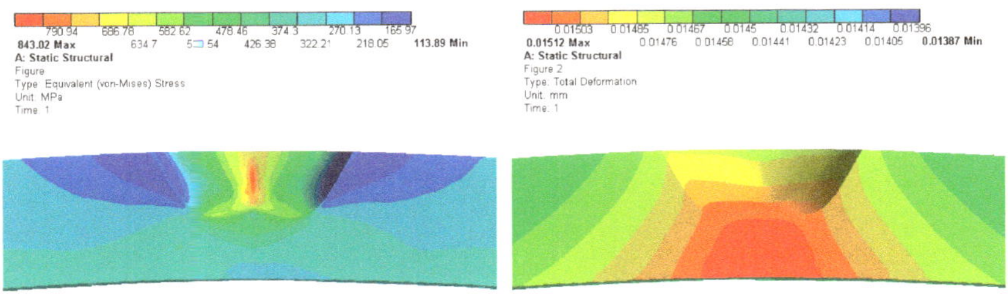

(a) (b)

Figure 22. Maximum principal stress (a) and strain (b) distribution around a 168 μm hemispherical shaped pit developed on a pressured thin-walled 350 μm tube specimen made of 316L SS; $\sigma_{0.02}$= 241 MPa, *E*= 190 GPa and *v*= 0.33. The internal pressure and the relation a/c were kept 110 bar and 0.67.

This plastic strain at pit bottom induces higher constrain than those observed just near to wall pit surface. Therefore, the pit to short crack transitions is done closer to the pit wall surface, as is shown in figure 22.

Location	Pit depth (μm)	Q	Hoop Stress (σ_1) (MPa)	Axial Stress (σ_2) (MPa)	Von-Mises Stress (MPa)	ΔG Hoop (J.m^{-2})	ΔG Axial (J.m^{-2})	ΔG Von-Mises (J.m^{-2})	ΔK$_I$ Hoop (MPa m$^{1/2}$)	ΔK$_I$ Axial (MPa m$^{1/2}$)	ΔK$_I$ Von Mises (MPa m$^{1/2}$)
Pit: near to the surface	168	1.397	1100	407	831	1722.15	235.77	982.84	18.09	6.69	13.66
Pit bottom			433	10.1	388	266.85	0.146	214.27	7.12	0.16	6.38
Pit: near to the surface	274	1.466	502	157	478	531.21	51.96	481.65	10.04	3.14	9.56
Pit bottom			436	12.3	375	400.72	0.32	296.44	8.72	0.24	7.50
Pit: near to the surface	291	1.587	206	565	643	789.36	81.02	609.47	12.25	3.92	10.76
Pit bottom			33.7	388	396	299.47	2.168	287.37	7.54	0.64	7.39

Table 4. Results for FEM Analysis of 3 pit depth modeled in a pressured thin-walled 350 μm tube specimen made of 316L SS; $\sigma_{0.02}$= 241 MPa, E= 190 GPa and v= 0.33. Q= Shape factor.

From figure 21 and table 3, it can be seen that for load mode I, when $\Delta K_{II}/\Delta K_{I}=0$, the value of ΔK_{Ith} is directly proportional to the aspect ratio. However, this relationship modifies as load mode II is dominant with respect to load mode I, ($\Delta K_{II}/\Delta K_{I}\sim3$) at an angle of 75°, from which the values a/c = 0.47 and a/c = 0.74 seem to have no effect on ΔK_{eq}. It has been reported that for short fatigue cracks under conditions of mixed load mode, the value of the ΔK_{th} can be substantially smaller than that of long cracks [62]. Thus, it can be assumed that the condition for crack growth initiation is reached at ΔK_{eq} values lower than for single mode loading (ΔK_{I}, ΔK_{II} or ΔK_{III}).

Tangential stress has a coupled influence with tension stress for angles from 30° to 55° (load modes I and II). For angles lower than 30, tension stress are favoured (load mode I) and become the dominant stress.

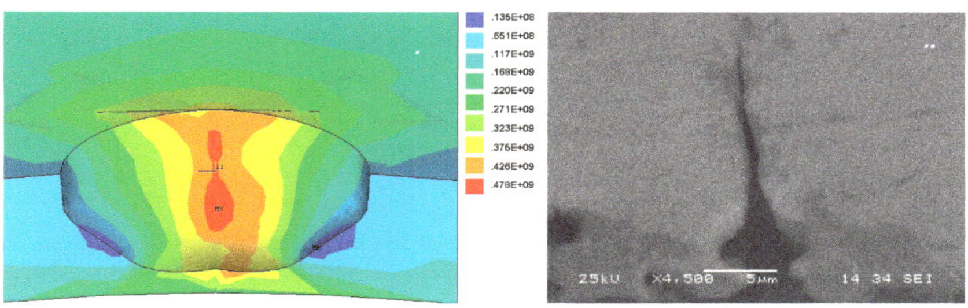

(a) (b)

Figure 23. (a) Maximum principal stress distribution at pit contour and (b) small crack emanating from a corrosion pit (Von-Mises: $\Delta K_{Isurface}$=9.56 MPa√m; $\Delta K_{Ibottom}$=7.50 MPa√m) , in a hemispherical shaped pit with a depth=291 μm and developed on a pressured thin-walled 350 μm tube specimen made of 316L SS; $\sigma_{0.02}$= 241 MPa, E= 190 GPa and ν= 0.33. The internal pressure and the relation a/c were kept 110 bar and 0.78.

The orientation of the stress in an angle relative to the pit-like crack plane generates a normal tension stress at the microscopic plane of the crack that can be even higher than the externally applied stress. This induces a stress singularity that makes the perpendicular tension stress component sufficiently high to activate the crack initiation from the pit. Based on the observations about the stress conditions around the pit-like cracks, the direction for crack initiation will be a function of the ratio, due to that the normal and tangential stresses vary with the orientation angle. Thus a crack will preferentially nucleate at sites of maximum principal normal stress as is indicated in figure 22 (a) and corroborated by figure 22 (b).

4. Conclusions

- From results obtained on corrosion fatigue tests using Electrochemical Noise and SRET, it was possible to determine the pit initiation time, pit activity and pit localization on specimen surface. The changes in dissolution current obtained from SRET measurements and the increment in the amplitude and frequency of potential transients observed through the Electrochemical Noise tests, gave semi-quantitative information of the early stages of corrosion fatigue damage and cracking. Despite the inability to spatially resolve either the pit or crack size or shape from SRET measurements in $FeCl_3$, the change detected in electrochemical activity can be related to the transition from localized metal dissolution to crack nucleation; which may involve a redistribution of current over the new crack surfaces. Tests on corrosion fatigue, performed by the authors, were aimed to establish a relationship between Electrochemical Noise and SRET results in order to determine in a semi-quantitative base, the localized corrosion rate and the pit to crack transition process.

- FEM analysis corroborated that the local stress conditions for crack initiation in mixed mode, in a semi-elliptic surface pit, overcome the minimum stress level or single threshold stress established from the fatigue limit concept. Superposition of cyclic shear (ΔK_{II} or $\Delta K_{III} > 0$) to cyclic tension can lower the mode I threshold stress intensity range, ΔK_I, below which crack growth is presumed dominant.
- Based on ΔG data, crack initiation will mainly depend on the pit-like crack a/c ratio (α) as well as orientation angle (θ). From theoretical results, it could be assumed that the condition for crack growth initiation at the surface zone ($\phi = 0^\circ$) are favoured when $\alpha \geq 0.56$ and θ ranges from 5° to 75°, being the most manifest when θ~45°.
- The alignment of corrosion pit with respect of the direction of applied nominal stress has a strong influence on the conditions for crack initiation. The crack initiation hot points are a function of a/c, θ and φ.

Author details

Narciso Acuña-González
Academic Vice-Chancellor Office, Anahuac-Mayab University, Merida, Mexico

Jorge A. González-Sánchez and Luis R. Dzib-Pérez
Centre for Corrosion Research, Autonomous University of Campeche, Campeche, Mexico

Aarón Rivas-Menchi
Faculty of Engineering, Anahuac-Mayab University, Merida, Mexico

5. References

[1] Miller K J (1987), Title. Fatigue Fract. Engng. Mater. Struct. 10 (1): p. 75.
[2] Sureh S (1998) Fatigue of Materials; second edition, Cambridge University Press. 679 p.
[3] Hagerup O, Strandmyr O (1998) Field experience with stainless steel materials in seawater systems. CORROSION-NACE 98, paper No. 707.
[4] Faimali M, Chelossi E, Pavanell G, Benedetti A, Vandecandelaere I, De Vos P, Vandamme P, Mollica A (2010) Electrochemical activity and bacterial diversity of natural marine biofilm in laboratory closed-systems. Bioelectrochem. 78: 30-38.

[5] Acuña N, Ortega-Morales B O, Valadez-Gonzalez A (2006) Biofilm colonization dynamics and its influence of austenitic UNS S31603 stainless steel exposed to Gulf of Mexico seawater. Mar. Biotechnol. 8: 62–70.
[6] Scotto V, Lai M E (1998) The ennoblement of stainless steels in seawater: a likely explanation coming from the field. Corros. Sci. 40:1007–1018.
[7] Mattila K, Carpen L, Raaska L, Alakomi H, Hakkarainen T, Salkinoja-Salonen MS (2000) Impact of biological factors on the ennoblement of stainless steel in Baltic seawater. J Indust Microbiol Biotechnol 24: 410–420
[8] Wang W, Wang J, Li X, Xu H, Wu J (2004) Influence of biofilm growth on corrosion potential of metals immersed in seawaters. Mater. Corros.. 55: 30–35.
[9] Gartland O P (1991) Aspects of testing stainless steels for seawater applications. Marine and Microbial Corrosion. p. 134
[10] Mansfeld F, Tsai R, Shih H, Little B, Ray R, Wagner P (1990) Results of exposure of stainless steels and titanium to natural seawater CORROSION NACE 90, paper No. 190
[11] Wei R P and Speidel M O (1972) Phenomenological aspects of corrosion fatigue, critical introduction. In: Corrosion Fatigue: Chemestry, Mechanics and Microstructure, NACE-2: 379-380.
[12] Jaske C E, Payer J H, Balint S V (1981) Corrosion Fatigue of Metals in Marine Environments. Springer – Verlag and Battelle Press. P.
[13] Kim K, Hartt W H (1995) Title. Journal of Offshore Mechanics and Artic Engineering, 117: p. 183.
[14] Akid R, Dmytrakh IM, Gonzalez-Sanchez J (2006) Fatigue Damage Accumulation: Aspects of Environmental Interaction. Materials Science, 42 (1): 42-53.
[15] Boukerrou and Cottis R.A.; (1993), Corrosion Science, 35, No. 1-4, p. 577.
[16] Acuña N (2005) Fatigue Corrosion Cracking of an Austenitic Stainless Steel Using Electrochemical Noise Technique. Anti. Corr. Meth. Mater. 52: 139-144.
[17] kid R., Dmytrakh I M, Gonzalez-Sanchez J (2006) Fatigue damage accumulation: the role of corrosion on the early stages of crack development. Corros. Engn. Science and Tech. 41 (4): 328-335
[18] Acuña-González N., García-Ochoa E., González-Sánchez J.; (2008), Assessment of the dynamics of corrosion fatigue crack initiation applying recurrence plots to the analysis of electrochemical noise data, Int. J. Fatigue 30:1211–1219.
[19] Horner D.A., Connolly B.J., Zhou S., Crocke L. and Turnbull A.; (2011) Novel images of the evolution of stress corrosion cracks from corrosion pits. Corrosion Science 53:3466–3485.
[20] Goswami T K, Hoeppner D W (1995) Pitting Corrosion Fatigue of Structural Materials. In: Chang C I, editor. Sun Structural Integrity in Aging aircraft. AD-Vol. 47. In: ASME, New York.
[21] Shreir L.L; (1976), "CORROSION Volume 1"second edition, Newnes-Butterworths, London U.K.
[22] Bockris J.O'M. and Reddy A.K.; (1970), "Modern Electrochemistry Vol. 2" Plenum Press, USA.

[23] Pourbaix M.; (1984), "On the Mechanisms, Teaching, and Research in Localised Corrosion", Journal of Electrochemical Society, Vol.131, No. 8
[24] Böhni, H.; (1987) " Localised corrosion in corrosion mechanisms", p. 285, Mansfeld and Dekker Eds
[25] Böhni H.; (1992), "Localised corrosion- Mechanisms and methods", Materials Science Forum Vols. 111-112, p. 401,Trans Tech Publications Switzerland.
[26] Williams D. E., Kilburn M. R., Cliff J., Waterhouse G.I.; (2010), "Composition changes around sulphide inclusions in stainless steels, and implications for the initiation of pitting corrosion", Corrosion Science 52:3702–3716
[27] Jones, D.; (1992), Principles and prevention of corrosion, Macmillan Publishing Co., N.Y., pg. 4.
[28] Marcus P., Olefjord I.; (1988), A Round Robin on combined electrochemical and AES/ESCA characterization of the passive films on Fe–Cr and Fe–Cr–Mo alloys, Corros. Sci. 28, 589–602.
[29] Mughrabi H., Wang R., Different K., and Essmann U., (1983); STP 811, American Society for Testing and Materials: p. 5
[30] Dawson, J.L., (1996), "ECN Measurement: The Definitive In-Situ Technique for Corrosion
[31] Legat, A. (1993), "EC Noise As the Basis of Corrosion Monitoring", Proceedings 12th International Corrosion Congress, NACE, Houston. 1410-1419.
[32] Dawson, J.L. et. al., (1989), "Corrosion Monitoring using ECN Measurements", Proceedings International Corrosion Congress, NACE, Houston, TX, Paper 89.
[33] Legat, A., Dolecek, V. (1995), "Corrosion Monitoring System Based on Measurement & Analysis of EC Noise", Corrosion Science, vol. 51 no. 4, pp. 295-307.
[34] Szklarska-Smialowska, Z; (1986), "Pitting Corrosion of Metals", NACE International, Houston, TX
[35] Benish, M.L., et.al. (1998), "A new electrochemical noise technique for monitoring the localized corrosion of 304 stainless steel in chloride-containing solutions", Proceedings International Corrosion Congress, NACE, San Diego, Paper 370
[36] Roberge, P.R. Wang, S. and Roberge, R. (1996); "Stainless steel pitting in thiosulfate solutions with Electrochemical Noise", Corrosion, vol. 52 no. 10, pp. 733-737
[37] Watanabe, Y. et.al., (1998), "Electrochemical noise characteristics of IGSCC in stainless steels in pressurized high-temperature water", Proceedings International Corrosion Congress, NACE, San Diego, Paper 129.
[38] Luo J., and Qiao L. J; (1999), "Application and evaluation of processing methods of electrochemical noise generated during Stress Corrosion Cracking", Corrosion, vol. 55 no. 9, pp. 870-876.
[39] González- Rodríguez, J. G., Salinas Bravo, V. M., García- Ochoa E. and Díaz Sánchez, A. (1997), "Use of electrochemical potential noise to detect initiation and propagation of stress corrosion cracking in a 17-4 PH steel", Corrosion, vol. 53 no. 9, pp. 693-699.
[40] González-Sánchez J. (2002); PhD thesis, Corrosion fatigue initiation in stainless steels: the scanning reference electrode technique, Sheffield Hallam University, Sheffield, UK.
[41] Djoudjou R., Lemaitre C., Beranger G.; (1993), Corrosion Reviews, Vol. 11, p. 157.

[42] Frankel G.S.; (1998) Pitting corrosion of metals – a review of the critical factors, Journal of the Electrochemical Society 145, 2186–2198
[43] Newman R.C., (2001);. Whitney W.R award lecture: understanding the corrosion of stainless steel, Corrosion 57, 1030–1041
[44] Sedriks A.J.; (1989), Corrosion - NACE, Vol. 45, p. 510.
[45] Sedriks A.J.; (1996), "Corrosion of Stainless Steels", second edition, John Wiley & Sons, Inc., USA
[46] Macdonald D.D.; (1992), J. Electrochem. Soc., Vol. 139, p.3434.
[47] Isaacs H.S. and Vyas B.; (1981), "Scanning Reference Electrode Techniques in Localised Corrosion", Electrochemical Corrosion testing, ASTM STP 727, p. 3, F.
[48] Bates S.J., Gosden S.R. and Sargeant D.A.; (1989), "Design and development of Scanning Reference Electrode Technique for investigation of pitting corrosion in FV 448 turbine disc steel", Materials Science and Technology, Vol. 5, p. 356
[49] Sargeant D.A., Hainse J.G.C. and Bates S.; (1989), " Microcomputer controlled scanning reference electrode apparatus developed to study pitting corrosion of gas turbine disc materials ", Materials Science and Technology, Vol. 5, p. 487
[50] Trethewey K.R., Marsh D.J. and Sargeant D.A.; (1996), "Quantitative Measurements of Localised Corrosion Using SRET", CORROSION-NACE 94, Paper no. 317.
[51] K.R. Trethewey, D.A Sargeant, D.J. Marsh and S. Haines, (1994), "New Methods on Quantitative Analysis of Localised Corrosion using Scanning electrochemical Probes", in Modelling Aqueous Corrosion: From individual Pits to System management, p. 417, eds. K.R. Trethewey and P.R. Roberge, Kluwer Academic Press.
[52] H.N. McMurray, S.R. Magill, and B.D. Jeffs (1996), "Scanning Reference Electrode Technique as a tool for investigating localised corrosion phenomena in galvanised steels", Iron and Steelmaking, Vol. 23, No.2, p 183
[53] H.N. McMurray and D.A. Worsley (1997), "Scanning Electrochemical Techniques for the Study of Localised Metallic Corrosion", in Research in Chemical Kinetics, Vol. 4, p. 149, Compton & Hancock eds. Blackwell Science Ltd
[54] Rosenfeld I L, Danilov IS (1967) Electrochemical Aspects of Pitting Corrosion. Corrosion Science, 7:129.
[55] Dzib-Pérez L., (2009), Ph.D. thesis, National Autonomous University of Mexico, Mexico.
[56] Acuña-González N.; (2001), Ph.D. Thesis, The effect of Micro-organisms on the Corrosion Fatigue Performance of Stainless Steel in Natural Seawater. National University of Mexico, Mexico.
[57] Akid R., Wang Y.Z., and Fernando U. S.; (1993) The Influence of Loading Mode and Environment on Short Fatigue Crack Growth in a High Strength Steel. Corrosion – Deformation Interactions (ed. T. Magnin and J. M. Gras), Les Editions de Physique, pp. 659-670.
[58] Hoeppner D.W.; (1979) "Model for Prediction of Fatigue Lives Based Upon a Pitting Corrosion Fatigue Process", Fatigue Mechanisms, Proceedings of an ASTM-NBS-NSF Symposium, J.T. Fong, De., ASTM STP 675, American Society for Testing and Materials, 1979 pp. 841-870.

[59] Lindley T.C., Mclntyre P. and Trant P. J.; (1982) "Fatigue Crack Initiation at Corrosion Pits," Metals Technology, Vol. 9, pp. 135-142.

[60] Kawai S., and Kasai K.; (1985) "Considerations of Allowable Stress of Corrosion Fatigue (Focused on the Influence of Pitting), "Fatigue Fracture of Engineering Materials Structure, Vol. 3, No. 2, pp. 115-127.

[61] Kondo Y.; (1983) "Prediction of Fatigue Crack Initiation Life Based on Pit Growth, "Corrosion Sciencie, Vol. 45, No.1, 1985, pp. 7-11.S. R. Novak, ASTM STP 801, p. 26

[62] Nalla RK, Campell JP, Ritchie RO; (2002) Fatigue Fract Engng Mater Strcut, 25

[63] Wei R. P; (1997) "Corrosion Fatigue – science and engineering" Conference of Recent Advances in Corrosion Fatigue", UK.

[64] Toyama K, Konda N; (1989) on International Conference of Evaluation of Materials Performance in Service Environments. Japan.

[65] Konda Y, and Wei R. (1989), on International Conference of Evaluation of Materials Performance in Service Environments. Japan.

[66] Raju IS, Newman JC Jr.; (1979) Engineering Fracture Mechanics; 11

[67] Newman JC Jr, Raju IS (1981) Eng Fracture Mech;15

[68] He M, Hutchinson JW. (2000), Eng. Fract. Mech. 65

[69] Acuña-González N, González-Sánchez J, Contreras E, Sauceda I (2009) Effect of Mixed Mode Loading Induced by Asymmetrical Stress upon Crack Initiation from Corrosion Pits. Advanced Materials Research. 65:39-46.

[70] ANSYS® Academic Research, Release 11.0, license number: 34216380, Flex ID: 18a905230e25 ANSYS, Inc.

Section 4

Fracture Mechanics Aspects of Power Engineering

Chapter 9

Methodology for Pressurized Thermal Shock Analysis in Nuclear Power Plant

Dino A. Araneo and Francesco D'Auria

Additional information is available at the end of the chapter

1. Introduction

The relevance of the fracture mechanics in the technology of the nuclear power plant is mainly connected to the risk of a catastrophic brittle rupture of the reactor pressure vessel. There are no feasible countermeasures that can mitigate the effects of such an event that impair the capability to maintain the core covered even in the case of properly functioning of the emergency systems.

The origin of the problem is related to the aggressive environment in which the vessel operates for long term (e.g. more than 40 years), characterized by high neutron flux during normal operation. Over time, the vessel steel becomes progressively more brittle in the region adjacent to the core. If a vessel had a preexisting flaw of critical size and certain severe system transients occurred, this flaw could propagate rapidly through the vessel, resulting in a through-wall crack. The severe transients that can lead the nuclear power plant in such conditions, known as Pressurized Thermal Shock (PTS), are characterized by rapid cooling (i.e., thermal shock) of the a part of the internal reactor pressure vessel surface that may be combined with repressurization can create locally a sudden increase of the stresses inside the vessel wall and lead to the suddenly growth of the flaw inside the vessel thickness.

Based on the long operational experience from nuclear power plants equipped with reactor pressure vessel all over the world, it is possible to conclude that the simultaneous occurrence of critical-size flaws, embrittled vessel, and a severe PTS transient is a very low probability event. Moreover, additional studies performed at utilities and regulatory authorities levels have shown that the RPV can operate well beyond the original design life (40 years) because of the large safety margin adopted in the design phase.

A better understanding and knowledge of the materials behavior, improvement in simulating in a more realistic way the plant systems and operational characteristics and a better evaluation of the loads on the RPV wall during the PTS scenarios, have shown that the analysis performed during the 80's were overly conservative, based on the tools and knowledge available at that time.

Nowadays the use of best estimate approach in the analyses, combined with tools for the uncertainty evaluation is taking more consideration to reduce the safety margins, even from the regulatory point of view. The US NRC has started the process to revise the technical base of the PTS analysis for a more risk-informed oriented approach. This change has the aim to remove the un-quantified conservatisms in all the steps of the PTS analysis, from the selection of the transients, the adopted codes and the criteria for conducting the analysis itself thus allow a more realistic prediction.

This change will not affect the safety, because beside the operational experience, several analysis performed by thermal hydraulic, fracture mechanics and Probabilistic Safety Assessment (PSA) point of view, have shown that the reactor fleet has little probability of exceeding the limits on the frequency of reactor vessel failure established from NRC guidelines on core damage frequency and large early release frequency through the period of license extension. These calculations demonstrate that, even through the period of license extension, the likelihood of vessel failure attributable to PTS is extremely low (≈10-8/year) for all domestic pressurized water reactors.

Different analytical approaches have been developed for the evaluation of the safety margin for the brittle crack propagation in the rector pressure vessel under PTS conditions. Due to the different disciplines involved in the analysis: thermal-hydraulics, structural mechanics and fracture mechanics, different specialized computer codes are adopted for solving single part of the problem.

The aims of this chapter is to present all the steps of a typical PTS analysis base on the methodology developed at University of Pisa with discussion and example calculation results for each tool adopted and their use, based on a more realistic best estimate approach.

This methodology starts with the analysis of the selected scenario by mean a System Thermal-Hydraulic (SYS-TH) code such as RELAP5 [2][3], RELAP5-3D [1], CATHARE2 [4][6], etc. for the analysis of the global behavior of the plant and for the evaluation of the primary side pressure and fluid temperature at the down-comer inlet.

For a more deep investigation of the cooling load on the rector pressure vessel internal surface at small scale, a Computational Fluid Dynamics (CFD) code is used. The calculated temperature profile in the down-comer region is transferred to a Finite Element (FE) structural mechanics code for the evaluation of the stresses inside the RPV wall. The stresses induced by the pressure in the primary side are also evaluated.

The stress intensity factor at crack tip is evaluated by mean the weight function method based on a simple integration of the stresses along the crack border multiplied by the weight function. The values obtained are compared with the critical stress intensity factor typical of the reactor pressure vessel base material for the evaluation of the safety margin.

2. Origin of the problem

The internal components of PWR vessels that are closest to the core (baffles, formers and core envelope in Solution Annealed 304 stainless steel [7], bolts in Cold Worked 316 stainless steel [8], etc.)[9] are highly irradiated; the most irradiated areas of some of these components may be exposed to doses reaching around 80 dpa after 40 years of operation. This neutron irradiation changes their microstructure and their mechanical properties, so they harden, lose ductility and toughness, suffer irradiation creep [9][10][11][12]. In addition, these changes seem to be the basis of increased sensitivity to stress corrosion [13].

Hardening (or embrittlement) starts at the nanometer level as the high energy neutrons are absorbed by the material causing lattice defects which cluster. The mechanisms proposed for the radiation damage are many, but on a fundamental level a single neutron scattering event can be considered. If a neutron of sufficient energy scatters off a nucleus, the nucleus itself is displaced. The atom associated with the nucleus finds itself embedded into the structure elsewhere in a high-energy, interstitial site. It is termed a self-interstitial as the matrix and interstitial atoms are in principle the same. The site the atom previously occupied is now empty: it is a vacancy. In this way, self-interstitial-vacancy pairs are formed.

Neutron scattering events are not isolated. On average, each displaced atom might then go on to displace further atoms, and likewise the neutron that caused the first displacement might go on to displace further atoms. This means that there is a local cascade of displacements, known as a displacement spike, within which there is a large amount of disorder in the structure.

Both the interstitial atoms and vacancies can diffuse through the lattice, but the interstitial atoms are more mobile. Both interstitials and vacancies are eventually removed from the lattice (when they reach sinks such as dislocations or grain boundaries). However, they are also always being generated by the neutron radiation. Thus steady-state populations of interstitials and vacancies are formed.

The majority of the self-interstitial quickly cluster to form small, disc-shaped features that are identical to small dislocation loops. Along with self-interstitial, these loops are very mobile. Diffusion of self-interstitial and loops within the cascade region causes additional recombination prior to their rapid long-range migration (unless they are strongly trapped by other defects or solutes). Although they are less mobile than the self-interstitial, vacancies also eventually diffuse.

In summary, displacement cascades produce a range of sub-nm clusters (defects, solutes, and defect-solute complexes) that directly contribute to irradiation hardening. Expressing damage exposure, or neutron dose, in terms of displacements-per-atom (dpa) partially accounts for the effect of the neutron energy spectrum on the generation of cascade defects and the net residual defect production scales with dpa.

The important thing to know, however, is that the neutron fluence causes the material to loose fracture toughness and in addition causes a shift upwards in the nil ductility transition temperature (RTNDT). It is this shift in the RTNDT value that is the heart of the problem. Consider Figure 1 which illustrates Charpy-V-Notch (CVN) results for ferritic steel before and after irradiation. The CVN test is an old approach to measure the fracture toughness of a material that is still in use today. Improved techniques are available but the CVN results can be used to clearly show the impact of neutron fluence.

The upper curve on Figure 1 represents the CVN energy curve for the un-irradiated specimens. The test measures the energy required to break a specific specimen at a given temperature. If the material is ductile (tough), the energy required to break the specimen is high. As the material is cooled, it loses fracture toughness (becomes more brittle) as illustrated by the curves in Figure 1. The upper curve shows relatively high fracture toughness for temperatures greater than 20° C. Since RPV temperatures are not expected to drop to this level, this material will remain tough during an overcooling event [14].

The risk is mainly associated to the presence of defect in the welding lines that face the core region, because at the edge of this defect an intensification of the stresses can occurs in case of a fast overcooling phase can be generated during the plant life by injection of cold water by the emergency systems.

From a safety point of view, the material properties of the RPV have to be regularly and carefully checked in order to evaluate the embrittlement level of the material by mean the analysis of specimens placed close to the inner RPV surface that can be analyzed by mean the classical CVN test.

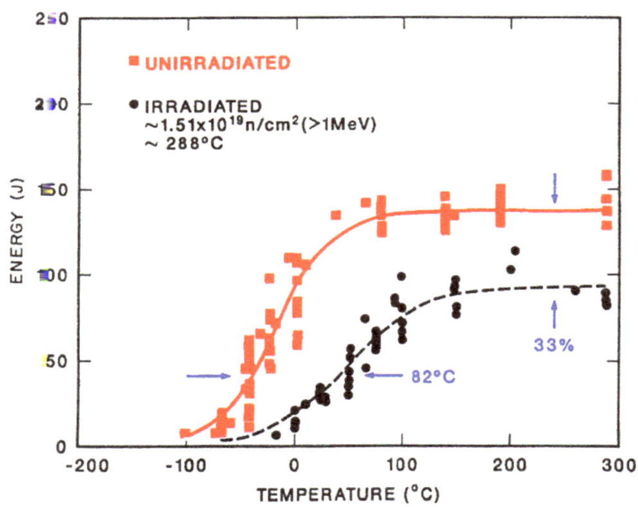

Figure 1. Effect of radiation damage on the CVN transition characterization of ferritic steels

Beside this process a complex analysis is required aimed at identifying those scenarios that can lead an over-cooling on the RPV internal surface where a crack flaw is supposed to exist.

The overcooling transients are usually very complex. It is often not possible to define in advance conservative or limiting conditions for all system parameters. Engineering judgment might not be sufficient to decide whether an accident under consideration is, by itself, a PTS event or along with other consequences can lead to a PTS event that may potentially threaten RPV integrity. Therefore thermal hydraulic analyses are often necessary for choosing, from a number of accidents, those initiating events and scenarios that can be identified as limiting cases within the considered group of events. The calculation period of a transient should be long enough to reach stabilized conditions or at least to overreach the critical time from the point of view of RPV integrity [16].

A wide variety of transients can contribute to the risk of vessel failure. These transients include reactor system overcooling attributable to a LOCA or a stuck-open primary side relief valve, a component failure that results in an uncontrolled release of steam from the secondary side (e.g., Main Steam Line Break (MSLB) or stuck-open secondary side relief valve), or a control system failure that results in overfilling the steam generators. Combinations of failures are also of concern and have to be taken into account during the analysis.

For a wide discussion of the PTS issue in PWR Nuclear Power Plant (NPP) see [15], hereafter only a short overview is reported.

Beside the main categories of transients highlighted, some phenomena were deemed to be most important to down-comer conditions during PTS events:

- natural circulation;
- Emergency Core Cooling Systems (ECCS) injection (mixing and condensation following ECCS injection);
- flow stagnation in case of primary system pressurization.

Natural circulation and flow stagnation are important because if loop mass flow continues (or restarts during a transient), warm water at the average coolant system temperature will be flushed through the reactor vessel down-comer, increasing the down-comer fluid temperature. In contrast, if the loop flow is stagnant, the cold ECCS water will not be mixed with water flowing from other parts of the reactor system and the down-comer temperature will be colder in comparison with the natural circulation case. Integral system response is important because the ECCS injection behavior (flow rates, timing, and to some extent temperatures) are functions of the overall system behavior. System pressurization is itself the primary phenomenon in the PTS analysis. The phenomena listed above were considered because of their potential impact on the down-comer conditions, in particular, the mixing phenomena occurring in the down-comer have the capability to mitigate the cooling effect of the cold water injected by the emergency systems [16].

3. UNIPI methodology for PTS analysis

The objective of the PTS analysis is to determine the safety margin for the RPV operability. The safety margin is obtained comparing the stress intensity factors at crack edge calculated for the identified spectrum of overcooling scenarios with the critical stress intensity factor of the RPV material obtained from the CVN test. In each one of the selected scenarios, there must be enough margin to be sure that the vessel can withstand the selected loads conditions.

University of Pisa developed a methodology concerning the use of a chain of codes to quantify this margin by mean a deterministic approach to the PTS issue, see [17] [18].

A preliminary analysis of the plant configuration and logics is required in order to identify the spectrum of the scenarios that lead to an overcooling of the down-comer region.

The methodology starts with the thermal hydraulic analysis of the Nuclear Power Plant (NPP) using a SYS-TH code such as RELAP5, CATHARE2, or equivalent, during a selected transient scenario. The goal of this step is to evaluate the plant response and to calculate the cooling load induced on the internal RPV wall surface by the Emergency Core Coolant (ECC) injection or by the cooling plug following a MSLB initiating event, to calculate the primary circuit pressure and to provide boundary conditions for the next step.

If the transient evolves in single phase, a more detailed analysis of the mixing phenomena occurring in the down-comer region can be performed by mean a CFD code. The result of this step needed for the PTS analysis is the temperature distribution inside the down-comer.

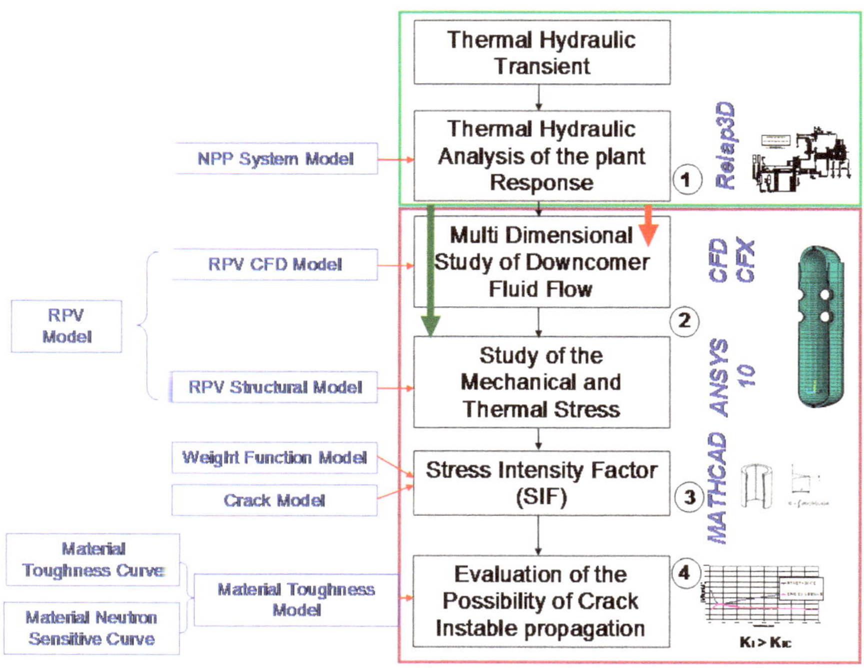

Figure 2. UNIPI Methodology for PTS Analysis

The thermal load to be applied to the FE model for the stress analysis can be extrapolate from the CFD result considering the temperature profile at the interface between down-comer fluid model and RPV wall. This step is accomplished using suitable subroutines developed for this purpose. The stresses due to mechanical load such self-weight, pretension in bolts and internal pressure are also accounted in the Ansys FE calculation. In the last step of the analysis, the Stress Intensity Factor (SIF) KI is calculated by means the Weight Function method, once the stresses generated by the loads identified before are known. The KI has to be compared with the critical SIF (KIc) of the material for the evaluation of the safety margin for the RPV operability. In the following paragraphs an example application to the methodology is provided.

In the next paragraphs a more detailed analysis will be provided for each steps of the analysis previously identified.

4. Analysis at system level

The thermal-hydraulic analysis has the task to determine accident sequences where the temperature differences in adjacent parts of the RPV-wall are large and last for a longer period of time. A catalogue on all relevant load conditions having the potential to cause such temperature differences has to be drawn up for each reactor plant individually, since each plant is characterized by system-specific equipment features.

Under the conditions described above, a high degree of detailing is required from a thermal-hydraulic analysis. Usually, the down-comer is being subdivided in thermal-hydraulic analyses into one or several vertically arranged flow areas, the so-called parallel channels. With regard to the accuracy required for a detailed brittle fracture analysis, this subdivision is not sufficient. This becomes obvious when considering the cooling mechanisms which have been observed in the corresponding test facilities.

Two different cooling mechanisms, the plume and stripe cooling, are considered in connection with the determination of thermal loads. Stripe cooling occurs when the cold-leg emergency core cooling takes place at a time when the water level in the down-comer has fallen below the opening of the cold-leg nozzle. Here, those loss of coolant accidents are taken into consideration which either show a sufficiently large leak cross section or where the emergency core cooling has limited availability.

The stripe cooling causes the biggest thermal load by far, because an only moderately heated cold water stripe of relatively small width cools down the RPV-wall with a large temperature difference to its surrounding area.

However, the significance of cold water stripes for the determination of thermal loads is restricted, since corresponding experimental analyses show that cold water stripes already become detached from the RPV-wall with a relatively low mass flow rate.

Depending on the constructive layout of the cold-leg nozzle, 10 kg/s are for example sufficient to detach a water stripe from the vessel surface (see Figure 3). The injection rate for a high-pressure emergency core cooling system in the conventional PWR NPP is higher than this value; therefore, the stripe cooling is most significant for the area of the cold-leg nozzle.

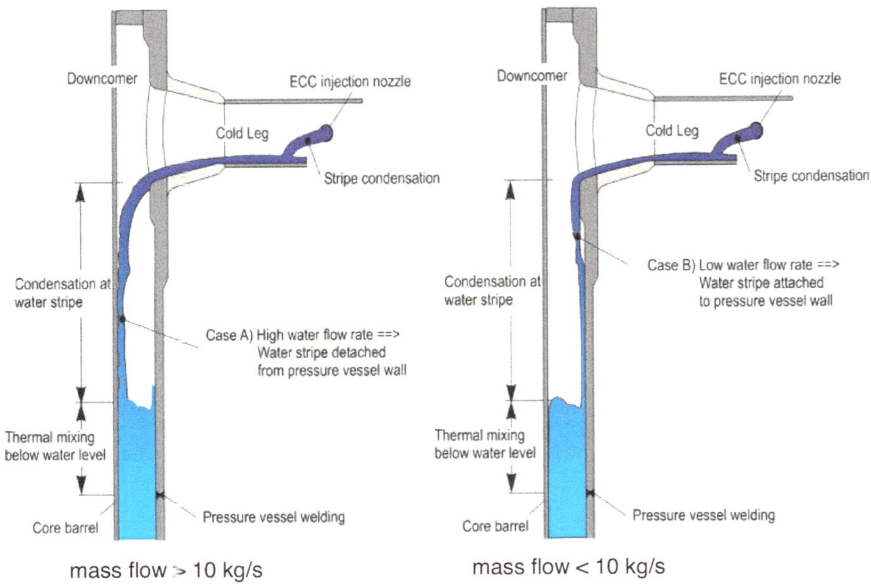

Figure 3. Water stripes in the down-comer with large and small mass flow rate mechanisms

The width of the cold water stripe depends on the flow velocity of the draining cold water stripe. Experimental investigations show that this velocity is determined by the flow phenomenon "critical flow". Dependent on the mass flow of the emergency cooling water, a water level in the cold-leg appears at which the flow velocity exactly corresponds to the critical flow velocity. It turns out that for the relevant mass flows stripe widths of about 10 to 30 cm are to be expected in the cold-leg nozzle. Within this width, the nozzle is cooled down locally.

Cold water plumes are formed if the cold-leg emergency coolant is injected into a down-comer filled with water. Such a situation arises, e.g., from loss-of-coolant accidents with a smaller leak cross section in the hot leg.

Since these plumes stay for a longer period until final mixing, i.e. about one to two hours, the temperature differences can act on the RPV-wall correspondingly long. In contrast to the stripe cooling, the concentration of cold emergency coolant here is lower. By admixture with the water surrounding the plume, there is a permanent exchange. The plume width is dependent on the injected mass flow and the exchange with its environment.

Such phenomena require a more detailed analysis at small scale level in order to better identify the position, the size of the overcooled region, the duration of the cooling phase and the cooling rate.

In order to reach this goal the use of the CFD code is envisaged taking the boundary conditions from the system thermal hydraulic code results. The only limitation in the use of such codes is the large computational resources required. Because of this, the analysis is restricted to the time period during the maximum cooling phase.

The boundary conditions to the CFD code from the system thermal hydraulic code are transferred following a scheme reported in Figure 4. The mass flow rates and the fluid temperature of the injected fluid in the cold legs and down-comer calculated by the SYS-TH code are imposed to the CFD model.

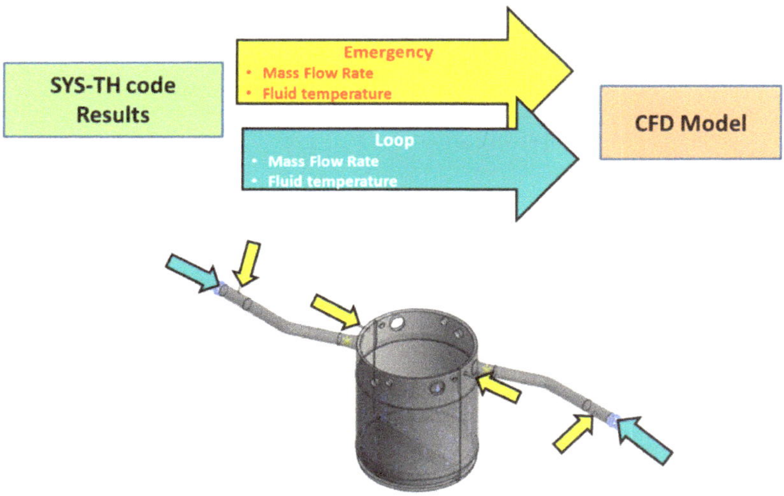

Figure 4. Coupling technique between SYS-TH and CFD codes

5. Analysis at small scale by means a CFD code

The second step of the methodology foresees the analysis at small scale by mean a CFD code for a more detailed calculation of the profile temperature inside the down-comer.

The CFD computer codes solve the Navier-Stokes equations optionally two or three-dimensionally, having the potential to reach the necessary degree of detailing.

With the increasing speed of modern computers, CFD techniques are becoming more widely used and may provide the best tool for computing the thermal fluid mixing. An example application is the emergency injection in the cold-leg by the emergency water reported in Figure 5.

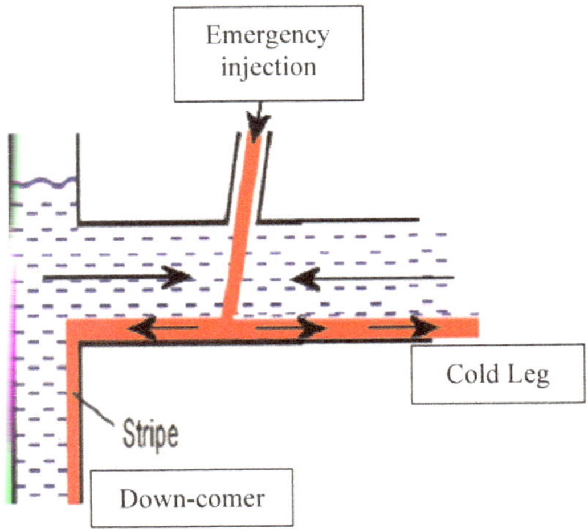

Figure 5. Mixing phenomena occurring during emergency injection

These approaches, however, still suffer from considerable user effect and the need for best practice guidelines continues. Challenges for the single phase CFD user include the turbulence modeling approach which must be able to adequately simulate the various mixing regimes which each have their own unique geometry and driving forces. In addition, the wall modeling approach is important. Mixed convection in the down-comer region is expected and typical CFD wall treatments do not account for this phenomenon.

All of the issues associated with CFD for the single phase PTS issue are compounded in the multi-phase problem by the relative immaturity of the multi-phase CFD techniques. For the near future, system analysis codes will still provide the overall system behavior and experimental results will provide the best source of information on the details of the multi-phase behavior related to PTS [14].

For reliable simulation of PTS related mixing processes the CFD methods must be validated to determine how well the CFD model, defined by the detail level of model geometry, the mesh and the used numerical and physical models can simulate the relevant physical processes and produce the needed data. The final target data of thermal hydraulic analysis are the pressure and temperature fields on structures needed as an input for structural analysis. The CFD model should be able to model the complex mixing and stratification processes in the cold legs and the down-comer of the pressure vessel as well as the heat transfer between fluid and structures accurately enough to reproducethis data. In the Figure 6 an example result of the CFD calculation of the temperature profile inside the down-comer is provided.

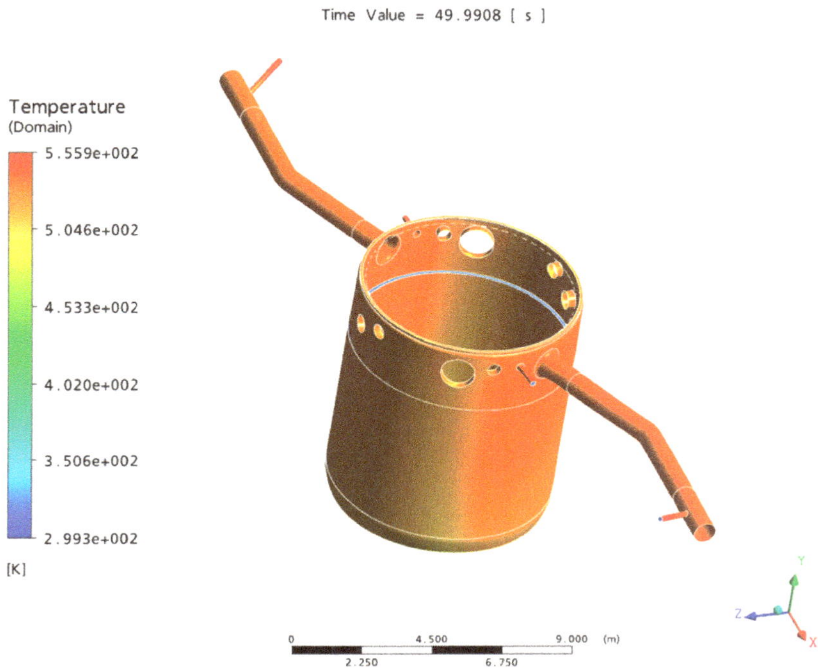

Figure 6. Example result of the down-comer CFD result

The temperature profile and pressure profiles calculated with the CFD code are the boundary condition for setting up the FE simulation for the evaluation of the stresses inside the RPV wall.

6. Structural mechanics analysis

A common approach in the calculation of the stress at the crack edge is the simulation of a part of the vessel obtained by mean symmetry consideration on the geometry and load condition where the crack is supposed to be. This approach in followed basically for saving computational resources, an example of this approach can be found in the ref. [19].

In the methodology developed at University of Pisa, the approach followed is to model the full geometry of the RPV without modeling the crack inside the wall. The reason of this approach is due to a more precise analysis of the local stress inside the RPV wall avoiding any simplification due to the fact that the phenomena occurring in a PTS scenario are intrinsically not symmetric. This choice is supported even from the adoption for the calculation of the stress intensity factor of the weight function method (described in the next paragraph) that needs the stress calculated in the undamaged structure (see Figure 7).

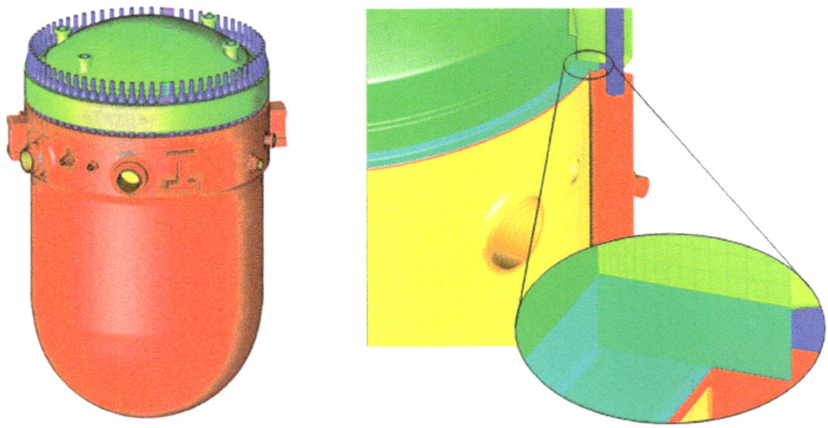

Figure 7. FE model of the RPV (example)

The thermal load calculated in the previous step by mean the CFD code has to be applied FE model. At this point, a problem arises because the CFD model is based on a finite volume mesh, while the structural mechanics model is based on FE mesh. Two techniques (shortly described hereafter) can be adopted, to transfer the temperature time history from one model to another:

- The first foresees the evaluation of the temperature time history of the FE RPV nodes internal surface interpolating the temperature of the CFD down-comer model closest nodes (external down-comer surface, Figure 8) and to calculate the profile inside the wall thicknesses solving the heat conduction equation by mean the FE code itself. The FE model implements all the material properties of the RPV for solving the thermal and stress calculation inside the thickness.
- A second approach foresees the modeling of the RPV wall and down-comer with the CFD model in order to solve the conductivity problem directly. This is called in literature as "conjugate heat transfer calculation". In this approach the temperature time trend in each of the nodes of the FE model is obtained interpolating the temperature of the closest nodes in the RPV CFD model. This second technique is more time consuming and requires a more sophisticated subroutines for performing the transfer (from RPV-CFD model to RPV-FE model) compared to the previous one because the number of nodes to manage is in the order of magnitude of millions.

An example of this transfer technique is reported in Figure 9.

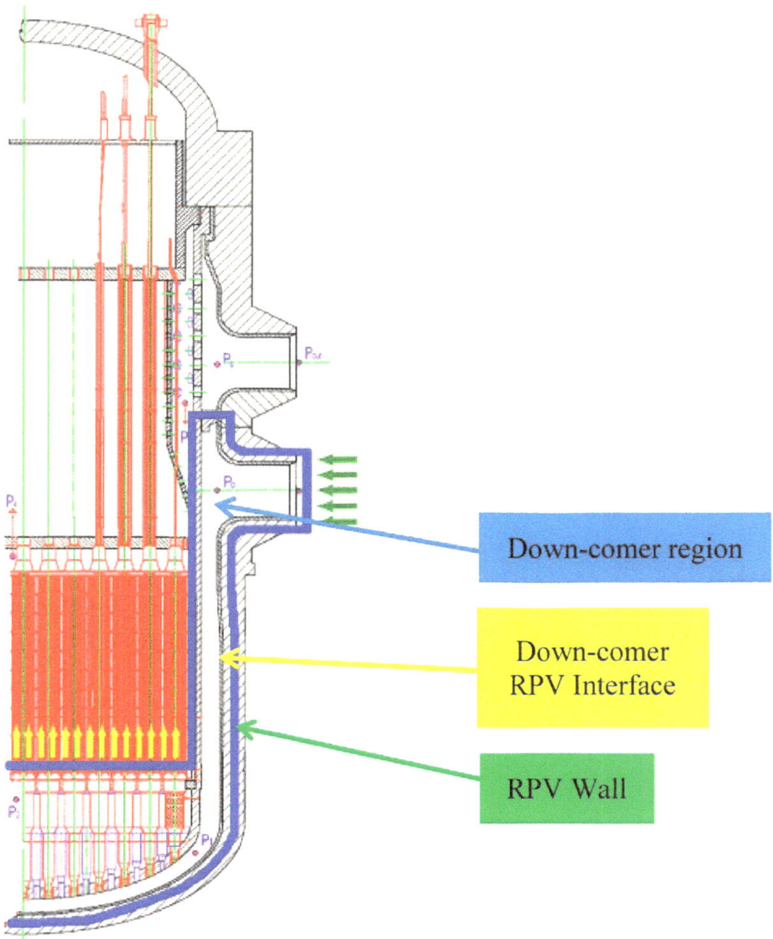

Figure 8. Section of the RPV

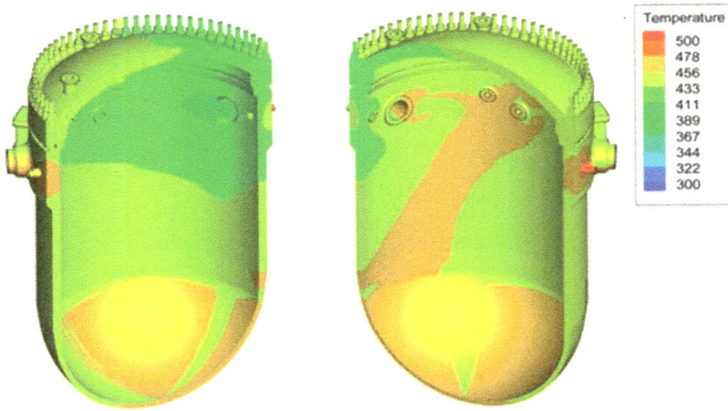

Figure 9. Temperature profile in the FE model of RPV

Once the pressure and the temperature profiles are implemented in the FE model, the stress analysis can be executed. An example result is shown in Figure 10.

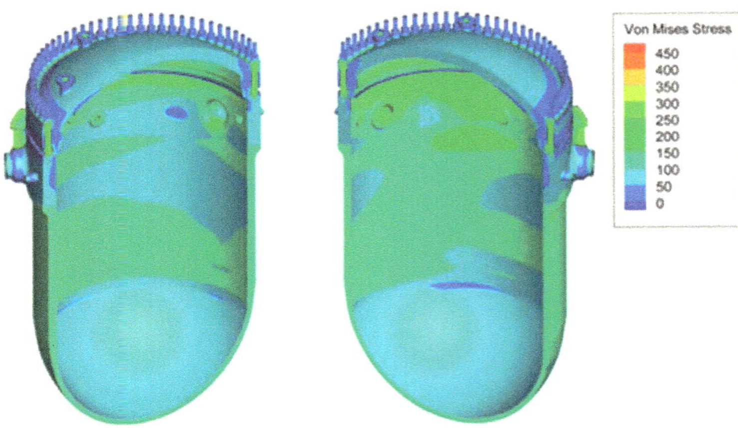

Figure 10. Von Mises stresses

The knowledge of the stress profile inside the RPV wall is the basis for the application of the Weight Function (WF) method that is described in the next paragraph for the calculation of the stress intensity factor.

7. Fracture mechanics

Most numerical methods require a separate calculation of the stress intensity factor for each given stress distribution and each crack length. The weight function procedure developed by Bückner [20] simplifies the determination of stress intensity factors. If the weight function is known for a crack in a component, the stress intensity factor can be obtained by multiplying this function by the stress distribution and integrating it along the crack length. The weight function does not depend on the special stress distribution, but only on the geometry of the component.

If $\sigma(x)$ is the normal stress distribution in the uncracked component along the prospective crack line of an edge crack (see Figure 11), the stress intensity factor is given by the expression (1):

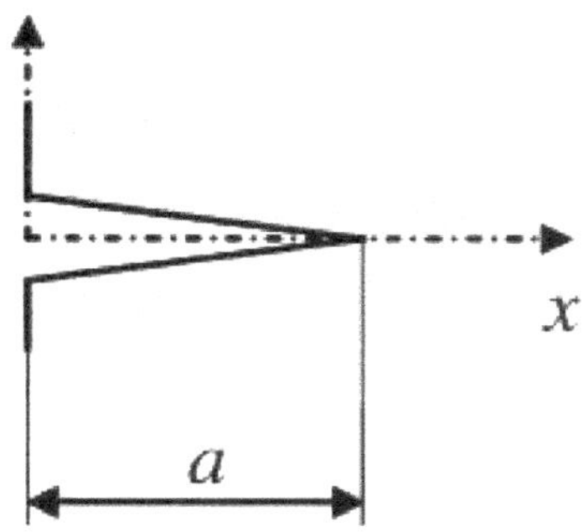

Figure 11. One dimensional crack model scheme

$$K_I = \int_0^a \sigma(x) h_I(x,a) dx \tag{1}$$

The integration has to be performed over the crack length. The WF h(x,a) does not depend on the special stress distribution, but only on the geometry of the component.

The general procedures for the determination of weight functions are described below for the weight function component h_I. The relation of Rice (see [20]) allows to determine the weight function from the crack opening displacement $V_r(x,a)$ under any arbitrarily chosen loading and the corresponding stress intensity factor $K_{Ir}(a)$ according to:

$$h_I(x, a) = \frac{E'}{K_{Ir}(a)} \frac{\partial V_r(x,a)}{\partial a} \tag{2}$$

(E' = E for plane stress and E' = E/(1-ν^2) for plane strain conditions), where the subscript r stands for the reference loading case. It is convenient to use $\sigma_r(x) = \sigma_0$ = constant for the reference stress distribution.

One possibility to derive the weight function with eq. (2) is the evaluation of numerically determined crack opening profiles which may be obtained by Boundary Collocation Method (BCM) computations. For more details on this method see [21].

Once the weight function is defined it can be implemented in a simple spreadsheet program that can easily compute the integral for the stress intensity calculation taking into account the stresses evaluated in the undamaged structure by mean the FE code described in the previous paragraph.

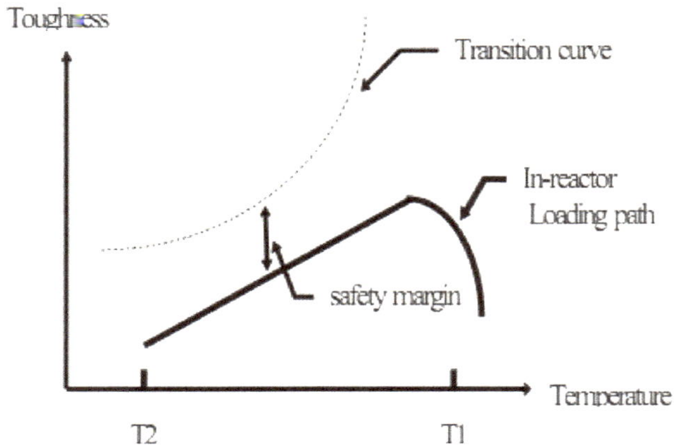

Figure 12. Safety margin evaluation for RPV brittle rupture

The results of the fracture mechanics analysis is a curve representing the value of the stress intensity factor in the selected scenario that is compared with the critical stress intensity factor curve of the material obtained from the analysis of the specimens by mean the CVN test.

In Figure 12 an example of this comparison is shown. The distance between the two curves gives the safety margin for the RPV operability.

Author details

Dino A. Araneo and Francesco D'Auria

GRNSPG, San Piero a Grado Nuclear Research Group, San Piero a Grado, Pisa, Italy

8. References

[1] INEEL, RELAP5-3D Code Manuals, vol. I, II, IV, and V, Idaho National Engineering and Environmental Laboratory, INEEL-EXT-98-00834, Rev. 1.1b, 1999.

[2] RELAP5 MOD 3.3 Code Manual, Volume IV "Models and Correlations", Nuclear Safety Analysis Division ISL 2006.

[3] W.L. Weaver, R.A. Riemke, R.J. Wagner and G.W. Johnsen. "The RELAP5 MOD3 code for PWR safety analysis". Proc. 4th International Topical Meeting on Nuclear Reactor Thermal-hydraulic, Karlsruhe, Germany. Vol. 2. 1991. pp. 1221-1226.

[4] G. Geffraye et al., "CATHARE 2 V2.5_2: a Single Version for Various Applications,"
[5] Proceeding of NURETH-13, Kanazawa City, Ishikawa Prefecture, Japan, Sept 27th - Oct 2nd, 2009.
[6] Barre, F. & Bestion, D. Validation of the CATHARE System Code for Nuclear Reactor Thermalhydraulics, CEA STR/LML/EM/95-347, Grenoble, France, (1995).
[7] IMS S.p.A. 1.4306 Reference Standard EN 10088.
[8] IMS S.p.A. 1.4401 Reference Standard EN 10088.
[9] J.C. Van Duysen, P. Todeschini, G. Zacharie. Effects of neutron irradiation at temperature below 500°C on the properties of cold worked 316 – a review. Effects of Radiation on Materials: 17th International Symposium, ASTM 1175 (1993) 747-776.
[10] G.E. Lucas. The evolution of mechanical property change in irradiated austenitic stainless steels Journal of Nuclear Materials 206 (1993) 287-305.
[11] G.E Lucas, M. Billone, J.E. Pawel, M.L. Hamilton. Implications of radiation-induced reductions in ductility to the design of austenitic stainless steel structures. Journal of Nuclear Materials 233-237 (1996) 207-212.
[12] P. Petrequin. Effect of irradiation on water reactor internals. AMES report n°11, COSU CT 94-074, June 1997, EUR 17694 EN.
[13] Scott P. A review of irradiation assisted stress corrosion cracking. J Nucl Mater 1994, 211:101.
[14] C. Boyd, US Nuclear Regulatory Commission, Office of Nuclear Regulatory Research, THICKET 2008_Session-IX_Paper_29.pdf THICKET 2008 University of Pisa May 5-9 2008.
[15] M. EricksonKirk, M. Junge, W. Arcieri, B.R. Bass, R. Beaton, D. Bessette, T.H.J. Chang, T. Dickson, C.D. Fletcher, A. Kolaczkowski, S. Malik, T. Mintz, C. Pugh, F. Simonen, N. Siu, D. Whitehead, P. Williams, R. S. Yin, NUREG-1806 Vol.1/Vol.2, "Technical Basis for Revision of the Pressurized Thermal Shock (PTS) Screening Limit in the PTS Rule (10 CFR 50.61)", August 2007.
[16] INTERNATIONAL ATOMIC ENERGY AGENCY, "Pressurized Thermal Shock in Nuclear Power Plants: Good Practices for Assessment Deterministic Evaluation for the Integrity of Reactor Pressure Vessel", IAEA-TECDOC-1627, IAEA, Vienna (2010).
[17] Araneo, D., "Procedura di analisi integrata di PTS per il RPV in un impianto nucleare WWER-1000/320 per mezzo dei codici accoppiati Relap5, Trio_U e Ansys", F. D'Auria, M. Beghini, D. Mazzini, Pisa 2003.
[18] L. Frustaci, "Analisi dello shock termico in un RPV tipo WWER1000 in condizioni di DEGB", F. D'Auria, M. Beghini, D. Mazzini, Pisa 2005.
[19] Myung Jo Jhung, Young Hwan Choi, Yoon Suk Chang1 and Jong Wook Kim2, The effect of postulated flaws on the structural integrity of RPV during PTS, Nuclear Engineering and Technology, Vol.39 No.5 October 2007.
[20] Bückner, H., A novel principle for the computation of stress intensity factors, ZAMM 50 (1970), 529-546.

[21] Rice, J.R., Some remarks on elastic crack-tip stress fields, Int. J. Solids and Structures 8(1972), 751-758.

[22] Fett T., Stress Intensity Factors T-Stresses Weight Functions, Schriftenreihe des InstitutsfürKeramikimMaschinenbau, IKM50, University of Karlsruhe 2008.

Section 5

Developments in Civil and Mechanical Engineering

Chapter 10

Fracture Analysis of Generator Fan Blades

Mahmood Sameezadeh and Hassan Farhangi

Additional information is available at the end of the chapter

1. Introduction

Critical gas turbine rotating component, such as turbine blades, compressor disks, spacers and cooling fan blades are subjected to cyclic stresses during engine start-up, operation and shut-down. The lifetime of these components are usually established on the basis of probabilistic crack initiation criterion for a known fracture-critical location (Koul & Dainty, 1992). Therefore, periodic inspections are carried out to detect the probable cracks and prevent suddenly fractures.

Shaft driven rotating fans are commonly utilized to provide the required cooling for generators. These fans circulate cooling gas, air or hydrogen, throughout the machine to maintain the electrical windings at safe operating temperatures. Cooling air is circulated in a closed cycle, in a way that after passage of air through rotor, it is heated and exhausted from top of the generator, which then passes through a cooler, which would cool it down using water flow. Cool air again flows towards rotor and by use of fans, which are installed on retaining ring at the generator sides, is blown around the rotor. Each fan is comprised of several blades, which have been separated by using spacers. In Fig. 1, overall plan of generator and air cycle is shown (Moussavi et al., 2009).

Failure of a rotating fan inside a generator will cause extensive damage. The stored rotational energy in a fan that lets loose will typically destroy the stator winding, sometimes damage the stator core and cause damage to other rotor components such as retaining rings, the rotor winding and possibly even the rotor forging (Moore, 2002). Fan blades are regularly inspected during overhauls by visual and dye penetrant inspections and are required to be replaced due to defects caused by crack, corrosion and impact.

This chapter reports the failure investigation of a rotating axial flow fan of the Iran Montazer-Ghaem-VI 123 MW capacity generator unit. The unit was equipped with two rotating fans, one at each end namely at the turbine side and the exciter side of the generator. The failed fan consisting of 11 blades was mounted on the generator-rotor at the turbine end, and had a total service life of about 41000 hours prior to the failure. The fan rotational speed was 3000 revolutions per minute (rpm) and the maximum operating temperature of the blades was 90°C.

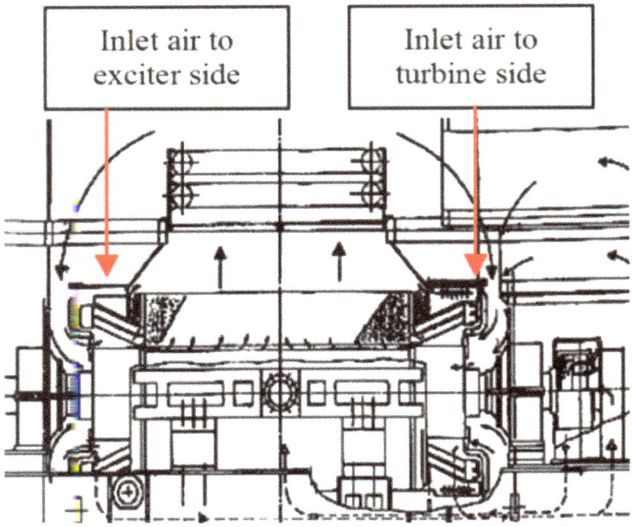

Figure 1. Generator diagram (Moussavi et al., 2009)

Initial investigation pointed out that three blades were fractured and several others were cracked just about 11 hours after resuming operation following the last major overhaul, causing extensive damage to the generator unit specially the stator windings. The failure of the blades was investigated using fractographic and microstructural characterization techniques as well as mechanical evaluations to identify the root cause of the failure. Two similar failures at this kind of fan that caused extensive damage to generator units have been reported from Iran. After some investigations, the corresponding company changed the mounting angle of blades from 19º to 14º to solve the problem of the fans (Iran Power Plant Repair Comany [IPPRC], 2003, 2004).

2. Experimental procedure

Visual inspections were taken on the generator parts especially on the fan blades and the effect of accident on them was studied. Three kinds of blades were found in the turbine casing after the accident: fractured blades, cracked blades and un-cracked blades. The failure was at the turbine side of the generator and according to the visual inspections, the fan blades at the excitor side were not damaged. Dye penetrant non-destructive test was used for detection of surface cracks on the blades. Chemical analysis of the fan blade material was conducted using optical emission spectroscopy. Brinell hardness measurements were taken on the sections prepared from the airfoils as well as on the base of all blades. All measurements were carried out using a 5 mm ball at a load of 1.23 kN. Longitudinal round tensile specimens were machined from the roots and tested according to ASTM E8M. Fatigue specimens were prepared from root of the fractured blades and rotary bending test was done in accordance with DIN 50113.

Longitudinal and transverse specimens were cut from the airfoils for scanning electron microscopy (SEM) and metallography. The metallography samples were prepared by using standard metallographic techniques and etched with modified Keller's reagent. The microstructure of the blade material was analyzed using an optical microscope and an Oxford MV2300 SEM equipped with an energy dispersive spectroscopy (EDS) facility. Fractographic studies were performed using visual examination. Following visual examination of the failed blades, portions of the fracture surfaces were cut for fractographic studies using SEM. All the specimens used for material characterization tests were prepared from the airfoils and bases of the fractured fan blades.

The stresses acting on the blades in steady state condition and at the time of final fracture were estimated using linear elastic fracture mechanics, finite element method (FEM) and fractographic results. At the end, a three-dimensional crack growth software was utilized to assess the crack growth rate and fatigue life in a simplified model of airfoils.

3. Results and discussion

3.1. Visual inspections

Visual inspections indicated that the accident has led to three different categories for the fan blades of the turbine side (Table 1). Dye penetrant testing revealed the cracked blades which did not completely fracture during the accident. A photograph of the fractured blades, labelled in according with their location on the fan is shown in Fig. 2.

All the examinations of mounting clearances, tightening and locking of the blades and the air guide mounting showed no defect also, there were no sign of foreign bodies in the turbine casing.

Category	Blade Number	Specification
Fractured	1, 8, 11	Blade completely fractured and airfoil separated from root
Cracked	2, 3, 7, 9	Including central or edge cracks
Un-cracked	4, 5, 6, 10	Without any surface crack

Table 1. Visual examination of turbine side blades

3.2. Materials characterization

3.2.1. Chemical composition

The chemical composition of the fan blades is given in Table 2. The closest standard aluminum alloy found in the literature is AA 2124 which is a wrought and heat treatable alloy (American Society for Metals [ASM], 1990). This alloy derives its strength mainly from second phase particles which are distributed in the matrix through a precipitation hardening process.

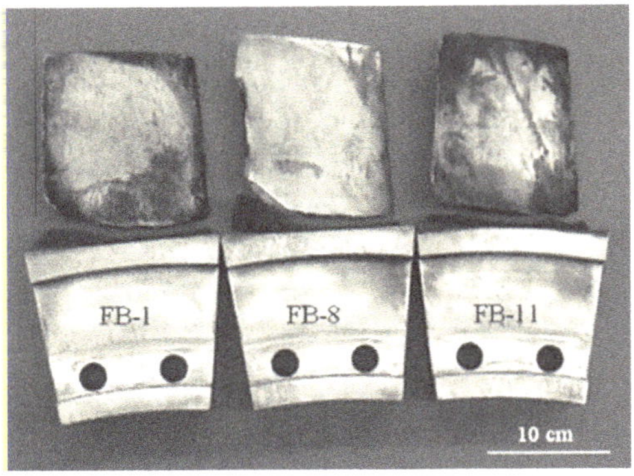

Figure 2. Fractured fan blades

Element / Alloy	Al	Cu	Mg	Mn	Si	Fe	Zn
2124	--Base--	3.8 – 4.9	1.2 – 1.8	0.30 – 0.9	0.20 max	0.30 max	0.25 max
Blade	Base	4.19	1.72	0.62	0.167	0.11	0.03

Table 2. Chemical composition of the fan blade material

3.2.2. Hardness

Brinell macrohardness measurements carried out on different sections of airfoils showed that the hardness was about 133 ± 5 HB, which was essentially uniform along various sections.

3.2.3. Tensile properties

The average values of yield stress, tensile strength, and elongation are 380 MPa, 510 MPa, and 22%, respectively.

The tensile properties and the hardness number of the blades material are all within the standard range reported for Aluminum alloy 2124 (ASM, 1990). The results indicate no degradation in the mechanical properties of the fan blades during service operation.

3.2.4. Fatigue test

The measured lifetime versus applied stress from the rotaty bending fatigue test is presented in Fig. 3. A SEM micrograph taken from the fracture surface of a tested specimen is shown in Fig. 4. Presence of several second phase particles on the surface is obvious. This kind of large particles can accelerate the fatigue crack initiation and affect the fatigue behaviour of the blade material.

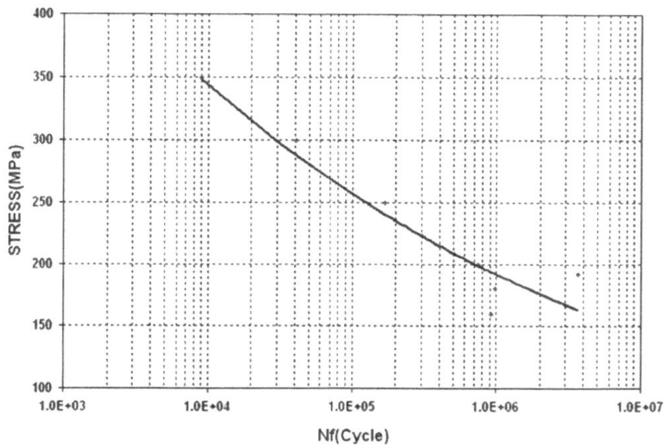

Figure 3. Fatigue test S-N curve

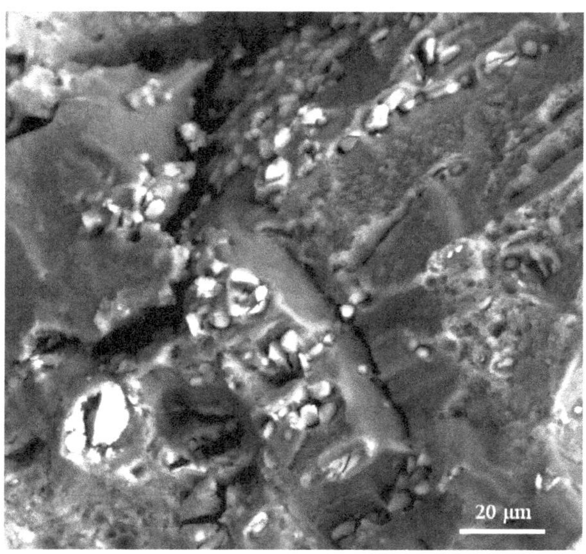

Figure 4. Fracture surface of a fatigue test specimen

3.2.5. Microstructure

Typical microstructures of the blade material in the longitudinal and transverse sections of a cracked airfoil are shown in Fig. 5. The microstructure consists of elongated grains and second phase particles in the longitudinal direction. Various types of second phase particles can be identified in the SEM micrograph, shown in Fig. 6. The large and elongated particle on the micrograph was subjected to EDS analysis. The composition of the particle contained iron, copper and aluminum which is consistent with the β-phase (Al_7Cu_2Fe) particles common to aluminum alloys (Merati, 2005).

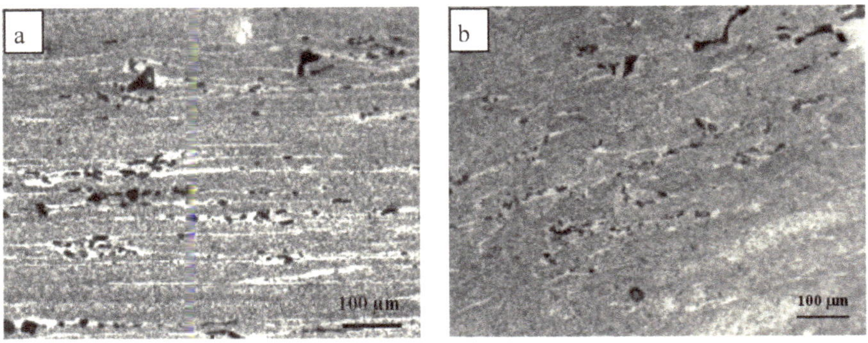

Figure 5. Microstructures of longitudinal (a) and transverse (b) sections of a cracked blade

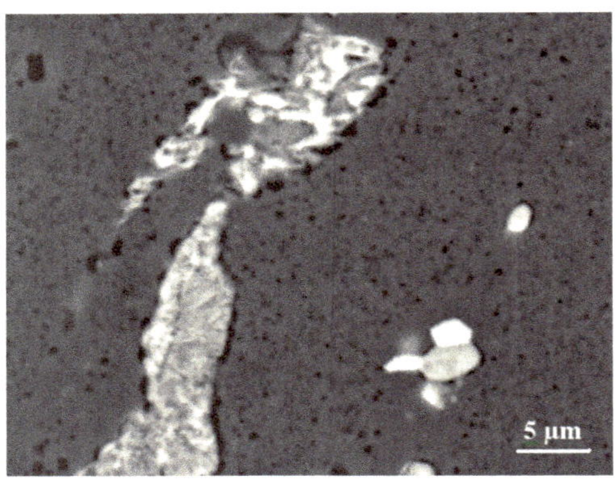

Figure 6. SEM micrograph showing large second phase particles

3.3. Fractography

The fracture location of the broken blades can be identified from Fig. 2. It can be seen that the fracture had occurred close to the transition radius between the blade airfoil and the blade root. All the fracture surfaces exhibited very similar macroscopic features. A representative fractograph of the fracture surface of blade No. 8 is shown in Fig. 7a. It is observed to consist of two distinct regions at low magnification, a semi-elliptical and smooth region which is oriented normal to the blade axis and exhibits a macroscopically brittle appearance, and an outer region with a rougher and more ductile appearance. The transition from semi-elliptical region to the outer region can be clearly identified in Fig. 7b. In this region the remaining cross section of the blade failed by tensile overload.

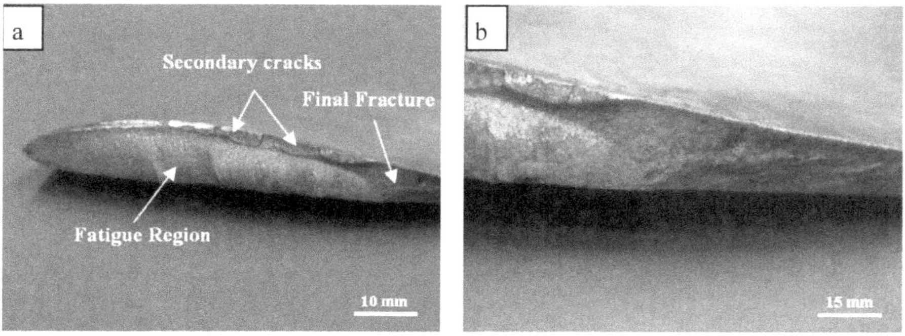

Figure 7. (a) Fracture surface of blade No. 8, (b) transition from fatigue to tensile overload fracture

Faint beach marks indicative of the progressive nature of crack growth in the semi-elliptical region can be seen in Fig. 8. Parallel microscopic fracture surface markings can also be observed in this region at higher magnifications in SEM micrograph as shown in Fig. 9. A schematic drawing of the fan with the location of damaged blades is presented in Fig. 10. Visual inspections and fractographic assessments revealed that the sequence of fracture for the broken blades has been blade No. 1, blade No. 8 and blade No. 11 respectively (Sameezadeh, 2005).

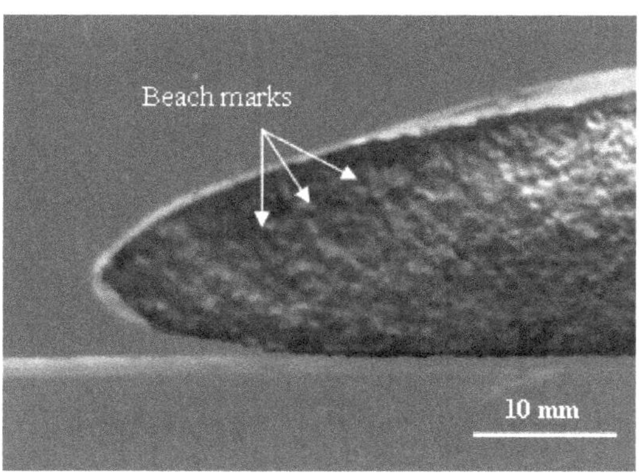

Figure 8. Beach marks on the fracture surface near the leading edge of the blade

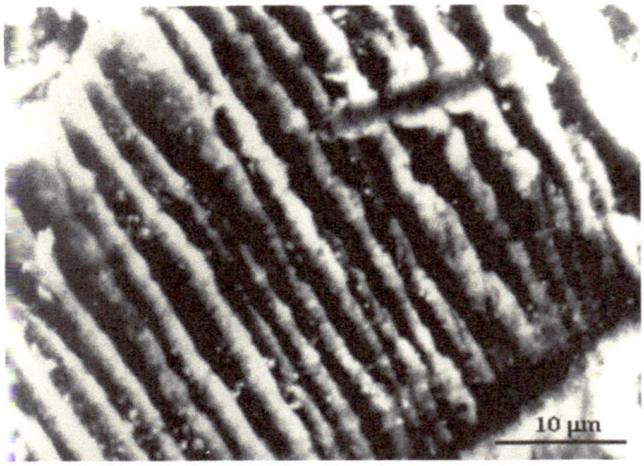

Figure 9. SEM fractograph showing parallel fracture surface markings

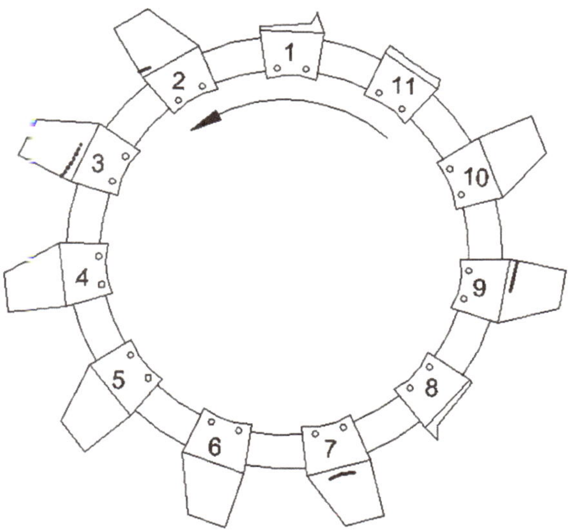

Figure 10. Schematic drawing of the fan with location of damaged blades

Table 3 shows the crack lengths which are measured on the surface of cracked blades. Schematic drawings of the fracture surfaces of three broken fan blades which show multiple crack initiation sites as well as the macroscopic crack growth paths are presented in Fig. 11. Also, For the first and the second fractured blades (blades No. 1 & No. 8), primary crack initiation sites are located on the concave side of the airfoils near the centre where the cross-sectional area is high. Primary cracks have coalesced during fatigue crack growth to form shallow semi-elliptical crack geometry and have propagated to reach the final cracks. In addition, several small semi-elliptical cracks are also shown to have initiated from the opposite convex side of the airfoils.

Blade No. / Crack location	2	3	7	9
Outer surface	24	3	70	78
Inner surface	79	90	-	80

Table 3. Measured lengths of surface cracks

Based on the fractographic observations, fatigue cracking is singled out as the primary fracture mechanism involved in the failure of the fan blades. Final fracture regions constitute only about 25–30 % of the fracture surfaces of the blades. Therefore, fatigue cracking of the blades can be considered to have occurred under high cycle fatigue conditions.

According to Fig. 11 the presence of shallow semi-elliptical cracks on both concave and convex sides of the airfoils is indicative of the influences of considerable bending stresses during crack propagation.

An SEM micrograph showing one of the typical primary crack initiation sites is presented in Fig. 12a. A secondary electron mode micrograph of this region revealed a large second phase particle with a length of about 100 μm at the crack origin as shown in Fig. 12b. The EDS spectrum of this particle identified that its composition is very similar to the intermetallic particle. Presence of the large particles at the origin of the crack have been shown to act as preferred fatigue crack nucleation sites in such alloys, despite the fact that they represent a small fraction of the particle population (Kung, 1979; Merati, 2005).

Crack propagation mode near the initiation region and throughout the fatigue fracture surface was predominantly transgranular and formation of secondary cracks was very limited. Examination of the airfoils near the crack initiation sites showed no apparent defects due to corrosion or foreign object impact.

Dye penetrant non-destructive testing that was carried out on the blades after the accident identified that some of the non-fractured blades have been cracked. The crack surface of one of these blades (blade No. 7) was disclosed as shown in Fig. 13 by cutting the remaining cross-section of the airfoil and opening carefully. In this figure the semi-elliptical fatigue region is obvious. The crack surface is covered by the penetrant material and a higher magnification view identifies the transition of the penetrant material out of the fatigue region that has different microscopic features, so it can be recognized as the final fracture zone. Therefore, the final fracture stage could not be completed because of the generator stoppage after the fracture of the three blades.

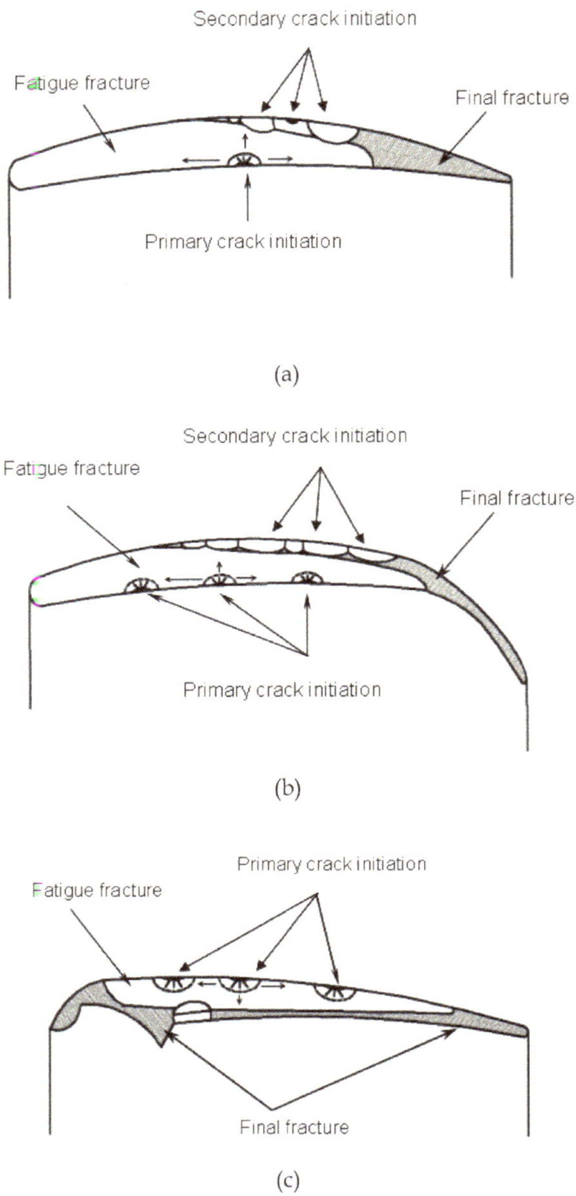

Figure 11. Schematic drawings of the fracture surfaces: (a) blade No. 1, (b) blade No. 8 and (c) blade No. 11

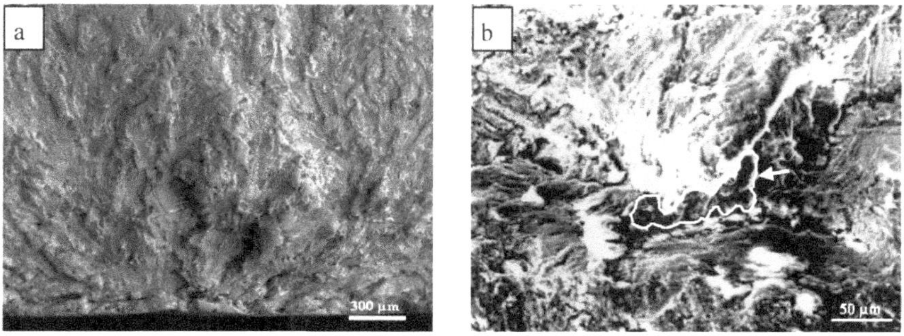

Figure 12. (a) Typical micrograph showing a primary crack initiation site and (b) a crack nucleating particle at the origin of the crack

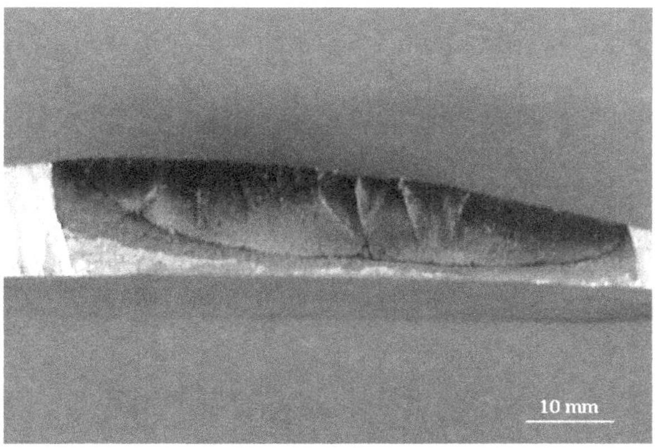

Figure 13. Opened crack surface of blade No. 7

Crack initiation and growth from the convex side and also decrease in the fatigue region in comparison with the fractured blades, prove that the fluctuation of the stresses and the maximum stress had been highly increased during the accident probably due to excessive vibrations.

3.4. Stress analysis of the blades

3.4.1. Steady state stresses

The most important stresses acting at the transition radius in the critical cross-section of the airfoils, under normal operating conditions, consist of a tensile stress component due to centrifugal forces and a bending stress component introduced by the action of the air flow pressure. The tensile stress depends on the rotational speed (N, 50 revolutions per second), mass of the airfoils (m, 0.51 kg), distance from the center of rotation (r, 0.615 m) and the cross-sectional area (A, 0.00155 m^2) as given by the following expression (Bleier, 1997):

$$\sigma_t = \frac{mr(2\pi N)^2}{A} \tag{1}$$

The tensile stress component due to the centrifugal forces is calculated from the above relationship to be about 20 MPa, which is essentially constant during operation.

A 3D finite element model was used to simulate the normal operation of the fan blades. The results revealed that the maximum total stress acting on the blade under normal condition was about 27 MPa which occurred close to the transition radius between the airfoil and the root where the cracks initiated in fractured blades (Fig. 14) (Ataei, 2006). The bending stress introduced by the action of the air flow pressure are less important (Cohen, 1987) and was estimated as about 7 MPa. According to the results, the steady state stresses are very low in comparison with the strength of the blade material therefore, it can be concluded that the fracture accident was happened during an abnormal condition.

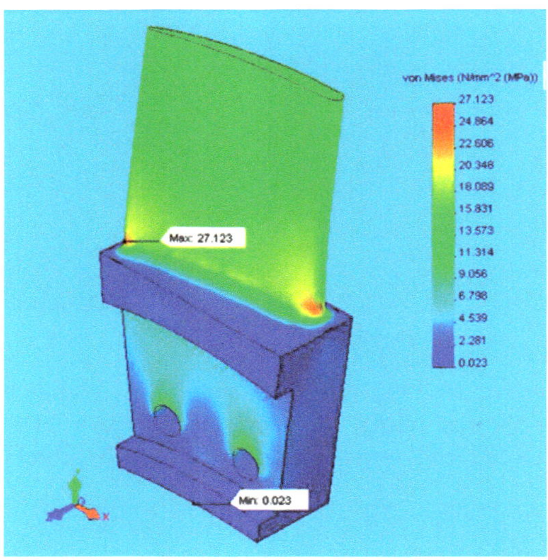

Figure 14. Finite element simulation of the fan blade, which shows the stress distribution under normal operating condition (Ataei, 2006)

3.4.2. Failure stresses

The low aspect ratio of primary fatigue cracks suggests that crack propagation was influenced by significantly higher bending stress levels than that caused by air flow pressure alone. Moreover, since mode I loading condition was dominant and the fatigue crack growth plane was normal to the tensile stress axis, individual components of stress intensity factors due to tension and bending can be added to obtain the total crack tip stress intensity. Accordingly, the failure stress at the time of final fracture and the magnitude of additional bending stresses which influenced the fatigue cracking process can be estimated by applying the superposition principle and fracture toughness data using the following equation (Anderson, 1995; Broek, 1995):

$$K_{IC} = K_t + K_b \quad (2)$$

$$K_{IC} = F\sigma_t\sqrt{\frac{\pi a}{Q}} + FH\sigma_b\sqrt{\frac{\pi a}{Q}} \quad (3)$$

where the fracture toughness of the material was taken as $31.9\ \text{MPa}\sqrt{\text{m}}$ from the literature (ASM, 1990) and the values for F, H and Q geometrical parameters can be calculated for each specific crack geometry under combined tension and bending stresses using Newman–Raju equations (Newman & Raju, 1984).

According to the fractographic results the final fatigue crack of the first broken blade (blade No.1) at the time of final fracture was assumed as a quarter-elliptical corner crack with $a = 12\times10^{-6}$ m and $c = 83\times10^{-6}$ m dimensions, thus the values for F, H and Q parameters in Eq. (3) were calculated as 5.02, 0.20 and 1.07 respectively. Using the above data and a 20 MPa tensile stress, the bending stress can be estimated by the following calculations:

$$K_t = F\sigma_t\sqrt{\frac{\pi a}{Q}} = 18.8\ \text{MPa}\sqrt{\text{m}} \quad (4)$$

$$K_b = HF\sigma_b\sqrt{\frac{\pi a}{Q}} = K_{IC} - K_t = 13.1\ \text{MPa}\sqrt{\text{m}} \quad (5)$$

$$\sigma_b = 69.5\ \text{MPa} \quad (6)$$

The bending stress at the time of fracture of the first broken blade was estimated as about 70 MPa and the total stress acting on the airfoil is thus about 90 MPa (adding a 20 MPa tensile stress). Subtracting the bending stress due to air flow pressure (7 MPa) from this total magnitude of bending stresses, an additional bending stress of 63 MPa is estimated to have influenced the failure of the fan blades. This significant magnitude of additional bending stress, probably caused by excessive vibrations, can explain the early initiation and rapid growth of fatigue cracks to final fracture in the fan blades, which occurred after only a short period of operation following the last overhaul.

The final fatigue crack of the second fractured blade (blade No. 8) is similar to blade No.1 and can be assumed as a quarter-elliptical corner crack too with $a = 11\times10^{-6}$ m and $c = 71\times10^{-6}$ m dimensions. According that, the values for F, H and Q parameters were calculated as 3.74, 0.24 and 1.07 respectively for this crack. With similar calculations as before the bending stress at the time of final fracture for blade No. 8 can be estimated as 115 MPa and the total stress acting on this airfoil is about 135 MPa.

Finally, for the third fractured blade (blade No. 11), the final fatigue crack shape is different and can be assumed as a semi-elliptical surface crack with $a = 11 \times 10^{-6}$ m and $c = 55 \times 10^{-6}$ m dimensions. After calculation of the crack geometry parameters (F=1.84, H=0.66, Q=1.10) and using the Eq. (3), the bending stress at the time of final fracture for this blade was estimated as 118 MPa. By adding a 20 MPa tensile stress, the total stress is thus about 138 MPa. The details of above calculations can be found elsewhere (Sameezadeh, 2005; Ataei, 2006). Table 4 shows the summary of stress analysis results for the fan blades. According to the estimated failure stresses for the fractured blades it should be noted that the significant magnitude of additional stresses acting on the blades and leading to the premature and catastrophic failure of the fan, possibly have been due to aerodynamical disturbances that have resulted in a state of resonant condition of vibration. Additionally, changes in blade installation conditions, such as the level of torque tightening applied to the fixing bolts, which can influence the blade natural frequency, may be regarded as a contributing factor to fan blade failure shortly after overhaul.

	Centrifugal tensile stress (MPa)	Total bending stress (MPa)	Additional bending stress (MPa)	Maximum stress (MPa)
Normal condition	20	7	0	27
Blade No. 1	20	70	63	90
Blade No. 8	20	115	108	135
Blade No. 11	20	118	111	138

Table 4. Summary of stress analysis results of the fan blades

3.5. Simulation of fatigue crack growth

Fatigue crack growth rates in a model of the airfoils, under the action of estimated loads, were computed using FRANC3D/BES crack propagation software. The FRANC3D (FRacture ANalysis Code for 3D problems)/BES (boundary element solver) software developed at cornell university, is capable of evaluating stress intensity factors (SIF) along 3D crack fronts. This software utilizes boundary elements and linear elastic fracture mechanics. The displacements and stress intensity factors are calculated on the crack leading edge to obtain crack propagation trajectories and growth rates. It was assumed throughout the calculations that linear elastic fracture mechanics conditions hold (Carter et al., 2000; Cornell Fracture Group [CFG], 2002).

A simplified 3D model of airfoils was created using a geometry pre-processor program called OSM (Object Solid Modeler). A boundary element model of the geometry, consisting of triangular and square elements was then meshed within the FRANC3D program, and the stresses were applied by the model boundary conditions.

To start the crack growth simulation in FRANC3D an initial crack was introduced into the model. The initial crack was assumed to be a semi-elliptical surface crack with $\frac{a}{c}=0.1$, based on previous fractographic findings. Initial crack length was computed from the threshold stress intensity factor range based on the equation (Hertzberg, 1989):

$$\Delta K_{\text{th}} = Y \Delta \sigma \sqrt{\pi a_0} \tag{7}$$

Taking ΔK_{th} from the literature and $\Delta\sigma$ values for the first and the second fractured blades based on the computed failure stresses and the positive portion of the stress cycle, and using the calculated values of Y from Newman–Raju equations (Newman & Raju, 1984), the initial crack lengths below which crack growth is arrested, a_0, were computed to be approximately 48×10^{-5} m and 22×10^{-5} m for the first and the second fractured blades respectively.

Based on the stresses applied by the model boundary conditions, the initial crack was grown in a series of crack propagation steps. Fig. 15a is a simplified 3D model of the airfoil that was meshed 2D (Boundary elements) in FRANC3D for simulating fatigue crack growth in fractured blades. Fig. 15b shows simulated fatigue crack in blade No. 8 after three steps of propagation.

Stress intensity factors at each step were calculated within FRANC3D and the fatigue crack growth curves were calculated using the Paris power law relationship given by (Broek, 1995):

$$\frac{\mathrm{d}a}{\mathrm{d}N} = C \Delta K^n \tag{8}$$

The empirical constants C and n were computed by fitting the near threshold crack growth data for AA 2124 aluminum alloy (Department of Defense, 1998) as $C = 9.5 \times 10^{-13}\ \text{m.}(\text{MPa}\sqrt{\text{m}})^{-\text{n}}$ and $n = 4.9$ for da/dN expressed in m.cycle^{-1} at a load ratio of $R = -0.1$ assumed in this case. The results of crack growth simulations in the first and the second fractured blades, where initial cracks are grown in small steps to their final dimensions are plotted in Fig. 16. These curves show the crack size as a function of the number of applied stress cycles. It can be seen that the number of cycles required to propagate initial cracks to their final dimensions for the first and the second fractured blades are just about 1.3×10^6 and 8×10^5 cycles respectively.

Stress calculations showed that the total stress acting on the third fractured blade (blade No. 11) is higher than for the two other fractured blades. Therefore, it can be assumed that the fatigue crack growth life of this blade is the shortest one and has the least portion on the total fatigue crack growth life of the failed fan. Thus, the simulation of blade No. 11 fatigue crack growth life was not performed and was ignored.

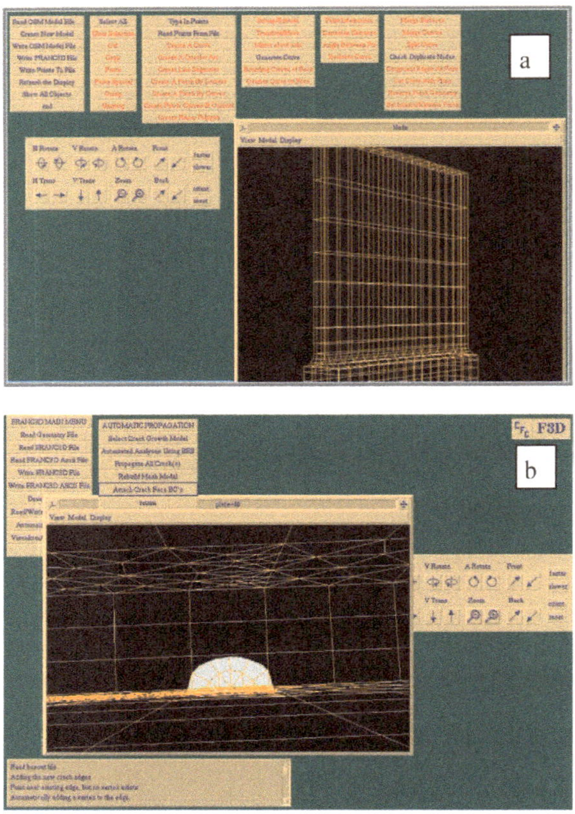

Figure 15. (a) A 3D model of the airfoil with 2D mesh for simulating fatigue crack growth in FRANC3D software and (b) simulated fatigue crack in blade No. 8 after three steps of propagation

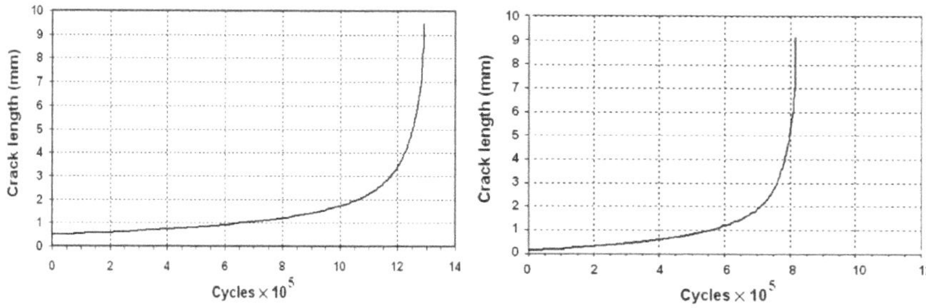

Figure 16. Crack length as a function of number of cycles in: (a) the first fractured blade (blade No. 1) and (b) the second fractured blade (blade No. 8)

As mentioned before, it is assumed that the excessive vibrations of the fan blades resulted from a resonant condition. Therefore, modal analysis was performed and natural frequencies and corresponding mode shapes were calculated. The results show that the first bending vibrational mode is a possible cause of the blades' failure. This first natural frequency calculated as 649 Hz (Ataei, 2006) So, the fatigue crack growth time for the first fractured blade with respect to the first natural frequency value (649 Hz) can be calculated as below:

$$t_{g1} = \frac{1.3 \times 10^6}{649 \times 3600} = 0.56 \ \ hour \tag{9}$$

and similarly for the second fractured blade:

$$t_{g2} = \frac{8 \times 10^5}{649 \times 3600} = 0.34 \ \ hour \tag{10}$$

Thus the total fatigue crack growth time is about one hour. This lifetime is within the 11 hours of actual operating period following the last overhaul, which ended with the failure of fan blades.

4. Conclusion

1. Based on fractographic observations, high cycle fatigue was identified as the primary cause of failure of fan blades.
2. Fatigue cracks initiated at large second phase particles.
3. Premature fatigue failure of the fan blades was caused by significant levels of additional bending stresses, probably due to excessive vibrations resulting from resonance.
4. The failure stress for the first fractured blade was estimated to be about 90 MPa, which is significantly higher than the normal operating stress (more than three times higher).

5. Based on crack growth simulations in the first and the second fractured blades, the total fatigue crack growth time was calculated as only about one hour, which is within the 11 hours of actual operating period following the last overhaul, ending with the failure of fan blades.

Nomenclature

A	cross-sectional area, m^2
a	crack depth, m
a_0	initial crack depth, m
c	length for corner cracks and half-length for surface cracks
a/c	crack aspect ratio
F	boundary correction factor on stress intensity factor for remote tension
H	bending multiplier on stress intensity factor for remote bending
K_{IC}	critical stress intensity factor, $\mathrm{MPa}\sqrt{\mathrm{m}}$
K_t	stress intensity factor due to tension, $\mathrm{MPa}\sqrt{\mathrm{m}}$
K_b	stress intensity factor due to bending, $\mathrm{MPa}\sqrt{\mathrm{m}}$
m	mass of blade, kg
N	rotational speed, revolutions per second
Q	crack shape factor
r	distance from center of rotation, m
t_g	fatigue crack growth time, hours
σ_t	tensile stress, MPa
σ_b	bending stress, MPa
$\Delta\sigma$	stress range, MPa
ΔK_{th}	threshold stress intensity factor range, $\mathrm{MPa}\sqrt{\mathrm{m}}$
ΔK	stress intensity factor range, $\mathrm{MPa}\sqrt{\mathrm{m}}$
$\frac{da}{dN}$	crack growth rate , m.cycle^{-1}
C	empirical material constant
n	empirical material constant

Author details

Mahmood Sameezadeh and Hassan Farhangi
University of Tehran, College of Engineering, Iran

Acknowledgement

The authors are grateful to Eng. H. Vahedi and Dr. E. Poursaeidi of the Iran Power Plant Repairs Company (IPPRC) for their support and stimulating discussions. The authors also acknowledge the helpful assistances of Eng. M. Vatanara and Eng. P. Ataei during the study.

5. References

American Society for Metals [ASM]. (1990), *ASM Handbook*, Vol. 2, 10th ed., ASM International, OH, ISBN o-87170-378-5

Anderson, T. (1995). *Fracture Mechanics Fundamental and Applications*, 2nd ed., CRC Press, ISBN 978-0849342608, Texas

Ataei, P. (2006). *Fatigue Fracture and Life Evaluation of Generator Rotor Fan Blades of a Thermal Power Plant*, MSc thesis, Islamic Azad University, South Tehran Branch, Iran

Bleier, F.P. (1997). *Fan Handbook, Selection, Application and Design*, McGraw-Hill, ISBN 978-0070059337, New York

Broek, D. (1995). *The Practical Use of Fracture Mechanics*, Kluwer Academic Publisher, ISBN 978-0792302230 ,OH

Carter, B.J.; Wawrzynek, P.A. & Ingrafera, A.R. (2000). Automated 3-D Crack Growth Simulation, International Journal for Numerical Methods in Engineering, Vol. 47, pp. 229-253, ISSN 0029-5981

Cornell Fracture Group [CFG]. (2002). *FRANC3D, Menu & Dialog Reference V2.2*, Cornell University, USA

Cornell Fracture Group [CFG]. (2002). *OSM, Menu and Dialog Reference V2.2*, Cornell University, USA

Department of Defense. (1998). *Metallic Materials and Elements for Aerospace Vehicle Structures*, MIL-HDBK-5H, New York

Hertzberg, R.W. (1989). *Deformation and Fracture Mechanics of Engineering Materials*, 3rd ed., Wiley, ISBN 978-0471012146, New York

Iran Power Plant Repair Company [IPPRC], (2003). *Failure Analysis Report of Iran-Neyshabour Unit 2B*, Iran

Iran Power Plant Repair Company [IPPRC], (2004). *Failure Analysis Report of Iran- Montazer-Ghaem Unit 4*, Iran

Koul, A.K. & Dainty, R.V. (1992). Fatigue fracture of aircraft engine compressor disks, In: *Handbook of Case Histories in Failure Analysis*, Vol. 1, K.A. Esaklul, pp. 241-245, ASM international, ISBN 978-0-87170-453-5, USA

Kung, C.Y. & Fine, M.E. (1979). Fatigue crack initiation and microcrack growth in 2024-T4 and 2124-T4 aluminum alloys, *Metallurgical and Materials Transactions A*, Vol. 10, pp. 603-610, ISSN 1073-5623

Merati, A. (2005). A study of nucleation and fatigue behavior of an aerospace aluminum alloy 2024-T3, *International Journal of Fatigue*, Vol. 27, pp. 33-44, ISSN 0142-1123

Moore, W.G. (2002). Several Case Histories of Generator Rotor Fan Failures and Methods of Repair, *Proceedings of international joint power generation conference*, ASME committee (Ed.), Phoenix, 2002

Moussavi, S.E.; Yadavar, S.M. & Jahangiri, A. (2009). Failure analysis of gas turbine generator cooling fan blades. *Engineering Failure Analysis*, Vol. 16, pp. 1686-1695, ISSN 1350-6307

Newman, J.C. & Raju, I.S. (1984). *Stress-intensity Factor Equations for Cracks in Three-dimensional Finite Bodies Subjected to Tension and Bending Loads*, NASA technical memorandum 85793, NASA, Virginia

Sameezadeh, M. (2005). *Analysis of Fatigue Failure of Generator Rotor Fan Blades*, MSc thesis, School of Metallurgy and Materials Engineering, University of Tehran, Iran

Chapter 11

Evaluating the Integrity of Pressure Pipelines by Fracture Mechanics

Ľubomír Gajdoš and Martin Šperl

Additional information is available at the end of the chapter

1. Introduction

Large engineering structures made with the use of sophisticated technology often include material defects and geometrical imperfections. These defects or imperfections do not exert their influence on the initial behaviour of structures designed in accordance with standard rules. Under the action of loading varying in time, however, they can reveal themselves in long-term operation by the initiation and growth of a fatigue crack from a defect root. Similarly, stress corrosion (SC) cracks can develop in a structure when there is an initial stress concentrator and the structure is exposed to both mechanical stress and a corrosion medium. A condition for the growth of a small fatigue crack is that the level of cyclic stress should be above the limit value given by barriers existing in a steel, and a condition for the growth of SC cracks is that the stress is greater than a certain limit value for a specific corrosion medium. It is important to pay due attention to the behaviour of cracks under various gas pipeline loading conditions in different environments, and to the influence of these conditions on the residual strength and life of the gas pipeline. The existence of a crack in the wall of a high-pressure gas pipeline mostly implies a shortened remaining period of reliable operation.

2. Theoretical treatment of cracks in pipes

At the present time, the manufacturing stage of pipes for gas pipelines includes sufficient flaw detection measures, and only products free of detectable material flaws are dispatched for operation. However, there are defects that are not revealed by the required inspection, and which manifest themselves during heavy-duty operation. The most dangerous defect is the occurrence of cracks – these are due to material defects that are difficult to reveal by a standard optical inspection. If the cracks are deep, and spread to a large extent, they can pose a threat to the pipeline operation. Using fracture mechanics it is possible to evaluate the threat that crack-like defects can pose to the pipeline wall, depending on whether a brittle, quasi-brittle or ductile material is involved. A model description of crack-containing systems, which relies on the stress intensity factor, K, can be used for brittle and quasi-brittle fracture, and also for subcritical fatigue growth, corrosion fatigue, and stress corrosion. In these cases, the surface crack is usually located in the field of one of the membrane tensile stress components, or in the field of bending stress, or in a combination of both. The extent of the plastic zone at the crack tip is small in comparison with the dimensions of the crack and the pipeline.

If the gas pipeline is made of a high-toughness material, the plastic strains become extensive before the crack reaches instability. Hence, some elasto-plastic fracture mechanics parameter, such as the J–integral or crack opening displacement, or a two-criterion method, should be employed to assess the threat that the crack poses to the pipeline wall. Although cracks of various directions may occur in the pipe wall, we will consider here only longitudinal cracks, because they are subjected to the biggest stress (hoop stress) in the pipe wall, and they are therefore the most dangerous (when we are considering the parent metal).

2.1. Stress intensity factor for a longitudinal through crack in the pipe wall

The first theoretical solution to the problem of establishing the stress intensity factor for a long cylindrical pipe with a longitudinal through crack under internal pressure was reported by Folias (Folias, 1969) and by Erdogan with Kibler (Erdogan & Kibler, 1969). They managed to show that the problem was analogous to that of a wide plane plate with a through crack. The only adjustment needed for transition to a pipe was to introduce a correction factor to multiply the solution for the plane plate. This factor, frequently referred to as the Folias correction factor and designated by symbol M_T, is only a function of the ratio $\lambda = c/\sqrt{Rt}$, where c is the crack half-length, R is the pipe mean radius, and t is the pipe wall thickness, and thus it depends only on the geometrical parameters of the crack and the pipe (Fig. 1).

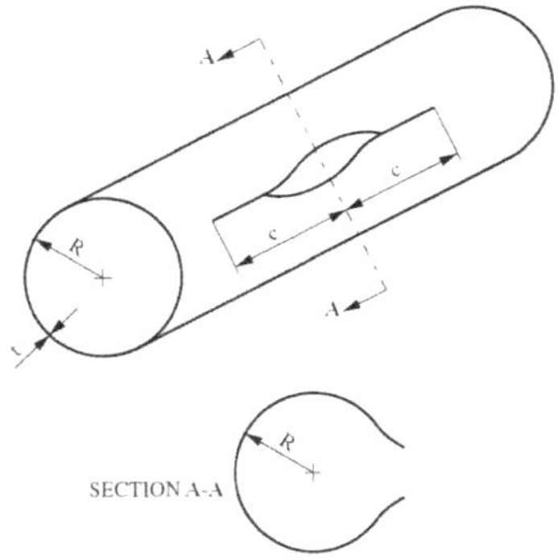

Figure 1. Deformation of a pressurized pipe in the vicinity of a longitudinal through crack

Several relations have been reported for determining the Folias correction factor. The following are the most frequently used at the present time:

The Folias relation (Folias, 1970):

$$M_T = \sqrt{1 + 1.255\lambda^2 - 0.0135\lambda^4} \tag{1}$$

The Erdogan et al. relation (Erdogan et al., 1977):

$$\begin{aligned} M_T &= 0.6 + 0.5\lambda + 0.4\exp(-1.25\lambda) \qquad \text{for} \quad \omega < 5 \\ M_T &= 1.761(\lambda - 1.9)^{0.5} \qquad \text{for} \quad \omega \geq 5 \end{aligned} \tag{2}$$

where $\omega = \sqrt[4]{12(1-\nu^2)}\lambda$

with ν denoting Poisson´s ratio

The following relation is the simplest:

$$M_T = \sqrt{1 + 1.61\lambda^2} \tag{3}$$

However, its validity is limited by the value $\lambda<1$.

If c is the half-length of a longitudinal through crack in the pipe, then the stress intensity factor of such a crack simply reads

$$K_I = M_T \sigma_\varphi \sqrt{\pi c} \tag{4}$$

where

$\sigma_\varphi = \pi D/2t$ is the hoop stress (D and t denoting pipe diameter and wall thickness, respectively), and

M_T is the Folias correction factor

2.2. Stress intensity factor for a longitudinal part-through crack

Various methods are used for analysing the problem of longitudinal semi-elliptical surface cracks in the wall of cylindrical shells (Fig. 2). As a 3D asymptotic solution to the stress intensity factor is virtually involved, the possibilities offered by accurate analytical procedures are confined to infinite or semi-infinite bodies. Solutions appropriate for finite bodies call for the application of approximate methods, such as the finite element method and the method of boundary integral equations, or various alternative methods (e.g. the weight function method).

The first solutions for semi-elliptical surface cracks in a plate subjected to uniaxial tension or steady bending were derived from solutions for an elliptical plane crack in an infinite 3D body. In order to account for the finite thickness of a body and the plastic zone at the crack tip, correction factors were introduced for the "front" surface and the "rear" surface of the body and for the plastic region at the crack tip (Shah & Kobayashi, 1973). However, solutions by different authors often showed rather considerable disagreement. Scott and Thorpe (Scott & Thorpe, 1981) therefore tested the accuracy of the solutions presented by various authors by measuring changes in the shape of a crack throughout its fatigue growth. They concluded that the best engineering estimation of the stress intensity factor for a part-through crack in a plate was provided by Newman's solution (Newman, 1973). An adjusted form of this solution for a thin-walled shell is given by:

$$K_I = \left[M_F + \left(E_{(k)}\sqrt{c/a} - M_F \right)\left(\frac{a}{t} \right)^s \right] \frac{\sigma_\varphi \sqrt{\pi a}}{E_{(k)}} M_{TM} \tag{5}$$

where

M_F is the function depending on the crack geometry (on the ratio a/c)

$E_{(k)} = \int_0^{\pi/2} \sqrt{1 - k^2 \sin^2\theta}\, d\theta$ is an elliptical integral of the second kind, k being $\sqrt{1 - \left(\frac{a}{c} \right)^2}$

s is the function depending on the crack geometry (the ratio a/c) and the relative crack depth (the ratio a/t)

$M_{TM} = \frac{\left(1 - \frac{a/t}{M_T} \right)}{(1 - a/t)}$ is the correction factor for the curvature of a cylindrical shell and for an increase in stress owing to radial strains in the vicinity of the crack root

In the last relationship, M_T is the Folias correction factor, determined by any of the relations (1) – (3). The functions M_F and s differ in form for the lowest point of the crack tip (point A in Fig. 2) and for the crack mouth on the surface of the cylindrical shell (point B in Fig. 2).

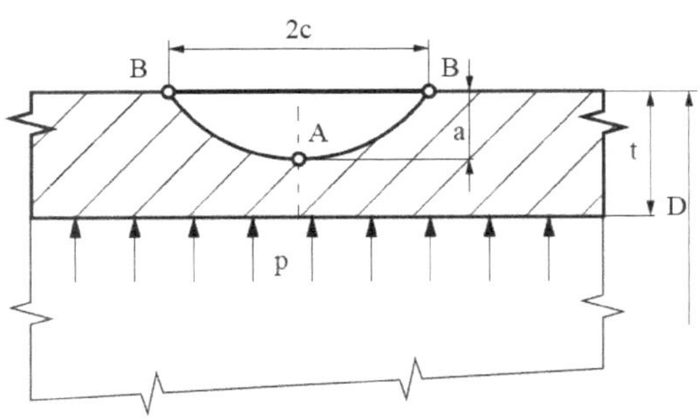

Figure 2. External longitudinal semi-elliptical crack in the wall of a cylindrical shell

2.3. Engineering methods for determining the J integral

2.3.1. *The FC method*

This method was proposed as the J_s method in Addendum A16 of the French nuclear code (RCC-MR, 1985). It stems from the second option for describing the transition state between ideally elastic and fully plastic behaviour of a material, i.e. from the function $f_2(L_r)$ of the R6 method (Milne et al., 1986). This function takes the form:

$$f_2\left(L_r\right)=\left(\frac{E\varepsilon_{ref}}{L_rR_e}+\frac{L_r{}^3R_e}{2E\varepsilon_{ref}}\right)^{-1/2} \tag{6}$$

where

$L_r = \sigma/\sigma_L$ (σ – applied stress, σ_L – stress at the limit load)
R_e is the yield stress
E is Young´s modulus
ε_{ref} is the reference strain corresponding to the reference (nominal) stress σ_{ref}

If we identify function $f_2(L_r)$ with function $f_3\left(L_r\right)=\left(\frac{J}{J_e}\right)^{-\frac{1}{2}}$ and express L_r as σ_{ref} / R_e and the elastic J integral J_e as K^2/ E', where $E' = E$ for plane stress state and $E' = E / (1-\nu^2)$ for plane strain state, we have:

$$J=\frac{K^2}{E'}\left(\frac{E.\varepsilon_{ref}}{\sigma_{ref}}+\frac{\sigma_{ref}{}^3}{2E.R_e{}^2.\varepsilon_{ref}}\right) \tag{7}$$

The stress σ_{ref} in the above equation is a nominal stress – i.e. a stress acting in the plane where the crack occurs. Taking into consideration the description of the stress-strain dependence by the Ramberg-Osgood relation (8) and adjusting Eq. (7), we obtain the J-integral in the form (9).

$$\frac{\varepsilon}{\varepsilon_0}=\frac{\sigma}{\sigma_0}+\alpha\left(\frac{\sigma}{\sigma_0}\right)^n \tag{8}$$

$$J=\frac{K^2}{E'}\left[A+\frac{0.5\left(\sigma/\sigma_0\right)^2}{A}\right] \tag{9}$$

where

$$A=1+\alpha\left(\frac{\sigma}{\sigma_0}\right)^{n-1} \tag{10}$$

In the above equations the stress σ_0 can be substituted by the yield stress R_e ; $\varepsilon_0 = \sigma_0 / E$; α, n – material constants

As a pipeline is a body of finite dimensions, stress σ in Eqs. (9) and (10) is a nominal stress – i.e. a stress acting in the plane where the crack occurs. Referring to the R6 method (Milne et al., 1986), this stress for a pipe containing a longitudinal part-through thickness crack may be written as:

$$\sigma = \frac{\sigma_\phi}{1 - \dfrac{\pi a c}{2t\left(t + 2c\right)}} \tag{11}$$

In eq. (11), $\sigma_\phi = \dfrac{pD}{2t}$ is the hoop stress, and the meaning of the symbols a, c, and t is clear from Fig.2.

2.3.2. GS method

The GS method was derived by Gajdoš and Srnec (Gajdoš & Srnec, 1994) on the basis of the limit transition of the *J*-integral, formally expressed for a semi-circular notch, to a crack, with the variation of the strain energy density along the notch circumference being approximated by the third power of the cosine function of the polar angle. If the stress-strain dependence is further expressed by the Ramberg-Osgood relation (8), with $\varepsilon_0 = \sigma_0 / E$, (α, n – material constants), we can arrive at Eq. (12)

$$J = \frac{K^2}{E'}\left[1 + \frac{2\alpha n}{\left(n+1\right)}\left(\frac{\sigma}{\sigma_0}\right)^{n-1}\right] \tag{12}$$

where σ is the nominal stress in the reduced cross-section of a body. For a pipe containing a longitudinal part-through thickness crack it may be determined by relation (11).

3. Consideration of the constraint

As mentioned above, the situation existing at the crack tip in conditions of small-scale yielding can be characterized by a single fracture parameter (e.g. K, J or δ). This parameter can be used as a fracture criterion, independent of geometry. However, single-parameter fracture mechanics fails in cases of developed plasticity, where fracture toughness is a function not only of the material, but also of the dimensions and the geometry of the specimen. It is well known from the theory of fracture mechanics that for small-scale yielding the maximum stress existing at the crack tip in a non-hardening material is about $3\sigma_0$, where σ_0 is the yield stress. Single-parameter fracture mechanics apparently does not apply to non-hardening materials under fully plastic conditions, because the stress and strain fields in the vicinity of the crack tip are affected by configurations of both the body and the crack. The situation is more favourable in hardening materials, where single-parameter fracture mechanics may approximately apply also for the developed plasticity, provided that the body maintains a high level of stress triaxiality.

The reported experimental studies suggest that the configuration of the specimen and the crack (the crack depth and the specimen dimensions, in particular) affect the fracture toughness in a brittle state. However, the fact that this configuration can also influence the R-curve of ductile materials is not so well known.

Generally, the bigger the dimension of the crack, the smaller the resistance of the material to fracture will be. The R-curve obtained on specimens with rather long cracks is, as a rule, below the R-curve obtained on specimens with rather short cracks. For this reason, standards require that the relative crack lengths be within a comparatively narrow range of values for valid values of fracture toughness J_{in}.

3.1. The J – Q theory

Some researchers dealing with fracture mechanics tried to extend the theory of fracture mechanics beyond the boundaries of the assumptions of single-parameter fracture mechanics, introducing other parameters to provide a more accurate characterization of conditions at the crack tip. One of the parameters is the so-called T-stress, which is a uniform stress acting axially (in the direction of the x-axis) in front of the crack tip in an isotropic elastic material loaded by the first mode, i.e. the opening mode, of the load. In this case, the stress field in front of the crack tip may be written as:

$$\sigma_{ij} = \frac{K_I}{\sqrt{2\pi r}} f_{ij}\left(\Theta\right) + T\delta_{1i}\delta_{1j} \tag{13}$$

The elastic T-stress heavily affects the shape of the plastic zone and the stress deep in this zone. T-stress values are linked with the stress biaxiality ratio, β, defined as

$$\beta = \frac{T\sqrt{\pi a}}{K_I} \tag{14}$$

It can be mentioned by way of illustration that the stress biaxiality ratio β equals -1 for a through crack in an infinite plate loaded by a normal stress applied far away from the crack plane. By implication, this remote stress, σ, induces a T-stress in the direction of the x-axis, whose magnitude is -σ. In an elastic case, positive values of the T-stress generally lead to a high constraint under fully elastic conditions, whereas a geometry with a negative T-stress leads to a rapid drop in the constraint as the load rises. For different geometries, the stress biaxiality ratio β can be used as a qualitative index for a relative constraint at the crack tip.

The so-called J – Q theory provides another approach to the extension of single-parameter fracture mechanics beyond the conditions of its validity. This theory aims to describe the stress field at the crack tip deep in the plastic zone. It is a well-known fact that if the small-strain theory is used, the stress field at the crack tip in the plastic zone can be described by a power series, in which the so-called HRR solution is the leading term (Hutchinson, 1968), (Rice & Rosengren, 1968). The other terms of higher magnitudes, when summed up, provide a difference stress field, which approximately corresponds to a uniform hydrostatic shift of the stress field in front of the crack tip. It has become customary to designate the amplitude of this approximate difference stress field with letter Q, according to its authors O´Dowd and Shih (O´Dowd & Shih, 1991). O´Dowd and Shih defined the Q parameter as:

$$Q \equiv \frac{\sigma_{yy} - \left(\sigma_{yy}\right)_{HRR\ or\ T=0}}{\sigma_0} \tag{15}$$

for $\Theta = 0$ and $\frac{r\sigma_0}{J} = 2$

The parameter is equal to zero (Q = 0) under small-scale yielding conditions, but it acquires negative values as the load (and in consequence the strain) grows. Classical single-parameter fracture mechanics assumes that fracture toughness is a material constant. However, the J-Q theory suggests that the critical value of the *J*-integral for a given material depends on the Q parameter – i.e. $J_c = J_c(Q)$ – and that fracture toughness is thus not some single-value quantity, but rather a function that defines the critical values of the *J*-integral and the Q parameter (Shih et al., 1993). Although the relation between critical J-integral values and the Q parameter shows a considerable scatter, the critical value of the *J*-integral tends in general to drop as the Q parameter increases in value.

The theory of single-parameter fracture mechanics assumes that the fracture toughness values obtained on laboratory specimens can be applied to a body. However, two-parameter approaches, such as the J-Q theory, reveal that the specimen must be tested at the same constraint as that of the body with a crack. In other words, the two geometries must have the same Q value at the moment of fracture, so that the corresponding critical values of the *J*-integral, J_{cr}, will be equal to each other. Since J_{cr} values are often scattered to a large extent, we cannot make a clear-cut prediction of this quantity. It is only possible to predict a certain range of plausible J_{cr} values for a given body or structure.

It should also be noted that the J-Q approach is only descriptive, and not predictive. This implies that the Q parameter quantifies the constraint at the crack tip, without providing any indication of the particular influence of the constraint on the fracture toughness. Two-parameter theories cannot be strictly correct as far as their universality is concerned, because they assume two degrees of freedom. Recent research into the influence of the constraint at the crack tip on fracture toughness indicates that geometries with a low constraint can in many cases be judged by a two-parameter theory, and geometries with a high constraint can be judged by a single-parameter theory (Ainsworth & O´Dowd, 1995).

3.2. Plastic constraint factor on yielding

A simple procedure based on the use of the so-called plastic constraint factor on yielding, C, can be applied to determine the fracture conditions in a thin-walled pressure pipeline. The factor is given by the ratio of the stress needed to obtain plastic macrostrains under constraint conditions to the yield stress at a homogeneous uniaxial state of stress (Gajdoš et al., 2004). The C factor can be expressed by the relation (16)

$$C = \frac{\sigma_1}{\sigma_{HMH}} \tag{16}$$

where σ_{HMH}, the Huber-Mises-Hencky stress, is put equal to the yield stress.

Let us now consider the state of stress at the crack tip in a thick-walled body, where the stress perpendicular to the crack plane, σ_1, and the stress in the direction of the crack, σ_2, are equal, and the stress in the direction of the thickness of the body, σ_3, is governed by the expression $\sigma_3=\nu(\sigma_1+\sigma_2)$. Then, based on the HMH criterion and assumed elastic conditions ($\nu \cong 0.33$), the plastic constraint factor $C \approx 3$. If the stress in the thickness direction, σ_3, falls between $2\nu\sigma_1$ and zero (thin-walled body), the value of the plastic constraint factor will range between $C = 1$ and $C = 3$. This data can be used to assess the fracture conditions in gas pipelines with surface part-through cracks, employing a C-factor which has to be experimentally determined. After the C factor has been determined, the value of $C\sigma_0$ would be used instead of the yield stress σ_0 in relations for calculating the J-integral. The C factor was experimentally investigated at the Institute of Theoretical and Applied Mechanics of the Academy of Sciences of the Czech Republic in the framework of a broader research project on the reliability and operational safety of high pressure gas pipelines. Fracture conditions were investigated on five pipe bodies, made of steels X52, X65 and X70, with cycling-induced cracks. Data on the pipe bodies that were used, the cracks in the walls, and the mechanical and fracture-mechanical material properties of the bodies are given in Table 1.

Material	X 52	X 65	X 65	X 70	X 70
D (mm)	820	820	820	1018	1018
t (mm)	10.2	10.7	10.6	11.7	11.7
c (mm)	50	100	100	127	115
a (mm)	7.0	7.7	7.0	6.7	7.1
a/t	0.686	0.720	0.660	0.573	0.607
a/c	0.14	0.077	0.07	0.053	0.062
p (MPa)	8.05	9.71	9.86	9.86	9.55
$p/p_{0.2}$	1.034	0.750	0.769	0.800	0.775
σ_0 (MPa)	313	496	496	536	536
α	2.40	5.34	5.34	5.92	5.92
n	6.25	8.45	8.45	9.62	9.62
C	2.1	2.4	2.4	2.0	2.07
J_{cr} (N/mm)	487	432	432	439	439
$-T/\sigma_0$	0.672	0.575	0.544	0.606	0.611
$-Q$	0.667	0.591	0.546	0.648	0.651

Table 1. Summary of data on the assessment of the fracture behaviour of model pipe bodies

The rows in the table show the following data (top to bottom): body diameter D, body wall thickness t, half-length of a longitudinal part-through crack c, crack depth a, relative crack depth a/t, aspect ratio a/c of a semi-elliptical crack, fracture pressure p, ratio of fracture pressure p and pressure $p_{0.2}$ corresponding to the hoop stress at the yield stress, yield stress in the circumferential direction of the body σ_0, Ramberg-Osgood constant α, Ramberg-Osgood exponent n, plastic constraint factor C, J-integral critical value J_{cr}, determined as J_m (corresponding to attaining the maximum force at the "force – force point displacement" curve), T-stress to yield stress ratio T/σ_0, and the Q parameter. Values of σ_0, α and n were derived from tensile tests, and the values of J_{cr} were derived from fracture tests run on CT specimens. Fracture pressure values p were read at the moment the ligament under the crack in the pipe body ruptured. Values for the plastic constraint factor on yielding, C, were determined on the basis of the J-integral in such a way that agreement was reached between the predicted and experimentally established fracture parameters for the given crack and fracture toughness of the material. The J-integral value was calculated using the GS method (Gajdoš & Srnec, 1994), on the one hand, and on the basis of the French nuclear code (RCC-MR, 1985), on the other.

It should be noted that in determining the C factor, the critical J-integral value established on CT specimens was considered – namely J_{cr} = 439 N/mm for steel X70, J_{cr} = 432 N/mm for steel X65 and J_{cr} = 487 N/mm for steel X52. It was found by a computational analysis of the CT specimens, employed to construct the R curve, that the Q parameter for these specimens was Q = 0.267. A comparison of this with the Q parameter for pipe bodies (Q ≈ -0.55 ÷ -0.65) reveals that the constraint in the CT specimens was much higher. This implies that the real fracture toughness – i.e. the critical value of the J-integral, J_{cr} – was higher in the pipe bodies. The real C factor for a cracked pipe body is lower, so that the J-a curve for a pipe body is steeper than the curve for CT specimens with a greater C factor (Gajdoš & Šperl, 2011). Due to this, the J-integral for the axial part-through crack reaches the corresponding higher fracture toughness (for a lower constraint) for the same crack depth as the J-integral with a higher C factor reaches lower fracture toughness (determined on CT specimens). The situation is illustrated in Fig.3.

The normalized T–stress values in Table 1 were obtained using the plane solution – i.e. a solution for a crack of infinite length oriented longitudinally along the pipe. The problem was solved at the Institute of Physics of Materials, Brno, by the finite element method. The solution consisted of two steps: (i) a corresponding FEM network was established and corresponding boundary conditions were formulated for each crack depth, (ii) the magnitudes of the stress intensity factor and the T-stress were calculated for each FEM network by means of the CRACK2D FEM system with hybrid crack elements. The Q parameter values were derived from the Q – T/σ_0 curves obtained by O'Dowd and Shih (O'Dowd & Shih, 1991), by modified boundary layer analysis for different values of the strain coefficient (Ramberg-Osgood exponent, n). Strictly speaking, the Q parameter values from Table 1 do not correspond accurately to the values for the examined cracks, because the T-stresses were not computed for real semi-elliptical cracks, but for cracks spreading along the entire length of the pipe body ($a/c \approx 0$). Nevertheless, due to the fact that the ratio of the depth to the surface half-length of the examined cracks (*a*/*c*) was close to zero (*a*/*c*=0.053÷0.14), we can assume that the differences between the real values of the *Q* parameter and the values listed in Table 1 will be small.

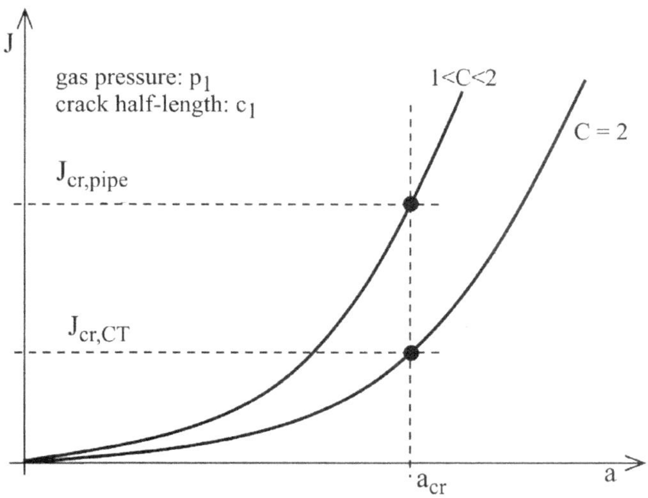

Figure 3. Schematic J-a dependence, (i) for a CT specimen, and (ii) for a pipe with an axial part-through crack

The figures shown in Table 1 provide an idea of the nature of the changes both in the plastic constraint factor, *C*, and in the *Q* parameter brought about by changes in the relative crack length, *a*/*t*. The diagrams shown in Figs. 4 and 5 can be obtained on the basis of the graphic representation of the pairs $C - a/t$ and $Q - a/t$.

These diagrams clearly show the trends of the changes in the two parameters with a change in the relative crack depth, *a*/*t*. It follows that, in the range of relative depths examined here (*a*/*t* = 0.57 to 0.72), the plastic constraint factor, *C*, and the *Q* parameter are a growing function of the relative crack depth, *a*/*t*, the $Q - a/t$ dependence being rather weak. Expressed simply (i.e. linearly), the following relations are involved:

$$C = 2.56\, a/t + 0.53 \tag{17}$$

$$Q = 0.32\, a/t - 0.83 \tag{18}$$

The high scatter of the *C* and *Q* values is (i) due to differences in the cross section dimensions of the DN800 and DN1000 pipes and (ii) due to different values of the strain exponent *n* in the Ramberg-Osgood relation, because the pipes were made of three different materials.

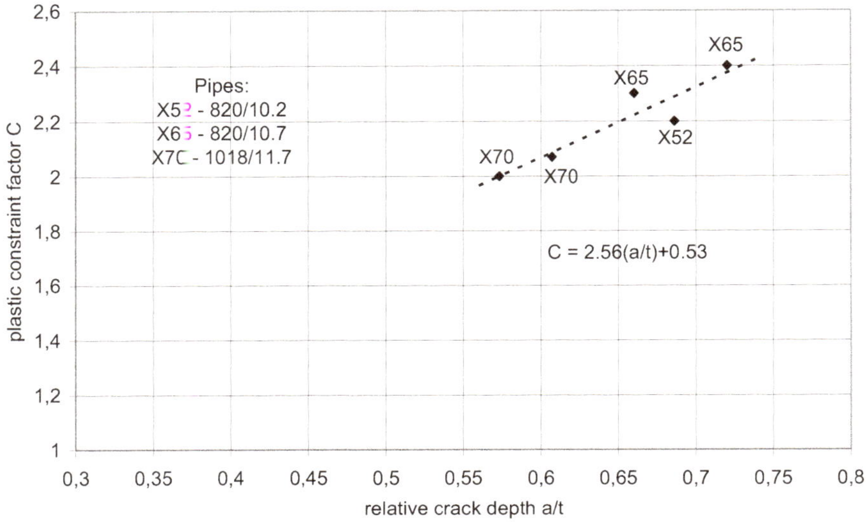

Figure 4. Plastic constraint factor, *C*, as affected by the relative crack depth, *a/t*

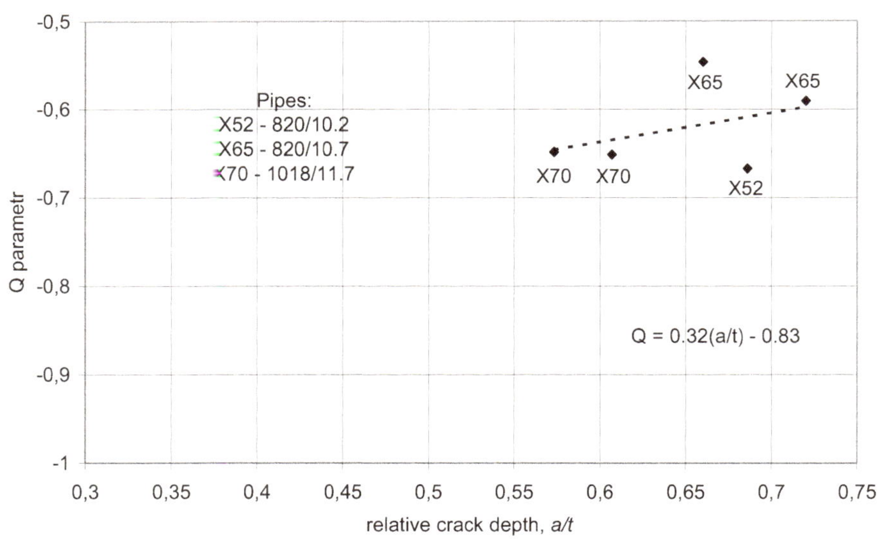

Figure 5. Parameter Q, as affected by the relative crack depth, *a/t*

Table 1 lists explicit values of Q and C for all examined cracks in the pipes that were used, and thus a graphic representation of the $C - Q$ relation can be plotted (Fig. 6). In the region where the established values of parameter Q for the examined pipe bodies are found, the $C - Q$ relation can be most simply described by the linear relation:

$$C = 3.4\,Q + 4.3 \tag{19}$$

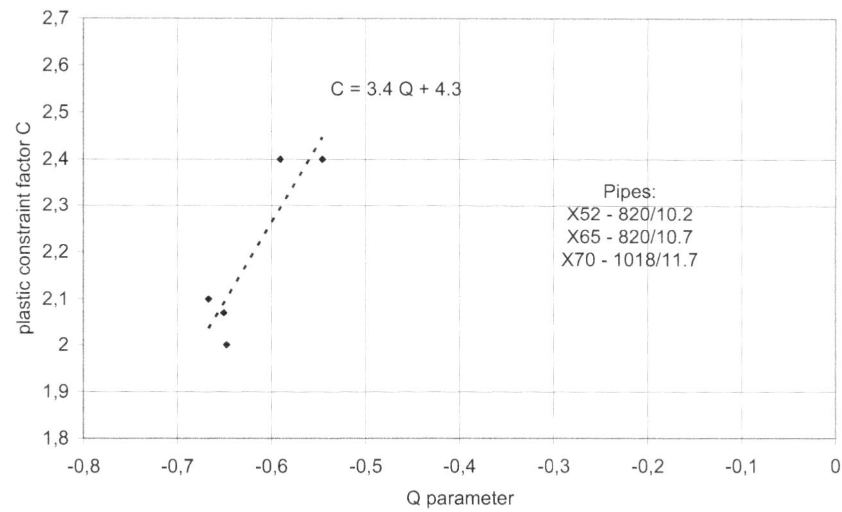

Figure 6. Dependence of the plastic constraint factor, C, on parameter Q

The relation implies that the plastic constraint factor, C, decreases with a decreasing value (increasing negative value) of parameter Q. The observed scatter of the experimental points is mainly due to inaccuracies of the T-stress estimate, which result from the substitution of the real conditions of cracks of certain lengths by the plane solution used in the task (crack along the entire length of the body).

4. Fracture toughness

If we are to evaluate the strength reliability and the remaining life of gas pipelines, we need to get an accurate picture of the properties of the material that the gas pipelines are made of. In the case of gas pipelines operated for different periods of time, we should be aware that the properties of the material of a used pipeline will be different from the initial properties.

In order to pass a qualified judgement on the reliability of a gas pipeline, we should know the true properties that the material displays at the time when the gas pipeline is being examined. The fracture properties can be characterized with sufficient generality by the fracture toughness, determined by quantities J_{in}, $J_{0.2}$, or J_m, where J_{in} is the so-called initiation magnitude of the J integral for a stable subcritical crack extension; $J_{0.2}$ is the J magnitude corresponding to the real crack extension Δa = 0.2 mm, and J_m is the magnitude of the J integral corresponding to attaining the maximum force at the "force – force point displacement" curve. We should point here to two aspects of fracture toughness that can be encountered when dealing with pressure pipelines. One of them is the effect of pipe band straightening, and the other is the effect of stress corrosion cracks on fracture toughness.

4.1. The effect of straightening

Fracture toughness tests are carried out with fracture mechanical specimens, e.g. single edge notched bend (SENB) specimens or compact tension (CT) specimens. Both types are plane specimens. When investigating the integrity of thin-walled pressure pipelines, we face the problem of ensuring the planeness of the semiproducts for manufacturing the fracture mechanical specimens. The only way is press straightening of pipe bands taken from the pipe that is under investigation. As a consequence of the plastic deformation that the semi-product undergoes during straightening, internal stresses are induced not only in the semi-product but also in the final specimens. Therefore there are still some doubts about the reliability of the fracture toughness characteristics obtained with straightened specimens. In order to verify this matter, Gajdoš and Šperl (Gajdoš & Šperl, 2012) carried out an experimental investigation of fracture toughness, as determined using press straightened CT specimens and curved CT specimens, manufactured directly from a pipe band, i.e. ensuring that their natural curvature and wall thickness were preserved.

The so-called curved CT specimens (see Fig. 7) to some extent simulate the stress conditions in the pipe wall upon loading by internal pressure. In order to apply a circumferential force on these specimens, we used a special testing rig, similar to that developed by Evans (Evans et al., 1995). The rig is shown in Fig. 8.

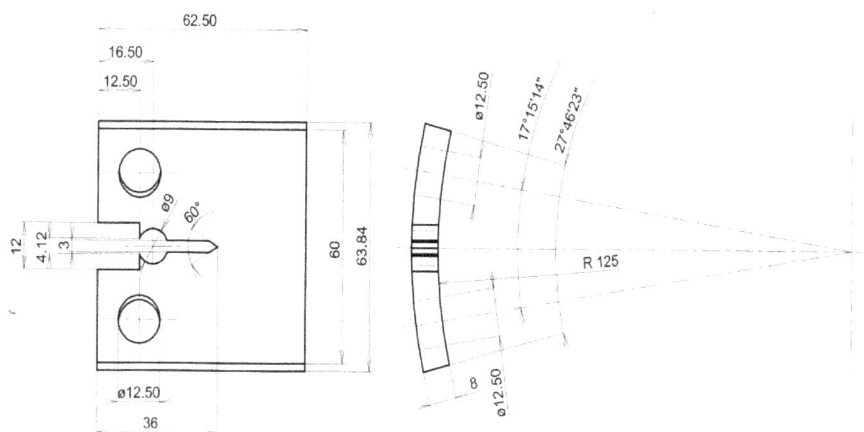

Figure 7. The shape and dimensions of the curved CT specimens

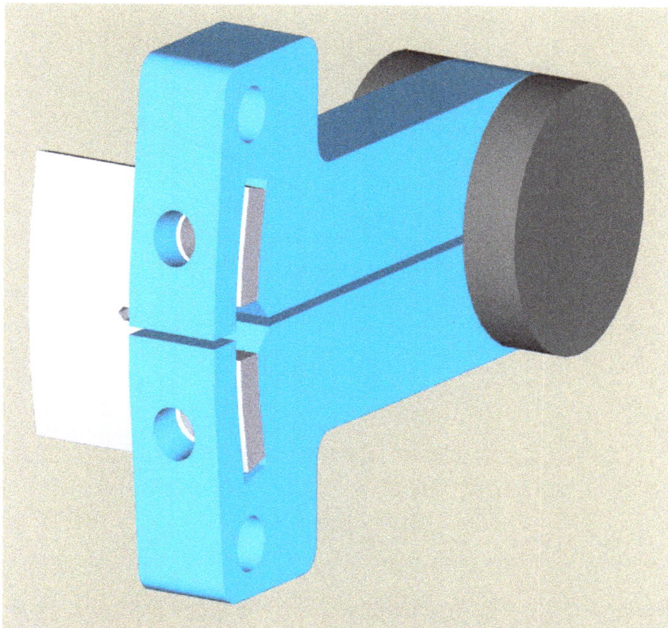

Figure 8. The testing rig for circumferential loading of a curved CT specimen

It is clear that the testing rig is tied with only certain cross – sectional dimensions of a pipe. In the case considered here, the dimensions corresponded to a pipe 266 mm in outside diameter and 8 mm in wall thickness. The material of the pipe was low-C steel CSN 411353. Static tests of the steel provided the following results: $R_{p0.2}$ = 286 MPa; R_m = 426 MPa; A_5 = 31%; Z = 54%. The Ramberg-Osgood constants had the following values: $\alpha = 6.23$; $n = 5.87$; σ_0 = 286 MPa.

First, fracture toughness tests were carried out by an ordinary procedure, as specified in the ASTM standard (E 1820-01, 2001), on CT specimens manufactured from a press-straightened band taken from the pipe.

The result in the form of an R-curve is presented in Fig. 9. One point (designated by a triangle) has not been included in the regression analysis because it was outside the valid area of the diagram. The positions of J_{in} and $J_{0.2}$ at the R-curve are clearly defined from the construction of the R-curve, the blunting line and the 0.2 offset line; the position of J_m is also indicated in the diagram, and it represents the mean of six values obtained on specimens where the maximum force was attained in loading the specimens. The R-curve determined by the least-square method is described by a power function (20):

$$J = 327.05\left(\Delta a\right)^{0.6406} \tag{20}$$

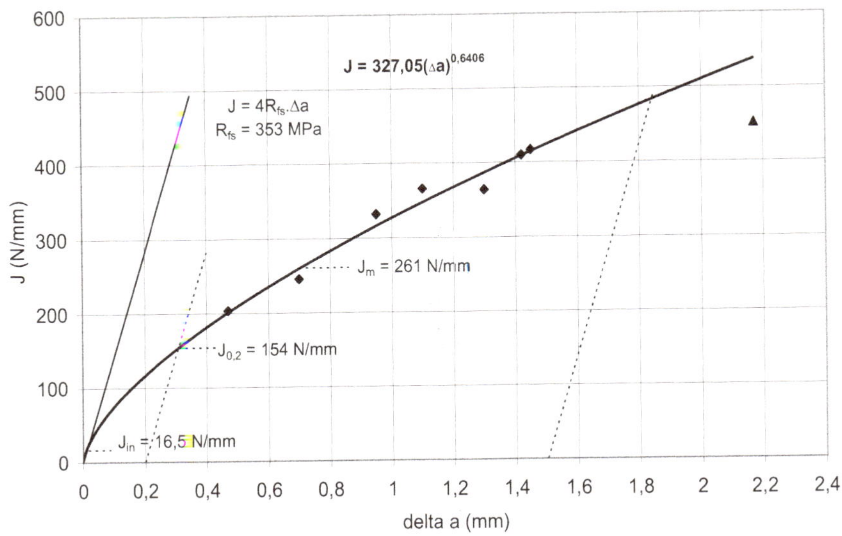

Figure 9. R curve for CT specimens manufactured from a press-straightened semi-product

Eight specimens were used for fracture-mechanical tests of curved CT specimens. Cracks were cycled up at a frequency of 4 Hz, using the testing rig, Fig. 8. The stress state in the inner side of a specimen was bigger than in the outer side, because of the bending moment induced by the out-of-axis action of the vertical component of the tangential force with regard to the intersection of the middle cylindrical area of the specimen with the symmetry plane of the specimen. For this reason, the growth rate of the fatigue crack was higher in the inner side than in the outer side.

This resulted in uneven length of the fatigue crack on the two sides of a specimen after finishing the cycling up. On one half of the specimens, a slant front of the starting notch was therefore made in such a way that the notch was 1 mm deeper on the outer side. By this operation, a much more even front of the fatigue crack was obtained. This is clearly demonstrated in Figs. 10 and 11, which show the fracture surfaces of specimens with a straight front and a slant front of the starting notch. In the two photographs, we can observe areas corresponding to the notch, fatigue, static crack extension and final break after the specimens were cooled down in liquid nitrogen.

On the basis of the finite element analysis and the compliance measurements made by Evans (Evans et al., 1995), it was concluded that the use of standard expressions for determining K factor will not cause error greater than 4% for curved CT specimens. By proceeding in the same way as in standard $J - \Delta a$ testing, an R-curve was obtained for curved CT specimens. It is described by a power function (21), and is presented in Fig. 12.

$$J = 278.21(\Delta a)^{0.525} \tag{21}$$

Figure 10. Fracture surface – straight front of the notch

Figure 11. Fracture surface – slant front of the notch

A comparison of the two R-curves shows that the decline of the R-curve obtained with the curved CT specimens is less than the decline of the R-curve obtained with plane, i.e. straightened, CT specimens. The higher decline of the R-curve with the straightened CT specimens is most probably connected with work hardening of a semiproduct during straightening. In the mathematical description of the R-curve of the curved CT specimens, not only the exponent but also the constant is less than for the standard R-curve. This means that the standard R-curve is situated above the R-curve of the curved CT specimens. However, the lower position of the R- curve for the curved CT specimens does not mean significantly lower magnitudes of the fracture toughness characteristics. For example, the J_m value is lower by 1.1%, the $J_{0.2}$ is lower by less than 3%, and the magnitude J_{in} is even higher than the respective characteristics for plane (straightened) CT specimens. In absolute units, the difference is 2.9 N/mm for J_m and 4.6 N/mm for $J_{0.2}$. There is a significant difference in J_{in}, namely 29.7 N/mm in favour of the curved CT specimens.

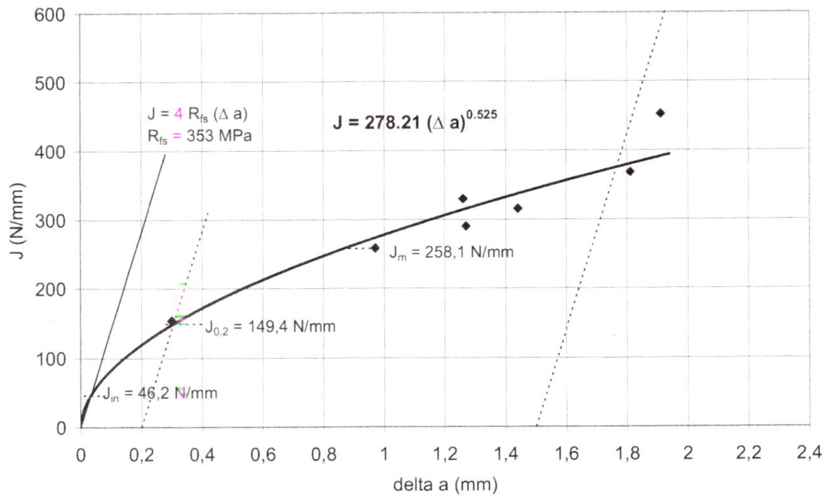

Figure 12. R-curve for curved CT specimens

By accounting the scatter of the results in the form of the J - Δa points, caused both by a natural process of subcritical crack growth and by inaccuracies in determining the J-integral and, in particular, the crack extension during monotonic loading of a specimen, it can be stated with a high level of reliability that the fracture toughness of a pipe material determined on straightened CT specimens is practically the same as the fracture toughness determined on curved CT specimens.

4.2. The effect of stress corrosion

(Gajdoš et al., 2011) investigated the stress corrosion fracture toughness of gas pipeline material, and compared it with fatigue fracture toughness. The material used for the investigation was a low-C steel according to CSN 411353 (equivalent to ASTM A519), containing 0.17% C, 0.035% P, 0.035% S. The test CT specimens were manufactured from a real pipe section cut out from a DN 150 gas pipeline 4.5 mm in wall thickness while it was being repaired after 20 years of operation. Before the CT specimens were manufactured, the pipe section was press straightened. Owing to the small thickness of the specimens (a low constraint), the fracture toughness values cannot be qualified to represent the real fracture toughness values. However, they can be used as a comparative measure of fracture toughness, thus enabling quantification of the effect of stress corrosion cracks on the apparent fracture toughness.

The CT specimens were first cyclically loaded by a routine procedure used in determining fracture toughness; the only difference was that the cycling was stopped when the growth of the fatigue crack reached approximately the magnitude $\Delta a_{FA} \approx 1.5$ mm. After that, the CT specimens were put into the stress-corrosion (SC) crack generator with an acidic solution according to the NACE Standard (NACE Standard TM0177, 2005). This solution consisted of 50 g NaCl (sodium chloride) + 5 g CH_3COOH (acetic acid) + 945 g H_2O, and during the generating process it was bubbled by H_2S (hydrogen sulphide). A constant force F of 3 kN was applied to the specimens. The corresponding level of the nominal stress (tension and bending) at the fatigue crack tip exceeded the yield stress $R_{p0.2}$ by about 25%. The crack length increment due to stress-corrosion Δa_{SC} was determined with the help of the relations for elastic crack-edge displacements at CT specimens. In total, three groups of CT specimens were prepared. The first group (A) was the reference group; the specimens from this group contained only the fatigue crack. The second group of CT specimens (B) contained specimens that were left freely in air at the indoor temperature for two weeks after being removed from the SC crack generator, and were then subjected to fracture toughness tests. The specimens from the third group (C) were tested immediately after they had been removed from the SC crack generator (the time difference between testing the first specimen and the last specimen being approximately 20 minutes).

The results confirmed that the fracture resistance of a component (given by the apparent fracture toughness) depends not only on the material of the component and on the crack tip constraint (the thickness of the wall of the component) but also on the origin of the crack (fatigue, stress corrosion), and thus on the corresponding crack growth mechanism. In contradiction with the opinion that low-C steels are not susceptible to stress corrosion cracking our results showed that under conditions specified in (NACE Standard TM0177, 2005) stress corrosion cracks can also be generated from fatigue cracks in low-C steels such as CSN 411353. Unlike a fatigue crack, the occurrence of a stress-corrosion crack in a component means a significant decrease in the fracture toughness characteristics while the crack is exposed to stress corrosion conditions, and a partial "recovery" of the fracture toughness when the stress corrosion conditions are removed. The results for all three groups of specimens are summarized in Fig. 13.

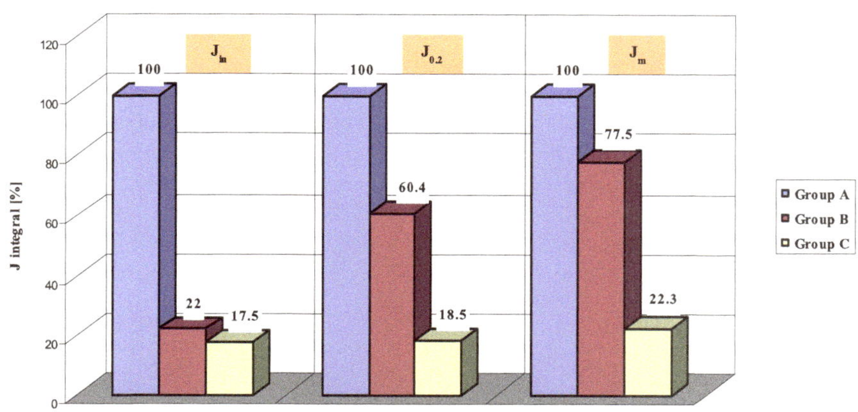

Figure 13. A bar chart of the J integral values for specimens of groups A, B and C

As this figure shows, the stress corrosion fracture toughness characteristics for the low-C steel CSN 411353 were lower than the fatigue fracture toughness characteristics by a factor ranging between 4.5 (J_m value) and 5.7 (J_{in} value). However, a two-week recovery period made it possible to recover their fracture properties to some extent, namely the J-integral J_m to almost 80%, the J-integral $J_{0.2}$ to about 60%, and the J-integral J_{in} to about 22% of the fatigue crack J-integral values. It follows from here that in evaluating the reliability of gas pipelines it is always necessary to examine the character of the cracks in the pipe wall, and in the case of stress corrosion cracks to take into account that the fracture toughness can be drastically lower than the values determined on specimens with cracks of fatigue origin.

5. Burst tests

An experimental verification of the fracture conditions of gas pipelines can be made most accurately on a test pipe body cut out of the gas pipeline to be examined. When deciding on the length of the test pipe body, we should bear in mind that the working length of the body (characterized by the absence of stress effects from welded-on bottoms) will be shorter by 2 x $2.5\sqrt{(Rt)} \approx 3.5\sqrt{(Dt)}$. It is usually sufficient for the distance between the welds of dished bottoms to be at least $3.5D$. This length permits a number of starting cuts to be placed axially along the length of the body. The cuts are made to initiate crack growth when the body is subsequently pressurized by a fluctuating pressure. The cuts can be made in several ways, one of which uses a thin grinding wheel. The smallest real functional thickness of such a wheel is about 1.2 mm, and the corresponding width of the cuts made with it is approximately 1.5 mm. Depending on the type of pipes of which gas pipelines are built (seamless, spirally welded, longitudinally welded), the starting cuts can be provided in the base material, in the transition region or in the weld metal, their orientation being axial, circumferential or along the spiral weld.

5.1. Preparation of test pipe bodies

It is appropriate to relate the surface length of the cuts to the wall thickness of the pipe body. Testing the body for the danger posed by so-called long cracks should be carried out with crack lengths not exceeding twenty times the wall thickness of the pipe body. The situation with the depth of the starting cuts is different. The depth of an initiated fatigue crack must be at least 0.5 mm along the whole perimeter of the cut tip, so that the cut with the initiated crack at its tip can be considered as a crack after the pipe body has been subjected to cycling. This value follows from the work done by Smith and Miller (Smith & Miller, 1977). If such a crack a_t in size finds itself in a notch root defined by depth a_v and radius of the roundness ρ (see Fig. 14), this configuration can be regarded as a surface crack a_e in depth, where

$$
\begin{aligned}
a_e &= \left(1 + 7.69\sqrt{\frac{a_v}{\rho}}\right) a_t \quad for \quad a_t < 0.13\sqrt{a_v \rho} \\
a_e &= a_v + a_t \quad for \quad a_t \geq 0.13\sqrt{a_v \rho}
\end{aligned}
\tag{22}
$$

It is evident that for $a_t \geq 0.13\sqrt{a_v \rho}$, a cut with a crack along the perimeter of the cut tip can be taken for a crack with a depth of $a_v + a_t$. For the cut width $2\rho = 1.5$–2.0 mm and the notch depth $a_v = 6$–10 mm (in relation to the wall thickness), we find that the fatigue increment of the size of the initiated crack, a_t, should be greater than about 0.5 mm.

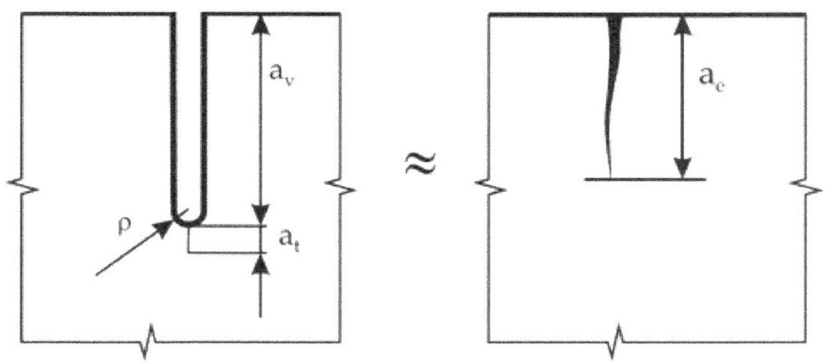

Figure 14. Substitution of a notch with a crack by the equivalent crack

As described in paragraph 3.2, three test pipe bodies, made of X52, X65 and X70 steels, were provided with working slits and so-called check slits, which were of the same surface length as the working slits but their depth was greater. These check slits functioned as a safety measure to prevent cracks that developed at the working slits from penetrating through the pipe wall. For illustration, a DN1000 test pipe body with a working length of 3.5 m is shown in Fig. 15. The check slits are denoted in Fig. 15 by a supplementary letter K. The material of the test pipe body is a thermo-mechanically treated steel X70 according to API specification. The pipe is spirally welded, the weld being inclined at an angle of $\varphi = 62°$ to the pipe axis. It is provided with starting cuts oriented either axially or in the direction of the strip axis (i.e. in the direction of the spiral) and then along or inside the spiral weld. The cuts differ in length ($2c$ = 115 mm or 230mm) and in depth (a = 5, 6.5, 7, and 7.5 mm). We are particularly interested in axial (longitudinal) slits situated aside welds, because these are sites where axial cracks will be formed in the basic material of the pipe.

Efforts were made in the fracture tests to keep the circumferential fracture stress below the yield stress, because the operating stress in gas pipelines is virtually around one half of the yield stress (and does not exceed two-thirds of the yield stress even in intrastate high-pressure gas transmission pipelines). Calculations reveal that in order to comply with this, the depth of the axial semi-elliptical cracks should be greater than one half of the wall thickness. Oblique cracks should be even deeper, as the normal stress component opening these cracks is smaller. If the crack depth is to have a certain magnitude before the fracture test is begun, the depth of the starting slit should be smaller than this magnitude by the fatigue extension of the crack along the perimeter of the slit tip. At the same time, we should bear in mind that the greater the fatigue extension of the crack, the better the agreement with a real crack.

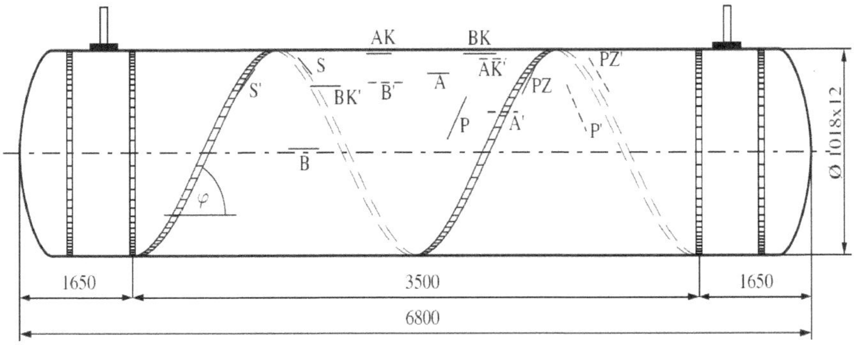

Figure 15. Test pipe body with the starting cuts marked

5.2. Prediction of fracture parameters

After the starting slits were made, the test pipes were subjected to water pressure cycling to produce fatigue cracks in the tips of the starting slits. The cycling was carried out in a pressurizing system, which included a high-pressure water pump, a collecting tank, a regulator designed to control the amount of water that was supplied and, consequently, the rate at which the pressure is increased in the pipe section. This was effected by opening by-pass valves.

In cycling the cracks, the water pressure fluctuated between p_{min} = 1.5 MPa and p_{max} = 5.3 MPa, and the number of pressure cycles was between 3 000 and 4 000. The period of a cycle was approximately 150 seconds. The cycling continued until a crack initiated in one of the check slits became a through crack. This moment was easy to detect, because it was accompanied by a water leak. By choosing an appropriate difference between the depths of the working slits and the check slits it was possible to obtain a working crack depth (= starting slit depth + fatigue crack extension) of approximately the required size. To run a test for a fracture, however, it was necessary to remove the check slit which had penetrated through the wall of the test pipe from the body shell and to repair the shell, e.g. by welding a patch in it.

After removing the check slit with a crack which penetrated through the wall, and repairing the shell of the test pipe, the pipe was loaded by increasing the water pressure to burst. The test procedure, which was common for all test pipes, will now be briefly described for the DN1000 pipe shown in Fig. 15. As the figure suggests, slits A, A′, B and B′ were oriented along the axis of the pipe. The nominal length of notches B, B′ was twice as long as notches A, A′, but notches B, B′ were shallower. As was mentioned above, the cracks at the slit tips were extended by fluctuating water pressure, and this proceeded until the cracks from the check slits (BK, BK′) grew through the wall and a water leak developed. Then the damaged parts of the shell were cut out, patches were welded in their place, and the test pipe was monotonically loaded to fracture at the location of crack B or B′. The burst of the test pipe at crack B is shown in Figs. 16 and 17 (as a detail). A part of the fracture surface is shown in Fig. 18.

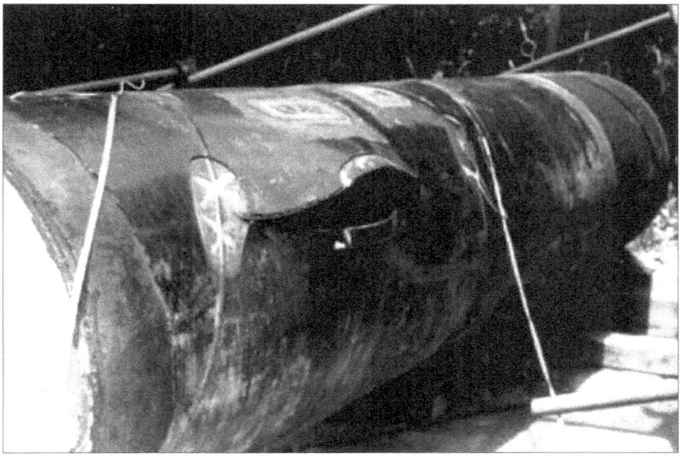

Figure 16. Burst initiated on slit B with a fatigue crack

Figure 17. Burst initiated on slit B – a detail

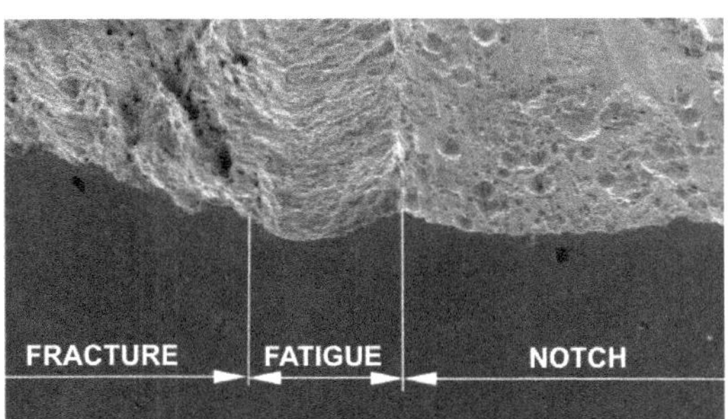

Figure 18. A part of the fracture surface of crack B (fatigue region ~ 2.4 mm)

Evidently, at the instant of fracture the crack spread not only through the remaining ligament, but also lengthwise. After removing the part of the pipe shell with crack B, a patch was welded in and the second burst test followed. Table 2 extracts from Table 1 the numerical values of the geometrical parameters, the J-integral fracture values, the Ramberg-Osgood constants, the fracture pressure and the fracture depth for cracks B and B′, respectively.

Characteristics	Crack B	Crack B′
CRACK DIMENSIONS		
half-length, c (mm)	115	127
depth in fracture, a_f (mm)	7.1	6.7
RAMBERG-OSGOOD PARAMETERS		
α / n /σ_0 (MPa)	5.92 / 9.62 /536	5.92 / 9.62 /536
FRACTURE TOUGHNESS		
$J_{cr} = J_m$ (N/mm)	439	439
FRACTURE PRESSURE		
p_f (MPa)	9.55	9.86

Table 2. Some characteristics referring to crack B and crack B′

It should be noted that Table 2 includes the Ramberg-Osgood constants for the circumferential direction of the test pipe, with the crack oriented axially in the pipe. This is because the stress-strain properties perpendicular to the crack plane are crucial in determining the J-integral for an axial crack. The stress-strain dependence in the circumferential direction should therefore be taken into account where an axial orientation of the crack is concerned. The most important fracture test results from the viewpoint of the fracture conditions are the magnitudes of the fracture pressure, p_f, and the fracture depth, a_f, for a given crack length $2c$. It follows from Table 2 that p_f = 9.55 MPa and a_f = 7.1 mm for crack B, and p_f = 9.86 MPa and a_f = 6.7 mm for crack B′. These values are also shown in the last two columns of Table 1.

Now let us predict the fracture conditions according to engineering approaches, and compare the prediction results with the real fracture parameter values (pressure, crack depth). The procedure for verifying the engineering methods for the predictions involves determining either the fracture stress for a given (fracture) crack depth, or the fracture crack depth for a given (fracture) pressure. To illustrate this, we select the latter case – i.e. determining the fracture depth of a crack for a given (fracture) pressure. Fig. 19 shows the J-integral vs. crack B depth dependences, as determined by the FC and GS predictions for the fracture hoop stress given by the measured fracture pressure. When using equations (9), (10), and (12) to determine J-integrals, the following parameters were used for the calculation: D = 1018 mm; t = 11.7 mm ; $p = p_f$ = 9.55 MPa; c = 115 mm; α = 5.92; n = 9.62; σ_0 = 2.07×536 = 1110 MPa (i.e. C = 2.07). Fig. 20 shows similar dependences for crack B′.

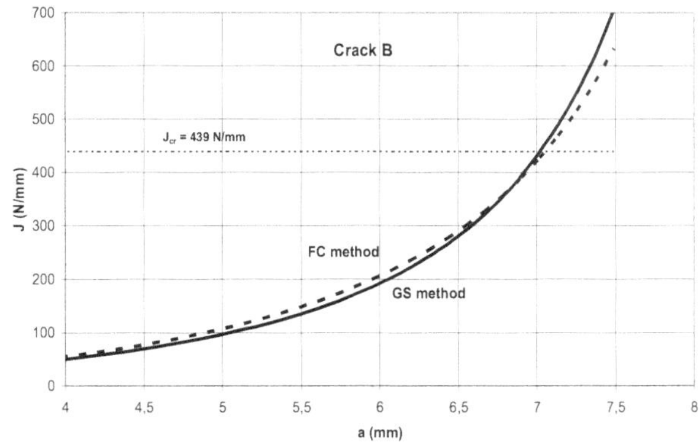

Figure 19. Prediction of the fracture depth for crack B ($p = p_f = 9.55$ MPa and C = 2.07)

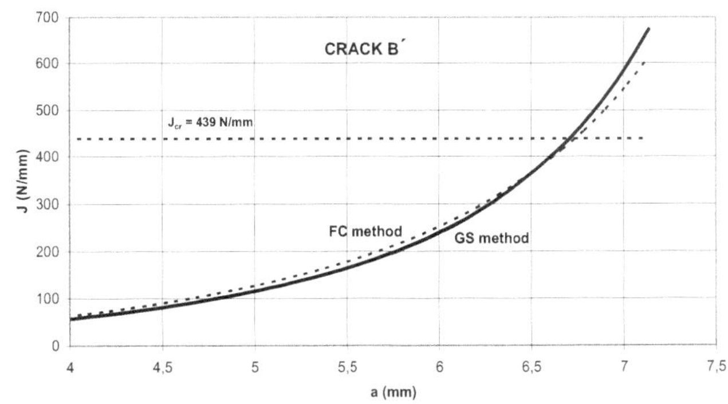

Figure 20. Prediction of the fracture depth for crack B´ ($p = p_f = 9.86$ MPa and C = 2.0)

The same computational parameters as those employed in the case of crack B were used in the equations to determine the J-integral according to the FC and GS methods, with the exception of the fracture pressure (p_f = 9.86 MPa), the crack half-length (c = 127 mm) and factor C (C = 2.0). As is evident from Fig. 19, the intersection of the straight line $J = J_{cr} = 439$ N/mm with the two $J - a$ curves gives the value $a_{cr} \approx 7.05$ mm, which is well consistent with crack depth B a_{cr} = 7.1 mm, established experimentally. Similarly, the intersection of the straight line $J = J_{cr} = 439$ N/mm with the J–a curves according to the FC and GS procedures in Fig. 20 shows the fracture crack depth a_{cr} to be virtually identical to the experimentally found fracture depth a_f = 6.7 mm. For other test pipes, namely DIA 820/10.7, made of X65 steel, and DIA 820/10.2, made of X52 steel, various magnitudes of the plastic constraint factor C were obtained to achieve good agreement of the geometric parameters at fracture with the experimental parameters. They are illustrated in Fig. 4. The conclusion can thus be drawn that very good agreement of the fracture parameter values predicted by the FC and GS engineering approaches with the values found experimentally can be achieved when using the plastic constraint factor on yielding, C, at the level $C = 2$. If a higher value of the C factor provides more precise results, the use of the value C = 2 will yield a conservative result.

6. Conclusion

A specific fracture-mechanics-based procedure for assessing the integrity of pressurized thin-walled cylindrical shells made from steels includes a theoretical treatment for cracks in pipes. On the basis of both experimental work and a fracture-mechanical evaluation of experimental results, an engineering method has been worked out for assessing the geometrical parameters of critical axial crack-like defects in a high-pressure gas pipeline wall for a given internal pressure of a gas. The method makes use of simple approximate expressions for determining fracture parameters K, J, and it accommodates the crack tip constraint effects by means of the so-called plastic constraint factor on yielding. Involving this in the fracture analysis leads to multiplication of the uniaxial yield stress by this factor in the expression for determining the J-integral. Two independent approximate equations for determining the J-integral provided very close assessments of the critical geometrical dimensions of part-through axial cracks. With the use of the crack assessment method, the critical gas pressure in a pipeline can also be determined for a given crack geometry.

The fracture toughness with which the J-integral is compared in fracture analysis is determined using fracture mechanics specimens (e.g. CT, SENB and others). Experiments made on press-straightened CT specimens and on curved CT specimens with a natural curvature, made from pipe 266/8 mm of low-C steel CSN 411353, showed that straightening a pipe band prior to the machining of CT specimens had a practically negligible effect on the fracture toughness characteristics ($J_{0.2}$, J_m). However, experiments with fracture toughness testing of specimens with stress corrosion cracks, formed by the hydrogen mechanism, showed a dramatic reduction of all fracture toughness characteristics in comparison with fracture toughness determined on specimens with fatigue cracks, e.g. the quantities J_{in}, $J_{0.2}$ and J_m dropped to 17.5%, 18.5%, and 22.3%, respectively. A partial "recovery" of fracture toughness characteristics was observed when the stress corrosion conditions were removed.

Author details

Ľubomír Gajdoš and Martin Šperl
Institute of Theoretical and Applied Mechanics, Academy of Sciences of the Czech Republic, Czech Republic

Acknowledgement

Financial support from Research Plan AV0Z 20710524 and from grant-funded projects GACR P105/10/2052 and P105/10/P555 are highly appreciated.

7. References

Ainsworth, R. A. & O'Dowd, N. P. (1995). Constraint in the Failure Assessment Diagram Approach for Fracture Assessment. *Transactions ASME – Journal of Pressure Vessel Technology*, Vol.117, pp. 260-267

ASTM Standard E 1820-01 (2001). *Standard Test Method for Measurement of Fracture Toughness*, 2001

Erdogan, F. & Kibler, J. J. (1969). Cylindrical and Spherical Shells with Cracks. *International Journal of Fracture Mechanics*, Vol.5, No.3, pp. 229-237

Erdogan, F.; Delale, F. & Owczarek, J. A. (1977). Crack Propagation and Arrest in Pressurized Containers. *Journal of Pressure Vessel Technology*, Vol.99, (February 1977), pp. 90-99

Evans, J. T., Kotsikos, G. & Robey, R. F. (1995). A method for Fracture Toughness Testing Cylinder Material. *Engineering Fracture Mechanics*, Vol.50, No.2, pp. 295-300

Folias, E. S. (1969). On the Effect of Initial Curvature on Cracked Flat Sheets. *International Journal of Fracture Mechanics*, Vol.5, No.4, pp. 327-346

Folias, E. S. (1970). On the Theory of Fracture of Curved Sheets. *Engineering Fracture Mechanics*, Vol.2, No.2, pp. 151-164

Gajdoš, Ľ. & Srnec, M. (1994). An Approximate Method for J Integral Determination. *Acta Technica CSAV*, Vol.39, No.2, pp. 151-171

Gajdoš Ľubomír et al. (2004). *Structural Integrity of Pressure Pipelines*. Transgas, 80-86616-03-7, Prague, Czech Republic

Gajdoš, Ľ. & Šperl, M. (2011). Application of a Fracture-Mechanics Approach to Gas Pipelines. *Proceedings of World Academy of Science, Engineering and Technology*. Vol.73, January 2011, pp. 480 - 487

Gajdoš, Ľ., Šperl, M. & Siegl, J. (2011). Apparent Fracture Toughness of Low-Carbon Steel CSN 411353 as Related to Stress Corrosion Cracks. *Materials and Design*, Vol.32, No.8-9, pp. 4348-4353

Gajdoš, Ľ. & Šperl, M. (2012). The Effect of Straightening a Curved Body on Its Fracture Properties (in Czech). In: *Proceedings of the 21st Colloquium "Safety and Reliability of Gas Pipelines"*, Prague, 2012

Hutchinson, J. W. (1968). Singular Behaviour at the End of a Tensile Crack Tip in a Hardening Material. *Journal of the Mechanics and Physics of Solids*, Vol.16, pp. 13-31

Milne, I.; Ainsworth, R. A.; Dowling, A. R. & Stewart, A. T. (1986). *Assessment of the Integrity of Structures Containing Defects*. CEGB Report No. R/H/R6 – Rev.3, Central Electricity Generating Board, London, U.K., (1986)

NACE Standard TM0177 (2005). *Laboratory Testing of Metals for Resistance to Sulfide Stress Cracking and Stress Corrosion Cracking in H_2S Environments*. Item No. 21212, 2005

Newman, J. C. (1973). Fracture Analysis of Surface and Through-Cracked Sheets and Plates. *Engineering Fracture Mechanics*, Vol.5, No.3, pp. 667-689

O'Dowd, N. P. & Shih, C. F. (1991). Family of Crack-Tip Fields Characterized by a Triaxiality Parameter – I. Structure of Fields. *Journal of the Mechanics and Physics of Solids*, Vol.39, pp. 898-1015

RCC – MR (1985). *Design and Construction Rules for Mechanical Components of FBR Nuclear Island*. First Edition (AFCEN–3-5 Av. De Friedeland Paris 8), (1985)

Rice, J. R. & Rosengren G. F. (1968). Plane Strain Deformation near a Crack Tip in a Power-Law Hardening Material. *Journal of the Mechanics and Physics of Solids*, Vol.16, pp. 1-12

Scott, P. M. & Thorpe, T. W. (1981). A Critical Review of Crack Tip Stress Intensity Factors for Semi – Elliptical Cracks. *Fatigue of Engineering Materials and Structures*, Vol.4, No.4, (1981), pp. 291 – 309

Shah, R. C. & Kobayashi, A. S. (1973). Stress Intensity Factors for an Elliptical Crack Approaching the Surface of a Semi – Infinite Solid. *International Journal of Fracture*, Vol. 9, (1973), pp. 133-146

Shih, C. F.; O'Dowd, N. P. & Kirk, M. T. (1993). A Framework for Quantifying Crack Tip Constraint. In: *Constraint Effects in Fracture*. ASTM STP 1171, American Society for Testing and Materials, Philadelphia, pp. 2-20

Smith, R. A. & Miller, K. J. (1977). Fatigue Cracks at Notches. *International Journal of Mechanical Sciences*, Vol.19, pp. 11-22

Structural Reliability Improvement Using In-Service Inspection for Intergranular Stress Corrosion of Large Stainless Steel Piping

A. Guedri, Y. Djebbar, Moe. Khaleel and A. Zeghloul

Additional information is available at the end of the chapter

1. Introduction

One of the important degradation mechanisms to be considered for alloyed steels, especially austenitic stainless steels, is stress corrosion cracking. This mechanism causes cracking in the material due to the combined action of corrosion and mechanical stresses. The purpose of this chapter is to apply probabilistic fracture mechanics (PFM) to the analysis of the influence of Remedial Actions on structural reliability. PFM provides a technique for estimating the probability of failure of a structure or component when failures are considered to occur as the result of the sub critical and catastrophic growth of an initial crack-like defect. Such techniques are inherently capable of treating the influence of non-destructive inspections. For these reasons, PFM is becoming of increased usefulness in analysis of the reliability of modem structures.

Various papers in the literature addressed the probabilistic failure analysis of components subjected to Stress Corrosion Cracking (SCC). Probabilistic failure analysis of nuclear piping of BWR plant was carried out by You and Wu (You Jang-Shyong, et al., 2002). Ting (Ting K, 1999) analyzed the crack growth due to Intergranular Stress Corrosion Cracking (IGSCC) in stainless steel piping of BWR plants. Zhang et al. (Zhang S et al., 2002) carried out experimental investigations to determine the time to crack initiation and crack propagation velocity for IGSCC in sensitized type AISI304 stainless steel in dilute sulphate solutions. From the statistical analysis results obtained by Zhang et al. (Zhang S et al., 2002), it was seen that the time of crack initiation follows an exponential distribution, where as the crack growth rate follows a Weibull distribution. Harris et al. (Harris et al., 1992) developed a computer code named PRAISE (piping reliability analysis, including seismic events) for estimating the probability of pipe leak age under SCC. Rahman (Rahman S. 1997) has developed another computer code named PSQUIRT (probabilistic see page quantification of upsets in reactor tubes) to determine the probability of leak age of piping made of stainless steel and carbon steel subjected to intergranular stress corrosion cracking (IGSCC) and corrosion fatigue. Failure probabilities of a piping component subjected to SCC was computed in (Priya C et al., 2005 and Guedri et al., 2009a , 2009b) using a Monte Carlo simulation (MCS) technique.

In this chapter, a piping component made of AISI 304 stainless steel is considered for the analysis, because this type of steel is highly susceptible to SCC. Empirical relations are used to model the initiation and early growth rate of cracks. Fracture mechanics concepts are used deterministically in this study. Although the general methodology recommended in the computer program Piping Reliability Analysis Including Seismic Events (pc-PRAISE) (Harris et al., 1996) is used for the modeling of initiation and propagation of cracks, the American Society of Materials (ASM) recommendations (American Society of Materials, 1996) are used for computing the stress intensity factors. The microstructural properties of the material and operating conditions like pressure and temperature show variations during the lifetime of the pipe. In order to account for these variations, degree of sensitization, applied stress, initiation time of cracks, crack growth velocity after initiation, and initial crack length are considered as random variables.

The present calculations build on past studies by the nuclear power industry: General Electric Company (GEC) (General Electric Company, 1982a, 1982b) and Nuclear Regulatory Commission (NRC) (Hazelton , 1988), which have addressed both Inter-Granular Stress-Corrosion Cracking (IGSCC) causes and preventive actions. A critical issue has been the difficulty of using Ultrasonic Testing (UT) to detect IGSCC. Pacific Northwest National Laboratory (PNNL) past research on Non-Destructive Examination (NDE) reliability has included systematic studies to quantify NDE effectiveness. The first statistically based data for the Probability of Detection (POD) of IGSCC in weld of stainless steel piping were generated based on a piping inspection round robin (Doctor et al., 1983). The resulting POD data related to crack size and other important variables covered the performance of several inspection teams participating in the round robin. The first part of this chapter describes the stress-corrosion cracking model used in the pc-PRAISE (Harris et al., 1981, 1986a, 1986b) for simulating the initiation and growth of IGSCC cracks. This model is based on laboratory data from IGSCC tests in combination with calibration of the model using field data from pipe-cracking experience. PNNL has improved on the prior calibrations (Harris et al., 1986b) by making adjustments to the modeling of plant loading/unloading cycles in addition to adjustments to residual stress levels. The crack detection data (POD curves) by use of the pc-PRAISE model are also described. A parametric approach was adopted in the present calculations to characterize IGSCC by a single damage parameter (D_{sigma}). This parameter depends on residual stresses, environment conditions, and degree of sensitization. In the second part, a matrix of calculations to address an example of large pipe size, materials, and service conditions was developed. The results of these calculations quantify the reductions in failure probabilities that can be achieved with various In-Service Inspection (ISI) strategies. The final subpart of this chapter presents conclusions regarding ISI effectiveness as a mitigation action to enhance the reliability of Boiler Water Reactor piping.

2. Methodology

2.1. Overview of the PRAISE code

The present section describes probabilistic fracture mechanics calculations that were performed for selected components using the PRAISE computer code. The calculations address the failure mechanisms of stress corrosion cracking and intergranular stress corrosion cracking for components and operating conditions that are known to make particular components susceptible to cracking. Comparisons with field experience showed that the PRAISE code predict relatively high failure probabilities for components under operating conditions that have resulted in field failures. It was found that modeling assumptions and inputs tended to give higher calculated failure probabilities than those derived from data on field failures. Sensitivity calculations were performed to show that uncertainties in the probabilistic calculations were sufficiently large to explain the differences between predicted failure probabilities and field experience.

The first version of PRAISE (Harris et al., 1981) was developed in the 1980s by Lawrence Livermore National Laboratory under contract to NRC, with the initial application to address seismic-induced failures of large-diameter reactor coolant piping. This version of the code addressed failures (small leaks and ruptures) associated with fabrication flaws in welds that were allowed to grow as fatigue cracks until they either caused the pipe to leak or exceed a critical size needed to result in unstable crack growth and pipe rupture. The next major enhancement to the code (Harris et al., 1986b) addressed IGSCC and simulated both crack initiation and crack growth. The enhanced code allowed for crack initiation at multiple sites around the circumference of a girth weld and simulated linking adjacent cracks to form longer cracks more likely to cause larger leaks and pipe ruptures. In the early 1990s a version of PRAISE (pc-PRAISE) was developed to run on personal computers (Harris and Dedhia, 1992). The mid-1990s saw the development of methods for risk-informed in-service inspection, for which there were many new applications of PRAISE. A new commercial version of PRAISE (win PRAISE) was made available by Dr. David Harris of Engineering Mechanics Technology that simplified the input to the code with an interactive front end (Harris and Dedhia, 1998). During this same time period, PNNL made numerous applications of PRAISE to apply probabilistic fracture mechanics to support the development of improved approaches to in-service inspection (Khaleel and Simonen, 1994a; 1994b; 2000; Khaleel et al., 1995; Simonen et al., 1998; Simonen and Khaleel, 1998a; 1998b). The objective of this work was to ensure that changes to inspection requirements could be justified in terms of reduced failure probabilities for inspected components. Other work at PNNL for NRC (Khaleel et al., 2000) involved evaluations of fatigue critical components that could potentially attain calculated fatigue usage factors in excess of design limits (usage factors greater than unity).The most recent upgrades to PRAISE (Khaleel et al., 2000) were developed to support these fatigue evaluations, with the upgrade consisting of a model similar to that for IGSCC but directed at predicting the probabilities of initiating fatigue cracks. This new model was used to develop the technical basis for changes to Appendix L of American Society of Mechanical Engineers (ASME) Section XI that addresses fatigue critical locations in pressure boundary components (Gosselin et al., 2005). The PRAISE code has been extensively documented, successfully applied to a range of structural integrity issues, and has been available since the 1980s as a public domain computer code. However, the code has not been maintained and upgraded in an ongoing manner. Upgrades have been performed to meet the needs of immediate applications of the code and as such have served to fill very specific gaps in capabilities of PRAISE.

Other probabilistic fracture mechanics codes for piping have been developed to calculate failure probabilities for piping. The SRRA code (Bishop 1997; Westinghouse Owners Group, 1997) developed by Westinghouse follows much the same approach as the PRAISE code, but is limited to failures associated with cyclic fatigue stresses considers only preexisting fabrication flaws. Fatigue crack initiation has been approximated by assuming a very small initial crack, but with only one initiation site per weld. Stress corrosion cracking is similarly treated by postulating a very small initial crack, and growing the crack according to user-specified parameters for a crack growth equation. The SRRA code includes an importance sampling procedure that gives reduced computation times compared to the Monte Carlo approaches used by PRAISE. Also the model can simulate uncertainties in a wide range of parameters such as the applied stresses. The European NURBIM (Brickstad et al., 2004) has looked at a number of codes including PRAISE as part of an international benchmarking study. Included were a Swedish code NURBIT (Brickstad and Zang, 2001), the PRODIGAL code from the United Kingdom (Bell and Chapman, 2003), a code developed in Germany by GRS (Schimpfke, 2003), a Swedish code ProSACC (Dillstrom, 2003) and another code (STRUEL) from the United Kingdom (Mohammed, 2003).

This review concluded that none of the other benchmarked codes provided capabilities significantly different than or superior to the capabilities of PRAISE. In any case, the predictions of all such codes are limited in large measure by the quality of the values that can be established for the input parameters, as well as the validation with service experience.

2.2. Stress corrosion cracking model

SCC is a corrosion mechanism that forms cracks in susceptible material in the presence of an aggressive environment and tensile stresses. SCC can be intergranular or transgranular in nature depending on the material, level of stress, and environment (ASME, 1998). Austenitic stainless steels are commonly used in power generating industries because of high ductility and fracture toughness (American Society of Materials, 1996).This type of steel is used in applications where corrosion resistance is an important characteristic.

The methodology recommended in pc-PRAISE for modeling IGSCC in pipe is followed in this study. PRAISE model the occurrence of IGSCC by considering it as a two-stage process, namely, (1) crack initiation and (2) crack propagation.

2.2.1. Crack initiation model

2.2.1.1. Time to initiation

Time to initiation of stress corrosion crack is considered as a function of damage parameter, D_{sigma}, which represent effects of loading, environment and material variables on IGSCC. The damage parameter is given by

$$D_{sigma} = f_1(material) \times f_2(environment) \times f_3(looding) \tag{1}$$

where f_1, f_2 and f_3 are given by

$$f_1 = C_1 (Pa)^{C_2} \tag{2}$$

where Pa is a measure of degree of sensitization, given by Electrochemical Potentiokinetic Reactivation (in C/cm^2).

$$f_2 = O_2^{C_3} \exp\left[C_4/(T+273)\right] \log\left(C_5 \gamma^{C_6}\right) \tag{3}$$

where O_2 is oxygen concentration in ppm, T is temperature in degrees centigrade, and γ is water conductivity in $\mu s/cm$. The loading term f_3 is considered to be a function of stress. For constant applied load case, f_3 is given by

$$f_3 = \left(C_8 \sigma^{C_9}\right)^{C_7} \tag{4}$$

where σ is stress in MPa. C_1 to C_9 are constants whose values depend on type of material. Values for these constants are presented in Table1.

Constant	C1	C2	C3	C4	C5	C6	C7	C8	C9
Value	23.0	0.51	0.18	−1123.0	8.7096	0.35	0.55	2.21×10^{-15}	6.0

Table 1. Numerical values of constants C_i used for predicting the initiation and propagation of SCC for AISI304 (Harris et al., 1992).

The time to initiation t_I for a given D_{sigma} is considered as a random variable following lognormal distribution. The mean and standard deviation of $Log\,(t_I)$ are given by:

Mean value of

$$Log(t_I) = -3.10 - 4.21 \log\left(D_{sigma}\right)$$

Standard deviation of

$$Log(t_I) = 0.3081 \tag{5}$$

2.2.1.2. Crack size at initiation

In pc-PRAISE, shape of surface crack initiated due to IGSCC is considered to be semi-elliptical (Fig.1), which is also consistent with shapes of stress corrosion cracks reported by Helie (Helie M, et al., 1996) and by Lu (Lu B T et al., 2005). Surface length of initiated cracks, ($l = 2b$), is assumed to be log normally distributed with a median value of 3.175*mm* and a shape parameter of 0.85 (Harris et al., 1996). Depth of initiated crack is taken to be 0.0254 *mm*.

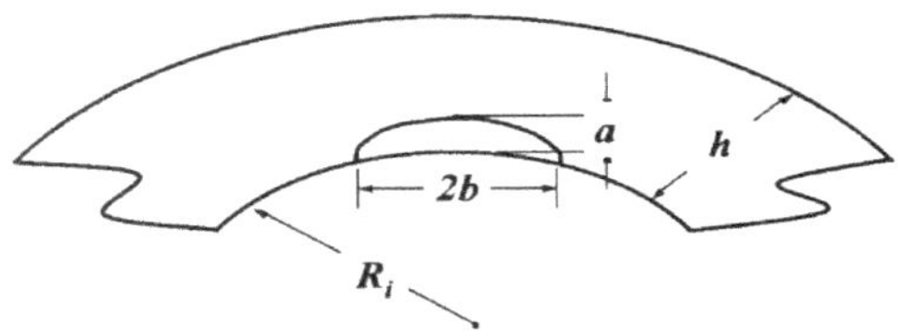

Figure 1. Geometry of the part-through circumferential crack considered

2.2.2. Crack growth model

The growth of very small cracks that have just initiated cannot be treated from a fracture mechanics standpoint (Andresen PL, 1994). Therefore, an initiation velocity is assigned to newly initiated cracks

$$Log(v_1) = J + G\log(D_{sigma}) \tag{6}$$

where J is normally distributed and G is a constant.

It can be noted that equation (6) is similar in form (power law) as that proposed by Helie (Helie M, et al., 1996), based on experimental observations. For AISI 304 austenitic stainless steel, J has a mean of 2.551 and standard deviation of 0.4269, and $G = 1.3447$ (Harris et al., 1996). From parametric studies carried out, a value of 5% coefficient of variation of $Log\ (v_1)$ was found to be critical to life of piping component (Priya C et al., 2005) , and hence this value is used in this study.

The procedure followed for transition from initiation to fracture mechanics crack growth rate in the present study is (Harris et al., 1996):

- Pre-existing cracks always grow at fracture mechanics velocity.
- Initiation velocity is always assigned to initiate cracks.
- At any given time, if fracture mechanics velocity (v_2) is greater than initiation velocity (i.e. $v_2 > v_1$) and depth of crack is greater than 2.54 mm, that particular crack grows at fracture mechanics velocity thereafter.
- If the stress intensity factor for a crack is negative, the crack will not grow.

Fracture mechanics based crack growth velocity, v_2 *(inches/year)*, is given by Harris (Harris et al., 1996):

$$Log(v_2) = C_{14} + C_{15} D_K \tag{7}$$

where D_K is the damage parameter given by

$$D_K = C_{12} \, Log\left[f_2(environment) \right] + C_{13} K \tag{8}$$

where K is stress intensity factor, C_{12}, C_{13} and C_{15} are constants and C_{14} is normally distributed.

For AISI 304 austenitic stainless steel, C_{12} = 0.8192, C_{13} = 0.03621 and C_{15} = 1.7935; mean value of C_{14} = -3.1671 and standard deviation of C_{14}=0.7260 (Harris et al., 1992).

From a probabilistic failure analysis of austenitic nuclear pipe against SCC, Priya in (Priya C et al., 2005) inferred that expressions given in PRAISE for computation of stress intensity factors for modeling crack propagation need modification. This modification has been introduced by using well-accepted expressions given in ASM Handbook (American Society of Materials, 1996), and with modified PRAISE approach, stochastic propagation of stress corrosion cracks with time has been studied. The expression for the stress intensity factor for a part-through crack in a hollow cylinder is given by

$$K = S \, F \sqrt{(\pi a)} \tag{9}$$

where S is the tensile or bending stress (assumed to be the value of applied stress in the study), F is a function of a/b, R/h, and a/h, where a is the crack depth, b is half the crack length, R is the inner radius of the cylinder, and h is the wall thickness. The values of F are calculated for length and depth directions separately using the tables given in ASM Handbook (American Society of Materials, 1996). Since these factors are applicable for values of a/b less than 1.0, the expression for stress intensity factor for part-through cracks in a finite plate is used when a/b becomes greater than 1.0. This expression is given by

$$K = \sigma F_s \sqrt{\left(\frac{\pi a}{Q} \right)} \tag{10}$$

where $\sqrt{Q}$ is the crack shape parameter which depends on a/b ratio and F_s is a function of a/b, a/h, and b/W and φ such that

$$F_s = \left[M_1 + M_2 \left(\frac{a}{h} \right)^2 + M_3 \left(\frac{a}{h} \right)^4 \right] G_1 \, f_\varphi \, f_w \tag{11}$$

where

$$M_1 = \left[1.0 + 0.04 \left(\frac{b}{a} \right) \right] \left(\frac{b}{a} \right)^{1/2} \tag{12}$$

$$M_2 = 0.2\left(\frac{b}{a}\right)^4 \tag{13}$$

$$M_3 = -0.11\left(\frac{b}{a}\right)^4 \tag{14}$$

$$G_1 = 1 + \left[0.1 + 0.35\left(\frac{b}{a}\right)\left(\frac{a}{h}\right)^2\right](1 - \sin\varphi)^2 \tag{15}$$

$$f_\varphi = \left[\left(\frac{b}{a}\right)^2 . \sin^2\varphi + \cos^2\varphi\right]^{1/4} \tag{16}$$

$$f_w = \sqrt{\sec\left[\left(\frac{\pi b}{W}\right)\sqrt{\left(\frac{a}{h}\right)}\right]} \tag{17}$$

φ is the parametric angle measured from the plate surface toward the centre of the crack (i.e. φ=0° is on the plate surface and φ=90° is at the maximum depth of the crack) and W is the width of the plate (assumed to be 50mm in the study). When the crack grows to become a through-wall crack, the expression for stress intensity factor used is

$$K = S\,F\,\sqrt{(\pi b)} \tag{18}$$

Here, F is given by the expression

$$F = A_1 + A_2\alpha + A_3\alpha^2 + A_4\alpha^3 + A_5\alpha^4 + A_6\alpha^5 + A_7\alpha^6 \tag{19}$$

where α is one-half of the angle that represents the crack tip-to-tip circumferential arc and the coefficients A_1 to A_6 are computed from the table given in ASM Handbook .

It has been noted that trend of distribution of crack depths at initial stages is in satisfactory agreement with relevant experimental observations reported in literature.

2.3. Multiple cracks

In materials subjected to IGSCC, many cracks would initiate successively and propagate simultaneously, and hence multiple cracks can be present in a given weld. The expressions, given in PRAISE, for determining statistical properties of t_I are mainly based on data from laboratory experiments on specimens about 50 mm long. Hence, these expressions are applicable to specimens of about 50 mm only. This is taken into account in PRAISE, by considering a given weld in the pipe to be composed of 50 mm segments adding up to length of weld. Initiation time for each segment is assumed to be independent and identically distributed.

2.4. Coalescence of crack

The multiple cracks that may be present can coalesce as they grow. Linkage of two cracks takes place if spacing between them is less than the sum of their depths. After coalescence of two cracks, the dimensions of modified crack are given by Eqn. (20).

$$\text{Length, } l = l_1 + d + l_2 \text{ and } \textit{Depth}, a = a_1 \textit{ or } a_2, \textit{which is greater} \tag{20}$$

where l_1 and l_2 are lengths of two cracks under consideration, a_1 and a_2 are crack depths and d is spacing between them.

The operating conditions and environmental conditions show variations during the lifetime of the power plant (Anoop M B et al., 2008). Also, there will be variations in micro-structural properties of the material of piping component. These variations should be taken into account while assessing the safety of the piping component against SCC. Various researchers have carried out studies on failure analysis against SCC in different types of components of power plants by considering different basic variables (such as those associated with material properties and applied loading) as random, (Harris et al., 1996; Ting K., 1999; Herrera ML et al., 1999; You J-S,2002; Priya C et al., 2005). However, safety assessment of nuclear power plant pipelines also involves information from expert judgment and/or data from in-service inspections.

2.5. Residual stresses

Residual stresses influence both crack initiation and propagation. The damage parameter D_{sigma} is a function of the stress, which consists of both the applied (service-induced pressure and thermal) and residual stresses. The crack-tip stress-intensity factor is given by

$$K = K_{ap} + K_{res} \tag{21}$$

where K_{ap} and K_{res} are the stress-intensity factors attributable to the applied stress and residual stresses, respectively.

The calculations reported here are concerned with the stress corrosion cracking behavior of large pipes (Outside Diameter > 508 *mm*). Residual stress is treated as a random variable in the Monte Carlo Simulation (MCS). Distributions of residual stress as a function of distance from the inner pipe wall were developed from experimental data for three categories of nominal pipe diameter. For large lines, residual stresses took the form of a damped cosine through the wall as based on data collected by GEC and Argon National Laboratory. Harris and Dedhia (Harris et al., 1996) documents the default pc-PRAISE inputs in detail for the complex pattern of residual stresses in large pipes. In summary, the inner surface had a mean tensile stress of 262 *MPa*. The through-thickness variation in stress had compressive stresses developing within the inner quarter-wall thickness and changing again to tension stress at greater depths.

2.6. Failure criteria

The part-through initial stress corrosion cracks considered can grow and become unstable part-through cracks or stable or unstable through-wall cracks. The stability of the part-through or through-wall crack is checked by comparing net-section stress with the flow stress of the material. The net-section stress criterion is applicable to very tough material, and the failure is due to the insufficient remaining area to support the applied loads (i.e. net-section stress due to applied loads becomes greater than the flow stress of the material). For leakage failure, the criterion was that of a crack depth equal to the pipe-wall thickness.

2.7. Numerical simulation

In this study, the stochastic evolution of cracks due to IGSCC in AISI304 austenitic stainless steel is simulated using MCS technique. The details regarding the random variables considered in this study are given in Table 2. N values of f_1 and f_3 are computed using the N values of Pa and generated using MCS. Using the values of f_1, f_2 (considered as deterministic), and f_3, N values of damage parameter are obtained. This represents N pipes with different damage parameters, which take into account the variation in microstructural properties of the material and the operating conditions. In order to take into account the variations in the time to initiation of stress corrosion cracks, the number of samples considered is $10N$. For each sample, n (the number of possible initiation sites in the pipe considering the weld length is divided into 25.4mm long segments along the circumference of the pipe) values of t_I are generated using MCS. In every time step, each one of the cracks is checked for initiation, values of initial crack depth and length are assigned to the initiated cracks and crack growth velocities are calculated. For all the initiated cracks in the 10^5 samples generated, crack propagation velocities are calculated based on initiation and fracture mechanics considerations as is appropriate. After each time step, the failure criterion, in section 2.6 is checked. Failure probability P_f is calculated as

$$P_f = \frac{N_f}{N} \tag{22}$$

where N_f is the number of failure cases and N is the total number of simulations.

Computer programs have been developed implementing the step-by-step procedure, in the form of flow charts, for computing failure probabilities using the net-section stress as failure criteria.

3. Calibration of model

The original calibration of the PRAISE model in (Harris DO et al., 1985) was redone by Khaleel et al. in (Khaleel MA. et al., 1995). The original calculations predicted substantial levels of material damage from loading and unloading events (i.e., complete start-up and shut-down of the plant) that used a model that applied strain-to-failure data from constant extension rate tests. A review of the damage model concluded that these predictions were extremely conservative and were inconsistent with more recent insights into stress corrosion cracking mechanisms. In the revised calculations, the loading/unloading events were decreased from once per year to once per 40 years, which essentially removed the contribution of these events to the calculated failure probabilities.

Outside diameter, (mm)	549
Wall thickness, (mm)	26.4
Welding residual stress, (MPa)	Randomized pc-PRAISE input values for large lines with adjustment of: f=0.75 Stress at ID: Mean= 262
Axial component of primary stress, (MPa)	4 to 86
Flow stress of piping material, (MPa)	Normal distribution Mean = 276 and Standard deviation= 29
Initial flaw distribution	Lognormal distribution Deterministic flaw depth=0.025 Mean flaw length= 3.2 Shape parameter=0.85
POD Curves	Three POD Curves as per table3
Frequency of inspection,(yr), (Time of initial ISI/ Frequency)	5/10, 4/4, 1/1
Range of D_{sigma} parameter	0.001 to 0.02
Crack initiation and growth	As per Eqs.1 to 10

Table 2. Input values for parametric calculation including the effects of in-service inspection

Fig. 2 compares the original PRAISE cumulative leak probabilities (Harris DO et al., 1985) and results obtained by Khaleel in (Khaleel MA. et al., 1995), for various values of residual stress adjustment factors and plant loading/unloading frequencies. In this case, the adjusted residual stress level used to limit the disagreement between predicted and observed leak probabilities was set at 75% of their original values. The resulting predictions had a much more rational basis and were in very good agreement with operational data for time periods beyond 6 years. The less satisfactory level of agreement for time periods less than 6 years can be attributed in a large measure to lack of observed failure events for the early periods of plant operation.

4. In-service inspection model

Performance demonstrations have addressed two measures of NDE reliability-Probability of Detection (POD) and flaw sizing accuracy. Both measures impact the ability of an inspection program to enhance the reliability of piping. The detection of growing flaws with a high level of reliability is an important step to ensure that in-service inspections will reduce the failure probabilities of the piping locations being inspected. However, such inspections impact piping reliability only if the detected flaws are correctly sized such that needed repairs or other corrective actions are performed. Significant under sizing of flaws can result in incorrect conclusions. Flaws can be incorrectly classified as benign if their measured sizes are less than the sizes of the governing flaw acceptance criteria used to dictate repair decisions.

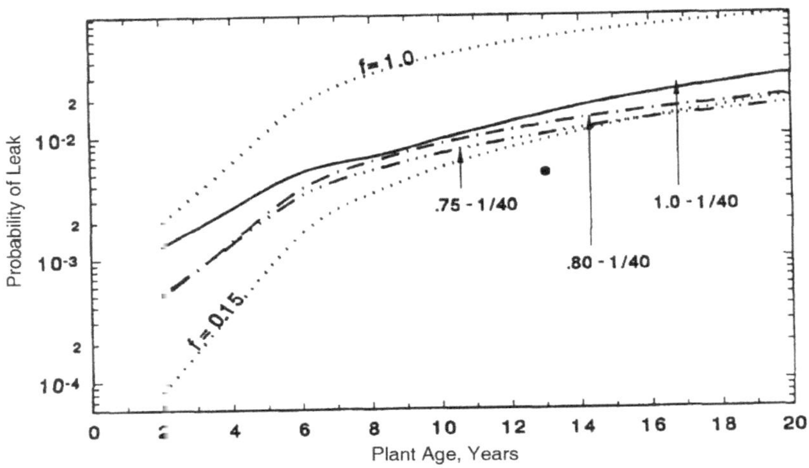

Figure 2. Field Observa ions of Leak Probabilities Compared with pc-PRAISE Results for Various Values of the Residual Stress Adjustment Factors and Plant Cycles (Large Diameter Pipes) (Khaleel MA. et al., 1995)

The procedures shown in Fig. 3 are applicable to a given location in a structure. The crack size distribution is combined with the non-detection probability to provide the post-inspection distribution. The manner in which the cracks that escape detection grow is then calculated by fracture mechanics techniques. The cumulative probability of failure at any time is simply the probability of having a crack at that time equal to or larger than the critical crack size. The crack size distribution at the time of the first ISI can be calculated. This pre-inspection distribution is combined with the non-detection probability to provide the post-inspection distribution. Fracture mechanics calculations then proceed up to the next ISI, at which time the procedures are again applied. Calculations of the failure probability for the general model are performed numerically because of the complexity of the fracture mechanics calculations of the growth of two-dimensional cracks as well as the complicated bivariate nature of the crack size distribution.

4.1. One-dimensional model

Consideration of one-dimensional cracks greatly simplifies the probabilistic fracture mechanics model. The two-dimensional surface of the initial crack size shown in Fig. 1 becomes a line that is a function of only one length, which will be taken to be crack depth a. Additionally, the fracture mechanics calculations are considerably simplified, because only one dimension of crack growth has to be considered. However, if one-dimensional cracks in finite bodies are considered, numerical calculations of subcritical crack growth are often still necessary. The procedures involved for one-dimensional cracks are depicted in Fig. 5. For simplicity, only the case of a pre-service inspection is shown in Fig. 5, and the critical crack depth, a_c is considered to be a constant.

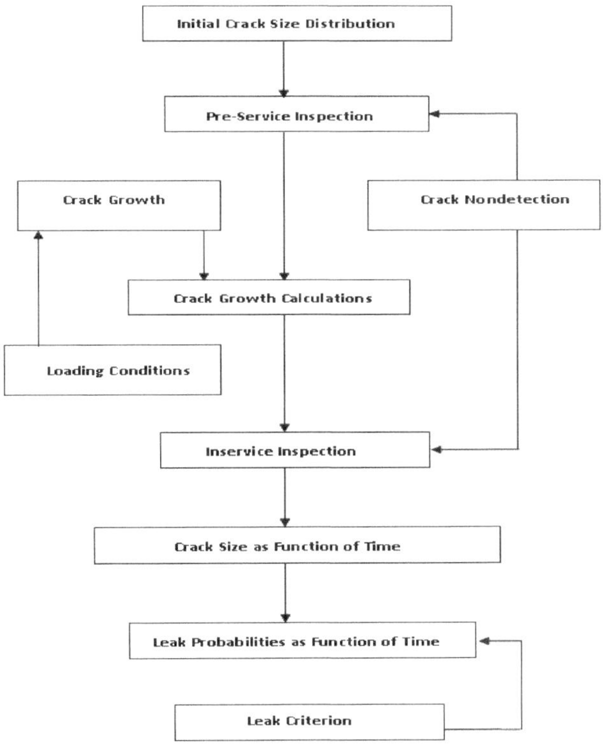

Figure 3. Computational procedure used in estimating leak probability for nuclear power plant piping

Let $P_0(a)$ be the probability density function of as-fabricated crack depth given that a crack is present. The as-fabricated complementary cumulative crack depth distribution is then given by

$$P_0(a > x) = \int_x^h P_0(y)dy \quad (23)$$

The density function and complementary cumulative distribution of crack depths following an inspection with nondetection probability $P_{ND}(a)$ are then given by

$$P_0^{'}(a) = P_0(a)P_{ND}(a) \quad (24)$$

$$P'_0(a > x) = \int_x^h P'_0(y)dy = \int_x^h P_0(y)P_{ND}(y)dy \tag{25}$$

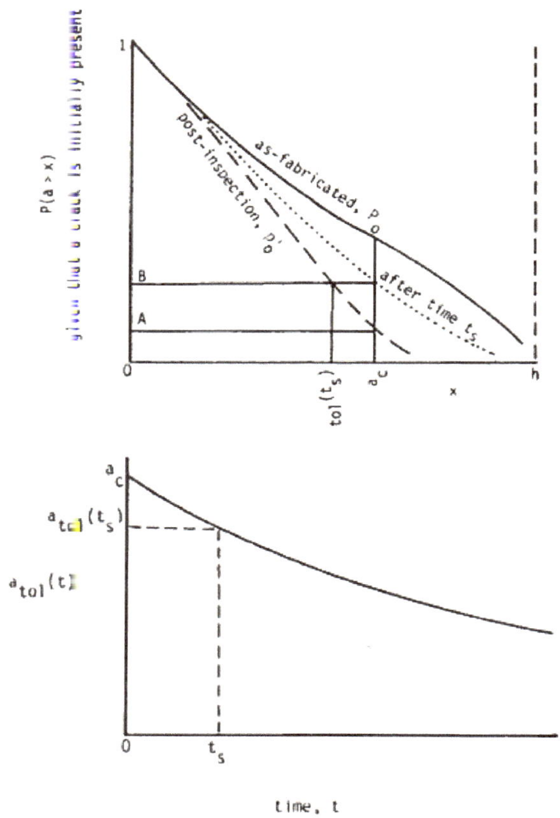

Figure 4. Procedures involved in calculating failure probability for one-dimensional crack problems (Harris DO et al., 1983).

The probability of failure at short time following the pre-service inspection is simply the probability of having a crack of depth greater than the critical value, a_c. Hence, for $t \sim 0$, the Point *A* in Fig. 4 gives the cumulative failure probability (given that a crack is initially present).

During succeeding time, the cracks that are initially present can grow due to subcritical crack growth, such as fatigue or stress corrosion cracking. Fracture mechanics calculations could be performed to determine the entire crack size distribution as a function of time, such as shown by the dotted line in Fig.4. The cumulative failure probability at or before t_s is then given by Point *B* in Fig. 4. Such a procedure requires extensive fracture mechanics calculations of crack growth in order to accurately define the crack size distribution at various times. An alternative procedure is to define $a_{tol}(t)$ to be the size of a crack at $t = 0$ (initially) that will just grow to a_c in time t (Fig. 4) schematically shows $a_{tol}(t)$. The probability of failure at or before t is then simply the probability of initially having a crack bigger than $a_{tol}(t)$. That is

$$P'_0(a > x) = \int_{a_{tol}(t)}^{h} P_0(y) P_{ND}(y) dy \tag{26}$$

Hence, once $a_{tol}(t)$ is known, the cumulative (conditional) failure probability is calculated in a straightforward manner. The failure rate, or hazard function, is equal to $(dP/dt)/(1-P)$. In the case where $P(t_f < t) << 1$, the $(1-P)$ term is nearly unity, and the failure rate is approximately given by the following

$$P(t_f < t) = \int_{a_{tol}(t)}^{h} P_0(y).P_{ND}(y) dy \tag{27}$$

$$P_f(t) = \frac{d}{dt}\left[P(t_f < t)\right] = -\frac{da_{tol}(t)}{dt} P_0(a_{tol}) P_{ND}(a_{tol}) \tag{28}$$

Consideration of the failure rate in conjunction with the tolerable crack depth simplifies analysis of the influence of ISI. The procedures for calculating the failure rate for an arbitrary schedule of in-service inspections are straightforward but lengthy. Details are provided in (Harris DO et al., 1983). The end result of the analysis is the following expression

$$\frac{P_f(t)\text{with inspection}}{P_f(t)\text{without inspection}} = \prod_{n=0}^{k} P_{ND}\left[a_n(t)\right] \tag{29}$$

The crack depth, $a_n(t)$, is shown schematically in Fig. 5 and is determined from fracture mechanics calculations. If $a_{tol}(t)$ is the depth of an initial crack that would just grow to a_c in time t, then a_n, is the depth this crack would have had at the arbitrary inspection times t_n.

The result expressed as Eqn. (29) shows that the relative benefit of ISI, when expressed as the ratio of failure rates with and without inspection, is independent of the initial crack size distribution. This ratio is dependent only on the nondetection probabilities and crack growth characteristics. The initial crack size distribution is currently ill-defined, and is a source of considerable uncertainty in probabilistic fracture mechanics analysis.

Another ill-defined parameter is the probability of initially having a crack in a body of volume V. This parameter is denoted as p^*. All failure probabilities just discussed were conditional on having an initial crack. To account for the possibility of not having a crack (that is, to make the results not conditional on having a crack to begin with) all those results are to be multiplied by p^*. This parameter is also not well defined, but cancels out in the relative benefit of ISI as expressed by Eqn. (29), because it appears in both the numerator and denominator.

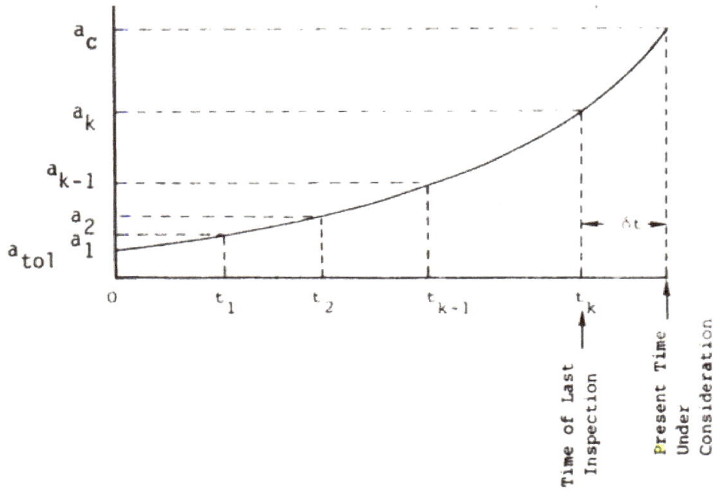

Figure 5. Schematic representation of a_k from known subcritical crack growth results and specified inspection (Harris DO et al., 1983).

Thus, the relative benefit of ISI is independent of both the conditional initial crack size distribution and the probability of having a crack. These two factors are currently sources of considerable uncertainty, and their absence from the relative benefit allows the influence of ISI to be assessed with greater confidence then if they were present in Eqn. (29). This result is for only one dimensional crack. The degree to which it is applicable to the more realistic case of two-dimensional cracks will be addressed in the following section.

4.2. Probability of detection curves

NDE techniques do not detect all cracks, but can be thought of as detecting cracks with a probability that depends on crack size. Detecting a crack depends on many factors, including the dominant degradation mechanism; the flaw location (e.g. surface or buried), orientation, and shape; material (e.g. stainless and ferritic steel); the inspector's training and experience; and the inspection method and procedure. In the literature, the POD has been used to define the capability of specific NDE techniques and inspection teams (Doctor et al., 1983), (Taylor et al., 1989) and (Heasler, 1996). Usually, however, the POD curve is given as a function of the defect size (depth) with all other factors governing detection assumed to be held constant for each given POD curve. In this paper, the POD curve is taken to be the measure of inspection reliability.

In general, the POD curve for a certain NDE method, procedure, and inspection team is estimated based on a statistical methodology using the experimental data, expert elicitation, or both data and elicitation. The characteristic POD curves for detection of IGSCC discussed here are based on expert judgement. Alternative inspection strategies for ultrasonic examination of welds in stainless steel piping were evaluated based on a range of POD curves. The approach was to establish three POD curves that represented widely differing levels of NDE performance. These curves were intended to bound the performance expected from inspection teams operating in the field. To establish the POD curves, an informal expert judgement elicitation was made using staff at PNNL with knowledge of NDE performance data from inspection round robins and of recent industry efforts in the area of NDE performance demonstrations.

It was recognized that a population of inspection teams operating under field conditions can exhibit a considerable range of POD performance, even though all such teams have successfully completed a performance demonstration. The basic premise in estimating POD curves for the present calculations was that all teams had passed the ASME Section XI Appendix VIII performance demonstration. The informal expert judgement also considered information and trends observed in the PNNL round-robin studies on UT inspection (Doctor et al., 1983). The NDE experts at PNNL were asked to define POD curves by estimating parameters for the following form of a POD function:

$$P_{ND} = \varepsilon + \frac{1}{2}(1-\varepsilon)\, erfc\left[\nu \ln\left(\frac{A}{A^*}\right)\right] \tag{30}$$

where P_{ND} is the probability of nondetection, A is the area of the crack, A* is the area of crack for 50% P_{ND}, ε is the smallest possible P_{ND} for very large cracks, and ν is the 'slope' of the PND curve. Based on measured performance for PNNL's mini round robin teams, a range of estimates for A* (crack area for 50% POD) was provided by the NDE experts. Harris et al. (Harris et al., 1981) assumed that the 'slope' parameter ν is 1.6. Several POD curves from PNNL studies were reviewed, and it was determined that a value of ν=1.6 was consistent with published curves. The slope was correlated to the detection threshold parameter A* and the value of ε was assigned to ensure that a smaller value of A* also implies a smaller value of ε. Three POD curves were selected:

1. Marginal performance: POD performance that is described by this curve would represent a team having only a small chance of passing an Appendix VIII performance demonstration.
2. Good performance: POD performance that is described by these curves corresponds to the better teams in the round robins.
3. Very good performance: this curve corresponds to a team that significantly exceeds the minimum level of performance needed to pass an Appendix VIII performance demonstration test.

Table 3 summarizes the input data for the above three POD curves. These particular curves assume that POD is a function of the crack depth as a fraction of pipe-wall thickness, independent of the actual wall thickness. Parameters indicated in Table 3 are considered appropriate to wall thicknesses of 25.4 *mm* and greater. Lower levels of POD as a function of a/h should be assumed for thinner-walled pipe. The parameter a* is the crack depth for 50% probability of non-detection and is related to A^* (i.e., $A^* = \pi/4\ D_B\ a^*$, where D_B is the beam diameter of the ultrasonic probe). The parameter a^* varies in the same manner as ε.

Inspection Performance Level	$a^*/h^{(+)}$	ε	ν
Marginal	0.65	0.25	1.40
Good	0.40	0.10	1.60
Very Good	0.15	0.02	1.60
(+) h is the wall thickness of the pipe.			

Table 3. POD Curve parameters for three performance levels

4.3. Factor of improvement

The effectiveness of an inspection strategy is quantified by the parameter 'Factor of Improvement (FI)', which is the relative increase in piping reliability due to a given inspection strategy as compared with the strategy of performing no inspection. The results of a systematic set of calculations are presented in this paper that address inspection effectiveness for operating stresses giving crack growth rates ranging from very low to very high. Inspection strategies are described that address three reference levels of ultrasonic inspection reliability, intervals between inspections ranging from 1 to 10 years, and in-service inspections. The FI is based on the cumulative leak probability that occurs over the 40-year design life of the plant, and is defined as the ratio

$$FI = \frac{\text{leak probability at 40 years without NDE}}{\text{leak probability at 40 years with NDE}} \tag{26}$$

5. Results

5.1. Predicted leak probability versus damage parameter

The reliability for a large number of welds and fittings in a piping system can be estimated quickly if the results of detailed Monte Carlo simulations are provided in a structured parametric format. D_{sigma} Values for various degrees of sensitization, different levels of applied stress, a conductivity equal to 0.2 $\mu S/cm$, different steady state temperature and different O_2 content are presented in Fig.6(a - d). The parametric calculations as presented below consisted of many pc-PRAISE runs that covered a range of leak probabilities from 1.0E-04 to 1.0. It was believed that D_{sigma} could serve as a suitable parameter to summarize results for calculated failure probabilities of stainless steel piping. Results presented later in this section show a good correlation between 40-year cumulative leak probabilities and D_{sigma}. Although the correlation tends to break down for smaller values of D_{sigma}, the parameter does provide a useful basis to generalize results for piping-leak probabilities (Fig.7).

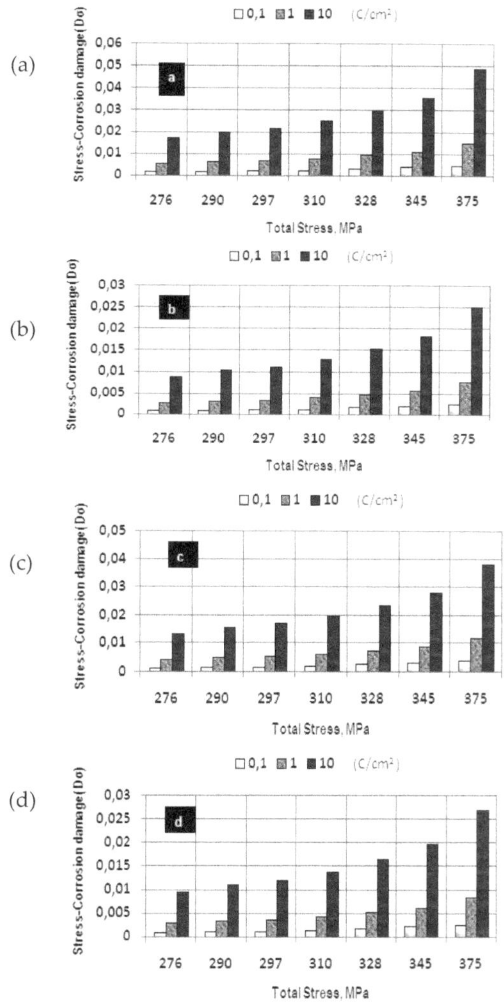

Figure 6. a: D_{sigma} Values for various degrees of sensitization, different levels of applied stress, a conductivity equal to 0.2 $\mu S/cm$ and a steady state temperature equal to 288°C, for O_2 content equal to 8ppm.
b: D_{sigma} Values for various degrees of sensitization, different levels of applied stress, a conductivity equal to 0.2 $\mu S/cm$ and a steady state temperature equal to 288°C, for O_2 content equal to 0.2ppm.
c: D_{sigma} Values for various degrees of sensitization, different levels of applied stress, a conductivity equal to 0.2 $\mu S/cm$ and a steady state temperature equal to 288°C, for O_2 content equal to 2ppm.
d: D_{sigma} Values for various degrees of sensitization, different levels of applied stress, a conductivity equal to 0.2$\mu S/cm$ and a steady state temperature equal to 140°C, for O_2 content equal to 0.2ppm.

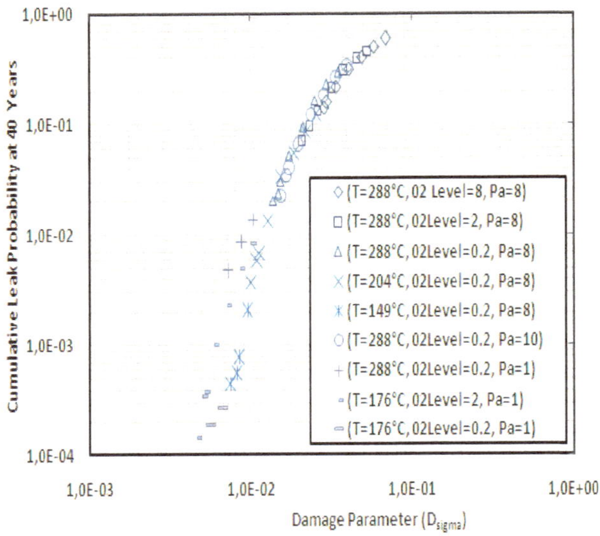

Figure 7. Cumulative Leak Probability over 40 Years as a Function of the Stress-Corrosion Damage Parameter for Various Temperatures, Oxygen Contents and degrees of sensitization

5.2. Improvement factors

The individual data points for improvement factors exhibited some data scatter, which was addressed by using a regression analysis to establish best fit curves (second order polynomial) in terms of improvement factors versus the logarithm (base 10) of D_{sigma} and the corresponding leak probabilities. For clarity, the plots below show only the best-fit curves rather than the individual data points. Figs. (8-12) show calculated improvement factors (over a 40-year design life).

Figs. (8-10) each addresses a specific inspection method (POD curve) with the individual curves corresponding to different inspection intervals ranging from 1 to 10 years. Figs. (10-12) have rearranged the curves to maintain a common inspection interval for each plot, with the individual curves corresponding to different POD capabilities. In general, it appears that a given inspection strategy (combination of POD curve and inspection interval) gives somewhat greater improvement factors.

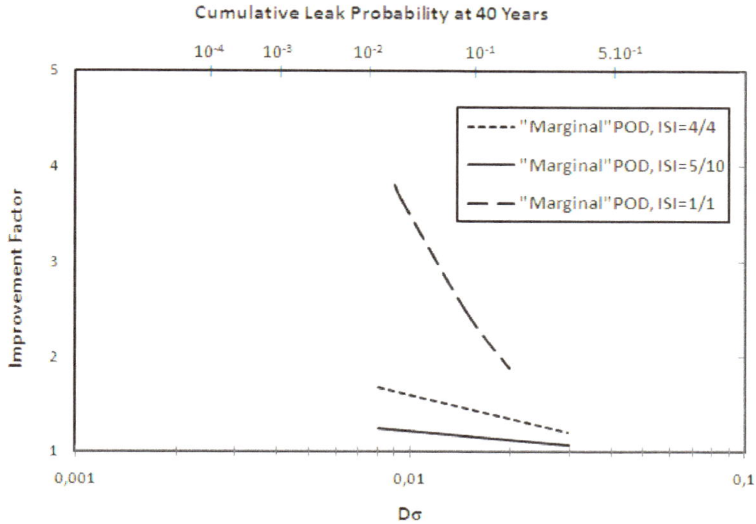

Figure 8. Improvement Factors versus D_{sigma} for "Marginal" POD Curve with various Inspection Intervals

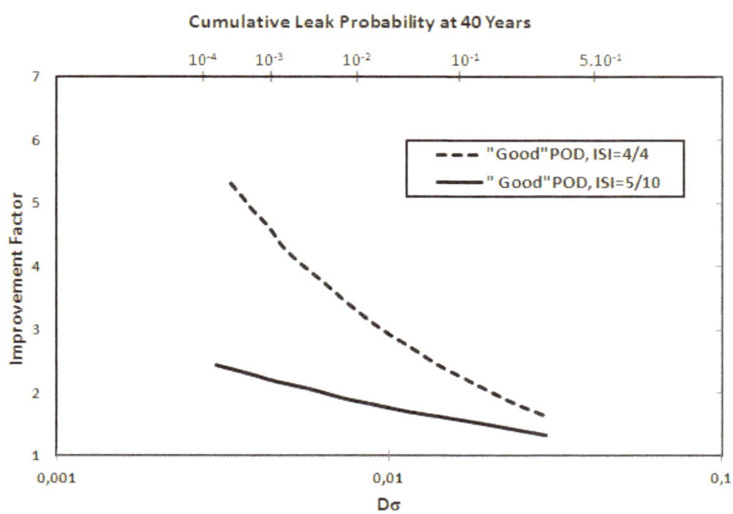

Figure 9. Improvement Factors versus D_{sigma} for "Good" POD Curve with various Inspection Intervals

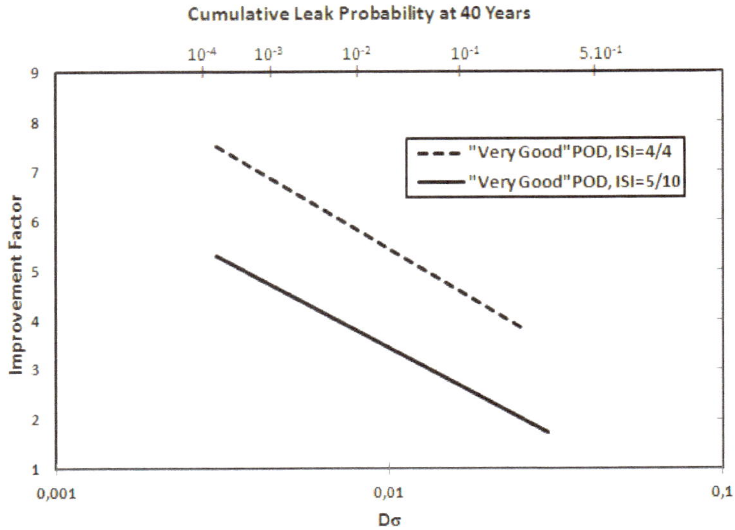

Figure 10. Improvement Factors versus D_{sigma} for "Very Good" POD Curve with various Inspection Intervals

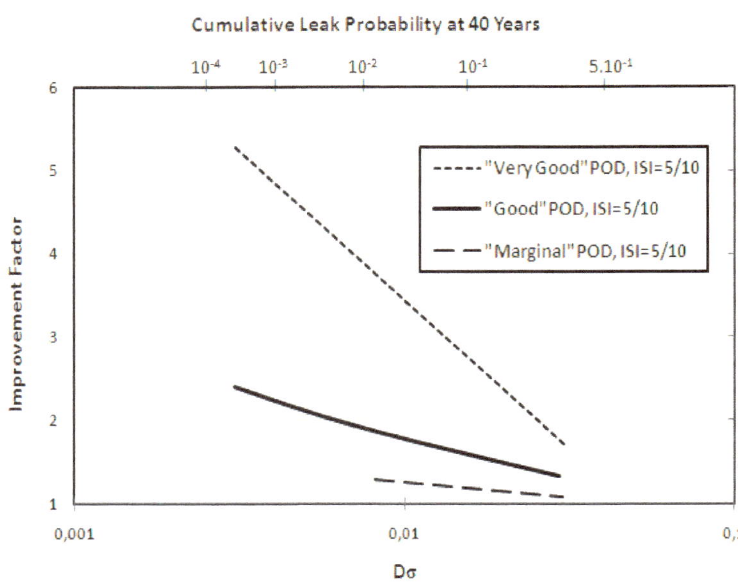

Figure 11. Improvement Factors versus D_{sigma} for 10-Year ISI Interval with various POD Curves

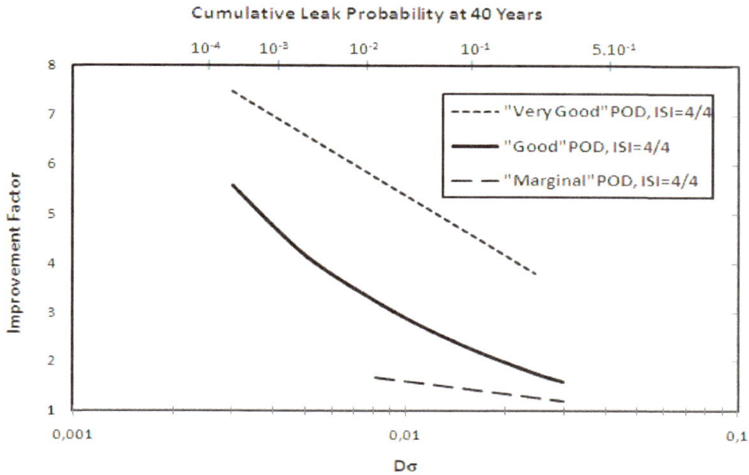

Figure 12. Improvement Factors versus D_{sigma} for 4-Year ISI Interval with various POD Curves

6. Conclusions

Stress corrosion crack growth can occur under constant loading conditions, and is therefore very different from crack growth driven by the cyclic loading. Our original probabilistic fracture mechanics model incorporated a simple model of stress corrosion cracking, based on the assumption that crack growth velocity in either the radial or circumferential direction is controlled by the value of stress intensity factor (for a given material and environment) at the crack tip. This model described crack kinetics by a simple functional relationship between crack growth rate and stress intensity factor. This probabilistic stress-corrosion cracking model was applied to assess the effect of various inspection scenarios on leak probabilities. This chapter has also discussed probability of detection curves and the benefits of in-service inspection in the framework of reductions in the leak probabilities for nuclear piping systems subjected to IGSCC. The results for typical NDE performance levels indicate that low inspection frequencies (e.g., one inspection every 10 years) can provide only modest reductions in failure probabilities. More frequent inspections appear to be even more effective. The greatest benefits are predicted for the "Very Good" NDE technology and procedures, for which an order-of-magnitude improvement on the leak probability can be achieved for an inspection frequency of once per year. The lower benefits of ISI for IGSCC compared to the benefits for fatigue crack growth can be explained in terms of long incubation periods for stress-corrosion cracking followed by a period of rapid crack growth.

Author details

A. Guedri
InfraRes Laboratory; University of Souk Ahras, Algeria
LEM3 Laboratory, UMR CNRS 7239, University of Metz, France

Y. Djebbar
InfraRes Laboratory; University of Souk Ahras, Algeria

Moe. Khaleel
Pacific Northwest National Laboratory, USA

A. Zeghloul
LEM3 Laboratory, UMR CNRS 7239, University of Metz, France

Acknowledgement

The authors would like to thank Dr. Khaleel Mohammad for his support and his invaluable guidance during the course of this work.

7. References

A. Guedri, A. Zeghloul and B. Merzoug (2009) Reliability analysis of BWR piping including the effect of residual Stresses. *International Review of Mechanical Engineering (I.RE.M.E.)*, pp.640-645, September 2009, Vol. 3, n. 5.

A. Guedri, B. Merzoug, Moe Khaleel and A. Zeghloul (2009) Reliability analysis of low alloy ferritic piping materials. *Springer Netherlands, Damage and Fracture Mechanics*. Failure Analysis of Engineering Materials and Structures, pp 33-42.

American Society of Materials (1996). *ASM handbook*: fatigue and fracture. USA: Materials Information Society International.

American Society of Mechanical Engineers (1998) *ASME Boiler and Pressure Vessel Code*, Materials, Section II Part D: Properties, (ASME, New York).

Andresen PL and FP Ford (1994) Fundamental Modeling of Environment Cracking for Improved Design and Lifetime Evaluation in BWRs." *International Journal of Pressure Vessels and Piping* 59(1-3): 61-70.

Anoop M B, Balaji Rao K, Lakshmanan N (2008) Safety assessment of austenitic steel nuclear power plant pipelines against stress corrosion cracking in the presence of hybrid uncertainties. *International Journal of Pressure Vessels and Piping*. 85(4): 238-247

Bell CD and OJV Chapman (2003) Description of PRODIGAL. *NURBIM Report D4/Appendix F*. Rolls-Royce plc., Derby, United Kingdom.

Bishop BA. (1997) An Updated Structural Reliability Model for Piping Risk Informed ISI, In *Fatigue and Fracture,* Volume 2, PVP Vol. 346, pp. 245-252. *American Society of Mechanical Engineers, New York.*

Brickstad B and W Zang. (2001) NURBIT Nuclear RBI Analysis Tool, A Software for Risk Management of Nuclear Components. *Technical Report No 10334900-1*. DNV, Stockholm, Sweden.

Brickstad B, OJV Chapman, T Schimpfke, H Schulz and A Muhammed (2004) Review and Benchmarking of SRM and Associated Software. *NURBIM Final Report* D4, Contract FIKS-CT-2001-00172. DNV, Stockholm, Sweden.

Dillstrom P. (2003) A Short Description of ProSACC. *NURBIM Report D4/Appendix G*. DNV, Stockholm, Sweden.

Doctor SR, FL Becker, PG Heasler and GP Selby (1983) Effectiveness of U.S. Inservice Inspection Technologies: A Round Robin Test." In *Proceedings of a Specialist Meeting on Defect Detection and Sizing*, Vol. 2 (CSNI Report No. 75 and EUR 9066 II EN), pp. 669–678.

General Electric Company. (1982a) The Growth and Stability of Stress Corrosion Cracks in Large Diameter BWR Piping, Volume 1: Summary. EPRI NP-2472, Vol. 1. *BWR Owners Groups and Electric Power Research Institute,* Palo Alto, California.

General Electric Company. (1982b) The Growth and Stability of Stress Corrosion Cracks in Large Diameter BWR Piping, Volume 2: Appendixes. EPRI NP-2472, Vol. 2. *BWR Owners Groups and Electric Power Research Institute,* Palo Alto, California.

Gosselin SR, FA Simonen, RG Carter, JM Davis and GL Stevens (2005) Enhanced ASME Section XI Appendix L Flaw Tolerance Procedure." In Proceedings of the PVP2005, 2005 ASME Pressure Vessels and Piping Conference, PVP2005-71100, July 17-21, 2005, Denver, Colorado. *American Society of Mechanical Engineers*, New York.

Harris DO and DD Dedhia (1998) WinPRAISE 98 PRAISE Code in Windows. *Engineering Mechanics Technology*, Inc., San Jose, California.

Harris DO, DD Dedhia and ED Eason (1986a) Probabilistic Analysis of Initiation and Growth of Stress Corrosion Cracks in BWR Piping." *American Society of Mechanical Engineers*, New York. ASME Paper 86-PVP-11.

Harris DO, DD Dedhia and SC Lu. (1992) Theoretical and User's Manual for pc-PRAISE, A Probabilistic Fracture Mechanics Computer Code for Piping Reliability Analysis. *NUREG/CR-5864.U.S.* Nuclear Regulatory Commission, Washington, D.C.

Harris DO, DD Dedhia, ED Eason and SD Patterson (1986b) Probability of Failure in BWR Reactor Coolant Piping: Probabilistic Treatment of Stress Corrosion Cracking in 304 and 316NG BWR Piping Weldments. *NUREG/CR-4792,* Vol. 3. U.S. Nuclear Regulatory Commission, Washington, D.C.

Harris DO, DD Dedhia, ED Eason, and SD Patterson (1985) Probabilistic Treatment of Stress Corrosion Cracking in Sensitized 304 Stainless Steel Weldments in BWR Piping. *Failure Analysis Associates Report to Lawrence Livermore National Laboratory,* Livermore, California.

Harris DO, EY Lim and DD Dedhia (1981) Probability of Pipe Fracture in the Primary Coolant Loop of a PWR Plant Volume 5: Probabilistic Fracture Mechanics Analysis - Load Combination Program Project 1 *Final Report. NUREG/CR-2189,* Vol. 5. U.S. Nuclear Regulatory Commission, Washington, D.C.

Harris, D. O. and Lim, E. Y., (1983) Applications of a Probabilistic Fracture Mechanics Model to the Influence of In-Service Inspection on Structural Reliability, *Probabilistic Fracture Mechanics and Fatigue Methods: Applications for Structural Design and Maintenance,* ASTM STP 798, J. M. Bloom and J. C. Ekvall, Eds. , American Society for Testing and Materials, pp. 19-41.

Hazelton W and WH Koo. (1988) Technical Report on Material Selection and Processing Guidelines for BWR Coolant Pressure Boundary Piping. *NUREG-0313,* Rev. 2. U.S. Nuclear Regulatory Commission, Washington, D.C.

Heasler PG and SR Doctor. (1996) Piping Inspection Round Robin. *NUREG/CR-5068, PNNL-10475.U.S. Nuclear Regulatory Commission,* Washington, D.C.

Helie M, Peyrat C, Raquet G, Santarini G, Sornay Ph (1996) Phenomenological modelling of stress corrosion cracking. *Intercorr/96 First Global Internet Corrosion Conference.*

Herrera ML, Mattson RA, Tang SS, Ahlgren CS. (1999) Probabilistic fracture mechanics analysis to justify in-service inspection intervals for the Helms penstock field welds. *Proceedings of Waterpower '99-hydro's future: technology, markets and policy,* Las Vegas, NV, July 6-9.

Khaleel MA and F Simonen. (1994a) The Effects of Initial Flaw Sizes and Inservice Inspection on Piping Reliability." In *Service Experience and Reliability Improvement: Nuclear, Fossil, and Petrochemical Plants,* PVP-Vol. 288, pp. 95-07. American Society of Mechanical Engineers, New York.

Khaleel MA and FA Simonen (1994b) A Parametric Approach to Predicting the Effects of Fatigue on Piping Reliability. In Service Experience and Reliability Improvement: Nuclear, Fossil, and Petrochemical Plants, PVP-Vol. 288, pp. 117-125. *American Society of Mechanical Engineers,* New York.

Khaleel MA and FA Simonen (2000) Effects of Alternative Inspection Strategies on Piping Reliability. *Nuclear Engineering and Design* 197:115-140.

Khaleel MA, FA Simonen, DO Harris, and D Dedhia (1995) The Impact of Inspection on Intergranular Stress Corrosion Cracking for Stainless Steel Piping." In Risk and Safety Assessments: Where Is the Balance? PVP-Vol. 296, pp. 411- 422. *American Society of Mechanical Engineers*, New York.

Khaleel MA, FA Simonen, HK Phan, DO Harris and D Dedhia (2000) Fatigue Analysis of Components for 60-Year Plant Life. *NUREG/CR-6674, PNNL-13227.* U.S. Nuclear Regulatory Commission, Washington, DC.

Lu B T, Chen Z T, Luo J L, Patchett B M, Xu Z H (2005) Pitting and stress corrosion cracking behaviour in welded austenitic stainless steel. *Electrochimica Acta* 50(6): 1391-1403

Mohammed AA. (2003) A Short Description of STRUEL. *NURBIM Report D4/Appendix H.* The Welding Institute, Cambridge, United Kingdom.

Priya C, Balaji Rao K, Lakshmanan N, Gopika V, Kushwaha H S, Saraf R K (2005) Probabilistic failure analysis of austenitic nuclear pipelines against stress corrosion cracking. *Proc. Inst. Mech. Eng. Part C: J. Mech. Eng. Sci.* 219(7): 607–626

Rahman S. (1997) A computer model for probabilistic leak-rate analysis of nuclear piping and piping welds. *International Journal of Pressure Vessels and Piping*; 70:209–21.

Schimpfke T. (2003) A Short Description of the Piping Reliability Code PROST. *NURBIM Report D4/Appendix C.* Gesellschaft fur Anlagen-und Reaktorsicherheit (GRS), Berlin, Germany.

Simonen FA and MA Khaleel (1998a) A Probabilistic Fracture Mechanics Model for Fatigue Crack Initiation in Piping." In Fatigue, Fracture and Residual Stress, PVP-Vol. 373, pp. 27-34. *American Society of Mechanical Engineers*, New York.

Simonen FA and MA Khaleel (1998b) Effects of Flaw Sizing Errors on the Reliability of Vessels and Piping." *ASME Journal of Pressure Vessel Technology* 120:365-373.

Simonen FA, DO Harris and DD Dedhia (1998) Effect of Leak Detection on Piping Failure Probabilities." In Fatigue Fracture and Residual Stress, PVP-Vol. 373, pp. 105-113. *American Society of Mechanical Engineers*, New York.

Taylor TT, JC Spanner, PG Heasler, SR Doctor and JD Deffenbaugh (1989) An Evaluation of Human Reliability in Ultrasonic Inservice Inspection for Intergranular Stress Corrosion Cracking Through Round Robin Testing." *Mater Eval* 47:338.

Ting K. (1999) The evaluation of intergranular stress corrosion cracking problems of stainless steel piping in Taiwan BWR-6 nuclear power plant. *Nucl. Eng. Design*; 191(2): 245-54.

You J-S, Wu W-F. (2002) Probabilistic failure analysis of nuclear piping with empirical study of Taiwan's BWR plants. *International Journal of Pressure Vessels and Piping.* 79(7):483–92.

Zhang S, Shibata T, Haruna T. (1997) Initiation and propagation of IGSCC for s ensitized type 304 stainless steel in dilute sulfate solutions. *Corrosion Science*; 39(9):1725–39.

Chapter 13

Interacting Cracks Analysis Using Finite Element Method

Ruslizam Daud, Ahmad Kamal Ariffin, Shahrum Abdullah and Al Emran Ismail

Additional information is available at the end of the chapter

1. Introduction

This chapter aims to introduce the concept of fracture mechanics and numerical approaches to solve interacting cracks problems in solid bodies which involves elastic crack interaction. The elastic crack interaction is a result of changes in stress field distribution as the applied force is given during remote loading. The main emphasis is to address the computational evaluation on mechanistic models based on crack tip displacement, stress fields and energy flows for multiple cracks. This chapter start with a brief discussion on fracture and failure that promoted by interacting cracks from industrial cases to bring the issues of how important the crack interaction behaviour is. The present fracture and failure mechanism is assumed to exhibit the brittle fracture. Thus, the concept of linear elastic fracture mechanics (LEFM) is discussed regarding the crack interaction model formulation. As the elastic crack interaction is concerned, the previous analytical and numerical solution of crack interaction are elaborated comprehensively corresponds to fitness-for-service (FFS) as published by ASME boiler and pressure vessel code (Section XI, Articles IWA-3330), JSME fitness-for-service code and BSI PD6493 and BS7910. A new computational fracture mechanics algorithm is developed by adopting stress singularity approach in finite element (FE) formulation. The result of developed approach is discussed based on the crack interaction limit (CIL) aspects and crack unification limit (CUL) in pertinent to the equality of two cracks to single crack rules in FFS. As a conclusion, the FE formulated approach was found to be at agreeable accuracy with analytical formulation and FFS at certain range of crack interval.

2. Fracture and failures by interacting cracks

This section provides the overview of failure cases in industries and related fitness-for-sevice (FFS) codes which used to assess any cracks or flaws that detected in structures. The works on solution models for FFS codes improvement in specific cases of interacting cracks also discussed.

2.1. Industrial failures

In this section, the fracture and failure by interacting cracks is explained by examples from industrial failures. Mechanical structures and components are designed with multiple stress concentration features (SCF) such as notches, holes, corners and bends. For example, welding and riveting in joining and fastening process have consumed to the increase of stress concentration factor. In every SCF, there is a critical point that experienced the highest concentrated stress field, named as multiple stress riser points (MRSP). Under multiple mode of loading and environmental effect, the interaction between SCF and MRSP tends to form multiple cracks in various types of cracks (e.g. straight crack, surface crack and curved crack) before the cracks propagate in various path to coalescence, overlapping, overlapping, branching and finally fracture in brittle manner.

Crack interaction that induced from MRSP has caused many catastrophic failures , for example in aircraft fuselage (Hu, Liu, & Barter, 2009), F-18 Hornet bulkhead (Andersson et al., 2009), rotor fault (Sekhtar, 2008), gigantic storage tanks (Chang & Lin, 2006), oil tankers (Garwood, 2001), polypropylene tank (Lewis & Weidmann, 2001) and the most recent is the fail of helicopter longerons (J. A. Newman, Baughman, & Wallace, 2010). The recent lab experimental work on multiple crack initiation, propagation and coalescence by (Park & Bobet, 2010) and metallurgical work by (J. A. Newman et al., 2010) supported the important role of crack interaction in fracture and failure. The above cases proved how crucial and important the research on crack interaction is.

To explain further, failure in aircraft is considered as an example. Al Alloy has been extensively used for the fabrication of fuselage, wing, empennage, supporting structures that involve many fastening and joining points. Under static, cyclic loading and environmental effect, the micro-cracks are initiated from MRSP. To certain extent, out of many factors, brittle failure may happen through catastrophic failure (Hu, Liu, & Barter, 2009). As the distance between MRSP is close, the interaction between cracks is become more critical. The fracture behavior due to interacting cracks as the distance between cracks is closed need more understanding. The conventional fracture mechanics may be insufficient to support

In this case, the advancement of computational fracture mechanic may contribute a lot in crack interaction research and increase the accuracy of failure prediction (Andersson et al., 2009). Most recent structural failure being reported is dealing with fuselage joints in large aircraft structures (Hu et al., 2009). The aircraft fuselage structure is made by 7050-T7451 aluminum alloy and designed to have multiple shallow notches that purposely used for lap joints. Due to environmental reaction and variable magnitude of operational loading, multi-site cracks are formed at MRSP. Therefore, the interaction of multi-site cracks is needed to be quantified for possible coalescence between cracks. The challenge is how to predict the accurate coalescence and fatigue crack growth for the multiple crack problems. Table 1 provides the list of industrial failures that originated from various kinds of multiple cracks. Under loading condition such as mechanical or thermal loading, it is observed that most of the cracks or flaws formation is start on the surface of the body rather than embedded inside the body.

Multiple Cracks	Industrial Failures	References
Collinear cracks	Failure of crack arresters-stiffeners in aircraft structures	(Isida, 1973)
Parallel and layered cracks	Failure of welded-bonded structures in composite structures for aircraft	(Ratwani & Gupta, 1974)
Collinear cracks and edge cracks	Catastrophic fracture accidents of turbine or generator motor	(Matake & Imai, 1977; Pant, Singh, & Mishra, 2011; Sekhtar, 2008)
Elliptical cracks	Failure in boilers	(O'donoghue, Nishioka, & Atluri, 1984)
Collinear and radial cracks	Failure of pressurized thick-walled cylinder	(Chen & Liu, 1988; Kirkhope, Bell, & Kirkhope, 1991)
Collinear and micro-cracks	Failure of ceramic material in heat exchangers and automobiles	(Lam & Phua, 1991)
Parallel cracks	Failure of aero-engine turbine engines coatings	(Meizoso, Esnaola, & Perez, 1995)
Parallel edge cracks	Failure of actuator piston rods	(Rutti & Wentzel, 1997)
Edge cracks	Brittle failure of Oil Tanker structures at welded joints	(Garwood, 2001)
Array of edge cracks	Failure of heat-checked gun tubes and rapidly cooled pressure vessels	(Parker, 1999)
Semi-elliptical surface cracks	Failure in pressure vessel and piping components	(Moussa, Bell, & Tan, 1999; Murakami & Nasser, 1982)
Collinear cracks and flat elliptical cracks	Multiple site damage in aircraft structures	(Gorbatikh & Kachanov, 2000; Jeong & Brewer, 1995; Jones, Peng, & Pitt, 2002; Milwater, 2010; Pitt, Jones, & Atluri, 1999)
Multiple flaws and surface cracks	Failure of nuclear power plant components	(Kamaya, Miyokawa, & Kikuchi, 2010; Kobayashi & Kashima, 2000)
Penny-shaped cracks	Brittle fracture of welded structures in pressure vessels	(Saha & Ganguly, 2005)
Offset collinear and layered cracks	Fracture and catastrophic failure in polymeric structures	(Lewis & Weidmann, 2001; Sankar & Lesser, 2006; Weidmann & Lewis, 2001)
Interface cracks	Failure in electronic packages and micro-electro-mechanical systems (MEMS)	(Ikeda, Nagai, Yamanaga, & Miyazaki, 2006)
Parallel cracks	Fracture in functional gradient materials (FGM)	(Yang, 2009)
Short cracks and micro-cracks	Fracture in bones	(Lakes, Nakamura, Behiri, & Bonfield, 1990; Mischinski & Mural, 2011; Ural, Zioupos, Buchanan, & Vashishth, 2011)

Table 1. Summary of industrial failures caused by multiple crack interaction

Crack interaction intensity exist in the form of stress shielding and amplification. The failure mechanism by cracks interaction may occurs under brittle, plastic and creep failure. In general, the interacting cracks problems that promotes the fracture and failure of structures are solved using the advancement of fracture mechanics. The fracture mechanics solution can be accomplished through analytical, numerical and experimental work. Based on the above individual approach or combination of them, typically the solution is represented by a model. The model may be defined based on the uncertainty input for the model such as crack geometry, loading and material properties. For crack interaction problems that randomness is relatively small, the deterministic analysis is the best to considered rather than probabilistic analysis. The model is suitable for any deterministic system response. However, when the randomness is relatively high, the model system response required a more robust solution, known as probabilistic approach.

Multiple cracks interaction can be defined into elastic crack interaction and plastic crack interaction that may be referred to theory of elasticity and plasticity, respectively. Under loading condition, the high stresses near crack tips usually accompanied by inelastic deformation and other non-linear effect. If the inelastic deformation and other non-linear effect are relatively small compared to crack sizes and other geometrical body characteristic, the linear theory is most adequate to address the crack interaction behavioral problems. Thus, the role of elastic driving force originated from crack tips can be translated into elastic crack interaction. Then, for every type of interaction, it may classified into interaction without crack propagation (EIWO) where the interaction occurs in the region of SIF $K < K_c$ and interaction with crack propagation (EIWI) occurs in the region of SIF $K \geq K_c$. In this case of EIWO, the quantification of crack extension is neglected. The EIWO becomes the main issue of interaction in present study since it is inadequate investigation on crack interaction limit and multiple to single crack equivalencies. The study on EIWO is typically measured the fracture parameter and its behavior based on SIF while EIWI more focused to evaluation and prediction of crack path, propagation, coalescence, branching and crack arrestment. Under mechanical or thermal loading condition, the generated interaction will varies depends on type of loading mode (e.g. Mode I and Mode II) being applied. The preceding sections outlined the related fitness-for-service (FFS) codes in pertinent to failures that caused by crack interaction.

2.2. Fitness-for-Service (FFS) codes

This section presents the guidelinea that have been published in fitness-for-service (FFS) codes. The related investigation works also discussed as the FFS codes are evaluated in different case of interacting cracks problems. ASME Boiler and Pressure Vessel Code Section XI (ASME, 1998, 2004) and API 579 (ASME, 2007) defines that the multiple cracks are assumed to be independent until or unless the following conditions are satisfied. In the case of two parallel cracks in solid bodies, if distance between crack planes d $d \leq 12.7$ mm , the cracks are treated as being coplanar. For coplanar cracks, if the distance between cracks s $s \leq 2 \times (\text{maximum of } \{a_i\}_{i=1,2})$. Indeed, the single enveloping crack is assumed equal to coplanar cracks if the condition of crack depth a and surface length c satisfy $a = \text{maximum of } \{a_i\}_{i=1,2}$, $c = c_1 + c_2 + s/2$. If the non-coplanar cracks in overlapped condition at $s < 0$, the cracks are assumed to be coplanar, the surface length c $c_{new} = c_1 + c_2 + s$. Thus, coalescence occur at $s = \max(2a_1, 2a_1)$. In United Kingdom, the engineering critical assessment (ECA) of potential or actual defects in engineering structures is codified into two prime standard; British Standard PD6495 (BSI, 1991) and Nuclear Electric CEGB R6 (R6, 2006). BSI PD6495 has replaced by BS7910 (BSI, 1997), and most latest (BSI, 2005). Original BSI PD6495 primarily concerned on assessment of defect welds. The PD 6495 used crack tip open displacement (CTOD) and SIF K based analysis while the BS7910 used CTOD, K or equivalent K that derived from *J*-integral. Both codes define the cracks are assumed to be independent until or unless the following distance between crack planes d satisfy $d \le 0.5(a_1 + a_2)$ and the cracks are treated as being coplanar. For coplanar cracks, if the distance between cracks s $s \le 2 \times (\text{minimum of } \{c_i\}_{i=1,2})$, a single enveloping crack is assumed and the coalescence occur at $s = 0.5(2c_1 + 2c_2)$. The R6 is an approach to upgrade the BSI PD6495 by Central Electricity Generating Board (CEGB) that focused on operating equipment at high temperature where the assumption of equal or greater fracture and possibility of plastic collapse together and fracture separately. The concept of failure assessment diagram (FAD) is introduced to occupy the need of fracture parameter of plasticity fracture. R6 provides a special form of *J*-integral analysis to impose the plastic collapse limit. Details of FAD can be found in (R6, 2006). JSME Fitness-for-Service Code provides no prescription for the interference between multiple cracks or flaws. In (JSME, 2000), in example of parallel offset cracks,the multiple cracks are replaced by an equivalent single crack based on the stage of detected cracks with satisfying the condition of $S \le 5\text{mm}(H \le 10\text{mm})$ and $S > 5\text{mm}(H < 2S)$ where S is relative vertical spacing and H is relative horizontal spacing. When the crack tips distance $S \le 0$ due to overlapped condition, the crack growth evaluation is considered about the coalescence stages. The guideline in JSME Code is based on experimental results. In (JSME, 2008), the interacting cracks are combined in crack growth prediction and the judgment is based on the relative spacing S and H at the initial condition. If the relative spacing at the beginning of the growth prediction meets the criterion, two cracks are combined when the distance S become zero during the crack growth.

2.3. Interacting cracks models for Fitness-for-Service

The solution for interacting surface or embedded cracks that based on FFS revision is limited in literature especially for interacting parallel edge cracks in finite body. Therefore, a review on available developed technique or models that related to FFS is presented in this section. The need of continues revision on FFS limitations based on ASME Boiler and Pressure Vessel Code Section XI and British Standard BS7910 started by the industrial failures in all major pressure vessels (Burdekin, 1982). The pressure vessels are designed and built to comply with ASME Boiler and Pressure Vessel Code Section XI and British Standard BS7910 codes but the failure occurrences are significantly high. Both codes can be expressed as

$$\begin{array}{ll} \text{ASME} & s \le 2 \times (\text{maximum of } \{a_1, a_2\}) \\ \text{BSI PD6493} & s \le 2 \times (\text{maximum of } \{c_1, c_2\}) \end{array} \quad (1)$$

To investigate the problem, (Burdekin, 1982) studied the interaction between the collapse and fracture in pressure vessels using the approach that successfully applied to nuclear applications. The approach applied the LEFM and EPFM using COD and *J*-integral based on single crack under bending condition. The study revealed the important of fracture mechanics as a tool for interaction in failures. This work can be considered as among the first work that put concern on the FFS codes.

Similarly, (O'donoghue et al., 1984) investigated the formation of elliptical cracks in aircraft and pressure vessel attachment lugs and identified the formation is due to stress risers and cracks interaction. Two equal coplanar surface cracks under Mode I loading are modeled using the proposed finite element alternating method (FEAM) in finite solid and the FE analysis results are compared to ASME Boiler and Pressure Vessel Code procedure. The interaction effects are defined by proposed magnification factor (normalized SIF) and the magnification factors seem to increase due to the increase interaction of two cracks and the depth of cracks. The Section XI of ASME Boiler and Pressure Vessel Code recommend that two interacting surface flaws in a pressure vessel should be modeled by a single elliptical crack that covers both flaws. It can be seen that SIF for single crack as proposed by FFS code are generally larger than those due to two interacting cracks as proposed by (O'donoghue et al., 1984). This trend of magnification factor shows that the ASME Boiler and Pressure Vessel Code in Section XI procedure will tend to underestimate the design life of multiple flaws structures. The ASME pressure vessel codes (ASME, 1998) and British Standard PD6495 (BSI, 1991) do not quantify the interaction between cracks especially in two close proximity cracks. At sufficiently close distance, the interaction may cause the increase of SIF. The exclusion of crack interaction may result with unrealistic SIF. Therefore, with the concern on the above standard guideline, (Leek & Howard, 1994) presents an empirical method to approximate the interaction factor of two coplanar surface cracks under tension and bending loading. The approximation approach resulted with good agreement with FE analysis using developed BERSAFE program , (Murakami & Nasser, 1982) and (J. C. Newman & Raju, 1981) within ± 5% discrepancy.

$$a = \text{maximum of } \{a_i\}_{i=1,2}\ ,\ \ c = c_1 + c_2 \quad (2)$$

$$(s / \bar{c}) \times (s / \bar{a}) > 3.38 \quad \text{and} \quad s / \bar{a} \ge 2.49 \quad (3)$$

Based on the (ASME, 2004) and (BSI, 1991) design code that expressed by Eq. (1), (Moussa et al., 1999) used FEM to analyze interaction of two identical parallel non-coplanar surface cracks subjected to remote tension and pure bending loads. The interaction factor as a function of stress shielding to cause overlapping in distance is studied using three dimensional linear finite element analyses. The formation of stress relaxation state is introduced near crack front, as a form of shielding effect at sufficient overlapping. *J*-integral is calculated based on models by (Shivakumar & Raju, 1992) and the interaction factor γ is defined as follows:

$$\gamma = K_{in} / K_{is} \tag{4}$$

where K_{in} and K_{is} are the SIFs with and without the influence of interaction, respectively. As conclusion, the interaction effect appears to diminish as the value of *s/c* approaches 2.0. The existing rules for re-characterization of interacting cracks as less conservative for high values of *s/c* and over-conservative as *s/c* is close to 2.0.

The existing (ASME, 1998; JSME, 2000) FFS combination rules provided no prescription for the interference between multiple cracks for corrosion fatigue. Therefore, (Kamaya, 2003) developed simulation model to extent the condition of coalescence rules in (JSME, 2000) for crack growth process using body force method (BFM). BFM is used to investigate the multiple cracks growth in stress corrosion cracking. Based on JSME code and the SIF value of coalescence behavior from experiments, the new SIF formulation is developed using BFM where focus is given to the interaction between cracks under various relative position and size. The crack propagation direction can be written as

$$\theta_{\max} = \mp \cos^{-1}\left(\left(3K_{II}^2 + K_I\sqrt{8K_{II}^2 + K_I^2}\right) / \left(9K_{II}^2 + K_I^2\right)\right) \tag{5}$$

where the sign in Eq. (5) positive in the case of $K_{II} / K_I < 0$ and negative in the case of $K_{II} / K_I > 0$. When the crack are close and overlapped, the crack interaction intensity between cracks is almost equivalent to single coalesced crack. The change of inner crack tips direction also found with little influence on the crack growth behavior. The relative crack length and position influenced the crack interaction intensity.

The combination rule in ASME Code is found to provide the relative large overestimation of the actual crack growth since the complex growth phenomena under interaction are summarized in simple combination rules. In order to reduce the conservativeness in existing code, (Kamaya, 2008b) proposed alternative assessment procedures based on the size of area and fatigue crack growth. Experimental analysis and testing is conducted using stainless steel specimens (A-H0S5 and B-H0S5) subjected to cyclic tensile loading. FE analysis is carried out to simulate the crack growth during coalescence. In the simulation, the automatic meshing was generated by command language in PATRAN and the SIF is derived from energy release rate obtained from virtual crack extension integral method using ABAQUS. The normalized SIF of Mode I, K_I is expresses as

$$F_I = K_I / \sigma_0 \sqrt{\pi a} \tag{6}$$

where σ_o denotes the applied tensile stress and the *a* is the maximum depth. As a result, the area of the crack face is concluded to be the predominant parameter for the crack growth of interacting cracks under test condition. The cracks of various shapes can be characterized as semi-elliptical cracks of the same area. In extension of parallel semi-elliptical cracks study, (Kamaya, 2008a) investigated the coalescence of adjacent cracks as a result of crack growth with the influence of crack interaction. The magnitude of interaction is represented by driving force of the crack growth (CGF), written as

$$W_m = \sum_{i=1}^{n-1} 0.5\left(D_p K_{I(i)}^{m_p} + D_p K_{I(i+1)}^{m_p} \right) \Delta g \tag{7}$$

where D_p and m_p are the material constant, $K_{I(i)}$ denotes the Mode I SIF of the i th node from $p = 0^o$. Δg and n are the distance of neighboring mode on the crack front and number of nodes. The CGF formulation proved that the interaction between surface cracks not only dependent on relative spacing but also the position of crack front. In the condition of $S > 0$, as the cracks overlapped, the stress shielding effect influenced the change of CGF. The most important, the study notified the cracks can be replaced with single crack of the same area when the relative spacing is sufficiently close, at crack spacing $H < a$. In regular inspection of pressure vessel components, the adjacent defects are found close enough. Under operational loading, the stress field around the crack tips will be magnified and accelerates the crack growth rate. This matter has been referred to current fitness-for-service (FFS) rules such as ASME Boiler and Pressure Vessel Code Section XI (ASME, 1998, 2004), API 579 (ASME, 2007), British Standard PD6495 (BSI, 1991) , BS7910 (BSI, 1997, 2005) and Nuclear Electric CEGB R6 (R6, 2006). The multiple interacting cracks are combined as single crack as the two cracks satisfy the prescribed criterion. As observed, this rule introduced unrealistic discontinuity in the process of crack growth due to the crack interaction is neglected. The evaluation of two interacting coplanar cracks in plates under tension is conducted by (Xuan, Si, & Tu, 2009) and creep interaction factor γ_{creep} is introduced by using C*integral prediction analysis and the FE analysis is executed using ABAQUS to verify the proposed approach. Creep interaction factor γ_{creep} is expressed as

$$\gamma_{creep} = \left(C^*_{Double} / C^*_{Single} \right)^{1/2} \tag{8}$$

In conclusion, the creep crack interaction represented by C*integral is affected by crack configuration (e.g. relative crack distance *c/d*, depth of crack *a/t* and location at crack front 2ø/π) and time dependent properties of material such as creep exponent *n*. The increasing crack aspect *a/c* resulted with no significant effect to C* integral.

Most recent, (Kamaya et al., 2010) used S-version finite element method to determine the SIF changes due to the interaction of stress field which caused variation in crack growth rate and cracks shape. The root of interaction problems is referred to (JSME, 2000, 2008) and (ASME, 2004, 2007) for the case of interacting dissimilar crack sizes. However, the effect of difference crack size or relative size effect is not taken into account in the aforementioned code. The results have shown that smaller cracks stopped growing when the difference in size of interaction was large enough. It means, the interaction effect on the fatigue life of the larger cracks was negligibly small. Moreover, the offset distance and the relative size were important parameter for interaction evaluation especially when the $S=0$ and the condition of crack spacing H/c_1 and cracks ratio c_2/c_1 must be considered most. In present study, the focus is given to determine the stage of crack interaction intensity is equal to single crack in a state of crack interaction limit (CIL) and unification (CUL) using finite element method.

3. Finite element analysis

3.1. Finite element analysis

The stress in the neighborhood of a crack tip in homogenous isotropic material exist in a form of square-root singular and there have been many special elements or singularity function based approach were described in details in (Banks-Sills, 1991). The square root singular stresses in the neighborhood can be modeled by quarter-point, square and collapsed, triangular elements for two dimensional problems, and by brick and collapsed, prismatic elements in three dimensions. Quarter-points square have been found to produce the most excellent results (Banks-Sills, 2010). The stiffness matrix of the element is evaluated using two-dimensional integral based on Gaussian quadrature approach. The plate is constructed with a consideration of singular element and assigned to both crack tips Ct_1 and Ct_2. It is because the high gradients of singular stress-strain and deformation fields are concentrated at both crack tips. The SIF calculation is limited to linear elastic problem with a homogeneous, isotropic material near the crack region.

3.1.1. Singularity stress field

The studies are conducted in a pure Mode I loading condition with specified material, Alloy 7475 T7351 solid plate in constant thickness, homogenous isotropic continuum material, linear elastic behavior, small strain and displacements, and crack surface are smooth. According to Westergaard method for single crack, Mode I K_I and Mode II K_{II} SIF can be expressed as:

$$\begin{aligned} K_I &= F / K_o = F(b/a, a/W)\,\sigma_0\sqrt{\pi a} \\ K_{II} &= F / K_o = F(b/a, a/W)\,\tau_0\sqrt{\pi a} \end{aligned} \tag{9}$$

where σ_0 is nominal stress, τ_0 is shear stress, W is width of specimen, a and b is the length of crack and crack interval, respectively. The work starts by determination of K_I and K_{II} using Eq. (9). The important issue which differs from single crack is the existence of cracks interaction in fracture analysis.

Consider two multiple edge crack of length a_1 and a_2 which occupies the segment of $0.05 \le a/W \le 0.5$ and $0.5 \le b/a \le 3.0$ in finite plate subjected to uniform equal stress σ along the y direction, as shown in Fig. 1(a) and (b). The SIF formulation is based on the creation of singular element at the crack tip based on quadratic isoparametric finite element developed in ANSYS evironment based on (Madenci & Guven, 2006), where the element is based on Barsoum (Barsoum, 1974, 1975), as depicted in Fig. 2 (a) and (b). The singularity is obtained by shifting the mid-side node the ¼ point close to the crack tip. To calculate the SIF, the elements are assumed to be in rigid body motion and constant strain modes. The master element mapping in Cartesian space is transformed into curvilinear space using Jacobian transformation which used to interpolate the displacement within the elements (Chandrupatla & Belegundu, 2002). The accuracy of special element has been addressed by (Murakami, 1976) where the crack tip nodal point is enclosed by a number of special element. In analysis, the size, number and compatibility of special elements really affect the accuracy. The special elements also defined as singularity function methods where stress singularity at crack tip is modeled. The condition of continuity between elements is the most important. By using singularity function method, Mode I and Mode II of stress intensity factor may be able to calculate with high accuracy (Shields, Srivatsan, & Padovan, 1992).

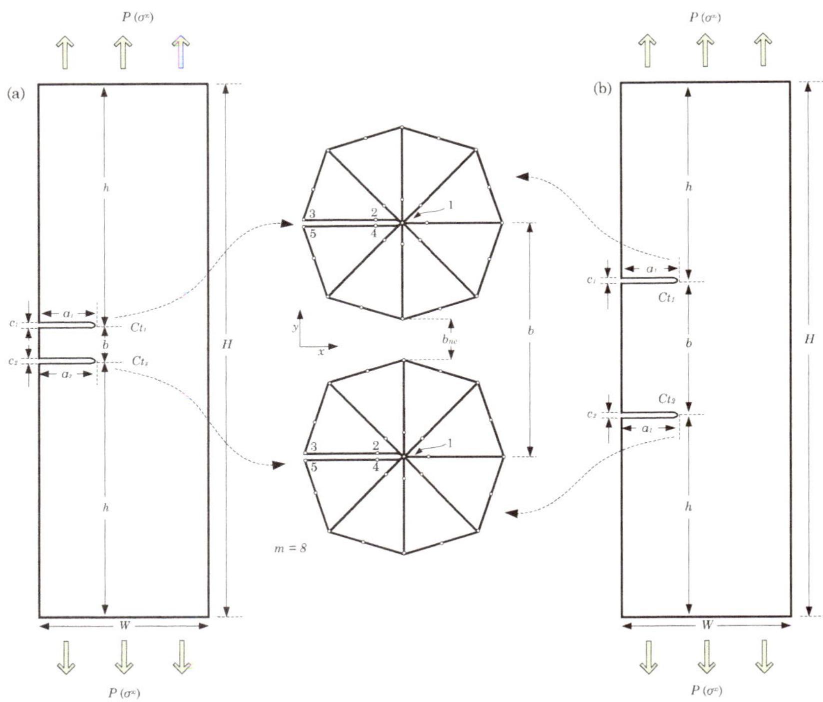

Figure 1. Barsoum singular element for (a) strong crack interaction and (b) weak crack interaction

According to Eq. (9), the shape correction factor can be converted to new elastic interaction factor

$$\gamma_{I,in,D} = K_I / K_o \tag{10}$$

$$\gamma_{II,in,D} = K_{II} / K_o \tag{11}$$

where γ_I and γ_{II} denotes to elastic interaction factor for Mode I and II fracture, respectively.

The SIF for Mode I and Mode II is determined using Displacement Extrapolation Method (DEM) by written APDL macro code in ANSYS (Madenci & Guven, 2006), and expressed as

$$K_I = \frac{E}{3(1+\nu)(1+\kappa)}\sqrt{\frac{2\pi}{L}}\left(4(v_2 - v_4) - \frac{(v_3 - v_5)}{2}\right) \tag{12}$$

$$K_{II} = \frac{E}{3(1+\nu)(1+\kappa)}\sqrt{\frac{2\pi}{L}}\left(4(u_2 - u_4) - \frac{(u_3 - u_5)}{2}\right) \tag{13}$$

where, E=Young Modulus, $\kappa = 3 - 4\nu$ for plain stress, $\kappa = 3 - 4\nu / 1 - \nu$ for plain strain, L is length of element, v and u are displacements in a local Cartesian coordinate system and υ is Poisson's ratio.

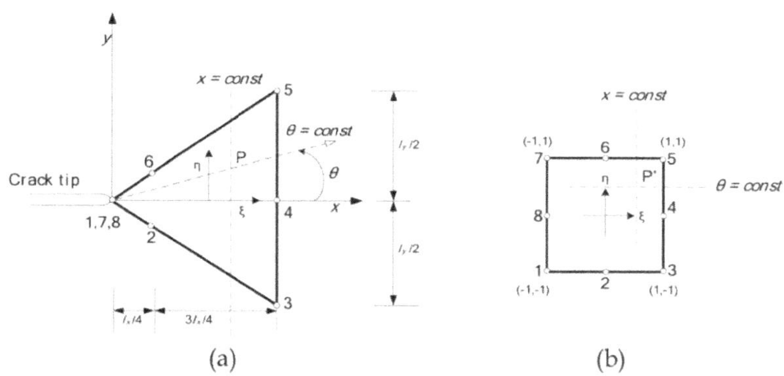

Source: (Barsoum, 1974, 1976; Henshell & Shaw, 1975)

Figure 2. (a) Eight nodes quadratic isoparametric elements (b) Parent element

4. Findings and discussion

Two parallel edge cracks interaction will mainly referred to shielding effect rather than amplification effect. The crack interaction is proportional to the magnitude of elastic interaction factor γ_I. The crack interaction will only exist at $b/a < 3$ (Z.D.Jiang, A.Zeghloul, G.Bezine, & J.Petit, 1990; Z.D.Jiang, J.Petit, & G.Bezine, 1990), the analytical formulation can be expressed as

$$K_I = K_o F_n(a / W, b / a) \tag{14}$$

in which

$$\gamma_{I,in,Ji} = F_n = F_2 = \frac{A_0 + A_1(a/w)^{1.5} + A_2(a/w)^4}{\sqrt{1-(a/w)^2}}$$
$$A_0 = 0.79 + 0.07(b/a) + 0.04(b/a)^2 - 0.011(b/a)^3 \quad (15)$$
$$A_1 = 1.74 + 2.84(b/a) - 1.44(b/a)^2 + 0.206(b/a)^3$$
$$A_2 = 6.02 + 2.19(b/a) - 3.26(b/a)^2 + 0.828(b/a)^3$$

and analytical single edge crack SIF reference by (Brown & Strawley, 1966),

$$K_{I,ref} = \sigma_0\sqrt{\pi a}\left(1.12 - 0.23(a/W) + 10.6(a/W)^2 - 21.7(a/W)^3 + 30.4(a/W)^4\right)$$
$$K_{I,ref} = \sigma_0\sqrt{\pi a}\left(f_{I,ref,BS}\right) \quad (16)$$

4.1. Crack interaction factor $\gamma_{I,in,D}$ comparison with analytical data $\gamma_{I,in,Ji}$

The mode I fracture of the elastic interaction factors $\gamma_{I,in,D}(\gamma_{I,in,D,ct1}, \gamma_{I,in,D,ct2})$ for crack interval ratio $b/a = 1.5 - 3.0$ and $a/W = 0.05 - 0.5$ are shown in Fig. 3-6. Overall, it can be seen that the interaction factor varies with the different a/W values where the crack interaction factor increases as the a/W increases and vice versa. The point of intersection also observed occurred for all the b/a values. For example, the plot of $\gamma_{I,in,D}(\gamma_{I,in,D,ct1}, \gamma_{I,in,D,ct2})$ against a/W at $b/a = 3.0$ are illustrated in Fig. 3.

The $\gamma_{I,in,D}$ prediction line is compared with the predicted result of single edge crack $f_{I,ref,BS}$. A general good agreement can be observed with a minimum difference 0.6% at $a/W = 0.1$ and maximum difference 5.39 % difference $a/W = 0.15$. In comparison, the present $\gamma_{I,in,D}$ has demonstrated more accurate prediction compared with $\gamma_{I,in,Ji}$ results. For example, in reference to $f_{I,ref,BS}$, at $a/W = 0.5$, the present prediction $\gamma_{I,in,D}$ is in difference of $\gamma_{I,in,D,ct1}$ 0.85 % and $\gamma_{I,in,D,ct2}$ 0.4%, while $\gamma_{I,in,Ji}$ is at 2.15% difference. In terms of the CIL point, the closer the crack interaction to $f_{I,ref,BS}$ of single crack, a more accurate CIL prediction can be achieved. It is noted that the $\gamma_{I,in,Ji}$ analytical expression was formulated using the numerical results of *J*-integral analysis. The formulation is unable to calculate the crack interaction factor for both crack tips and become the weakness of $\gamma_{I,in,Ji}$. Therefore, the present work of DEM has improved the existing *J*-integral analysis by improving the accuracy of CIL to predict fracture due to crack interaction.

Theoretically, the study of intersection point is most significant in identification of crack interaction limit (CIL) and crack unification limit (CUL). The intersection point of two cracks $\gamma_{I,in,D}$ with single crack $f_{I,ref,BS}$ justifies the realization of CIL at higher a/W and CUL at lower a/W. The CIL and CUL also differ with different b/a. From Fig. 3, the identified CUL is at $a/W = 0.1$ and CIL approximately at $a/W = 0.5$.

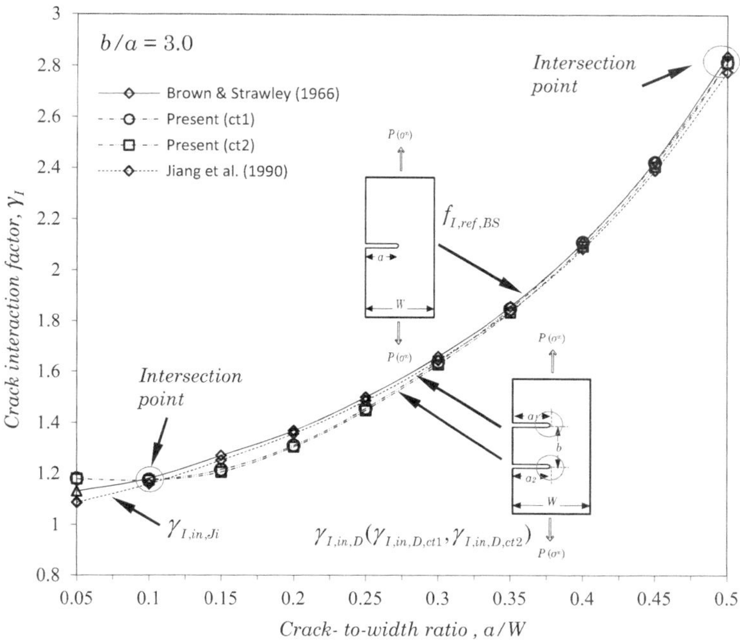

Figure 3. Variation of $\gamma_{I,in,D}$ against a/W for $b/a = 3.0$

Another significant improvement of present $\gamma_{I,in,D}$ is the moving intersection point as a/W decreases for every b/a, as shown in Fig. 3-6. From Fig. 4, the moving intersection point can be noticed moves from $a/W = 0.1$ to $a/W = 0.075$. The moving intersection point exhibits similar prediction trend of $\gamma_{I,in,D}$. The intersection point also can be denoted as the crack unification limit (CUL) point, which indicates the starting point of strong interaction region start approximately at $a/W < 0.07$.

The intersection point is also observed to move from $a/W = 0.075$ for $b/a = 2.5$, $a/W = 0.06$ for $b/a = 2.5$, $a/W = 0.05$ for $b/a = 2.0$ and $a/W < 0.05$ for $b/a = 1.5$, as shown in Fig. 4-6. It means that the CUL is not in a fix limit, it exist in dynamic condition which depends on crack interval ratio b/a. Conversely, the $\gamma_{I,in,Ji}$ prediction model overruled the FFS codes because it does not lead to a single independent or combined crack because of not having any intersection point. The intersection point could not be defined by $\gamma_{I,in,Ji}$ prediction model.

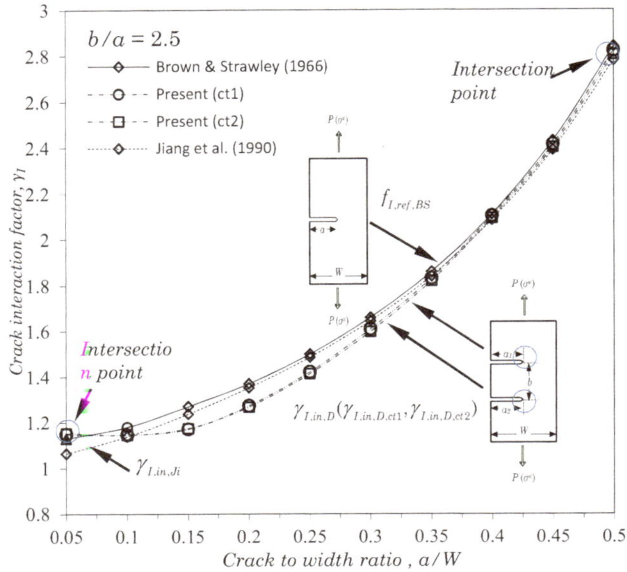

Figure 4. Variation of $\gamma_{I,in,D}$ against a/W for $b/a = 2.5$

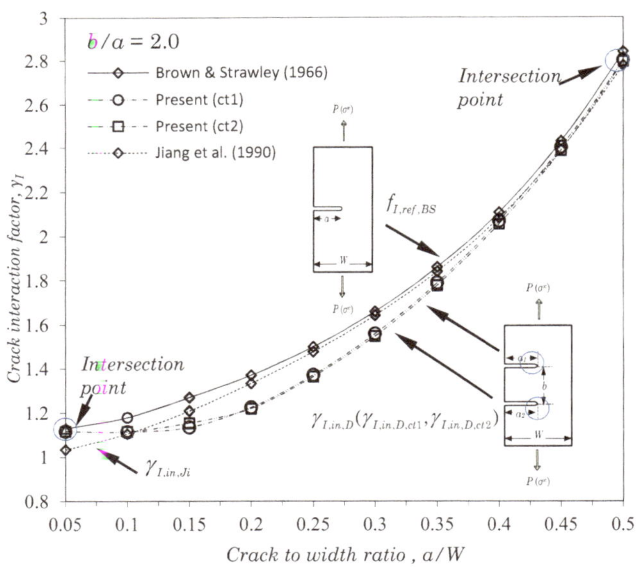

Figure 5. Variation of $\gamma_{I,in,D}$ against a/W for $b/a = 2.0$

It also can be seen that $\gamma_{I,in,Ji}$ prediction is unable to display the unification of crack interaction factor, which defined equivalent to $f_{I,ref,BS}$ of single crack. In this case, the $\gamma_{I,in,Ji}$ prediction is expected to encounter some numerical errors since at lower $a/W < 0.07$. In analysis, the ratio of crack length and width also define the critical stress field, if the path independent radius of *J*-integral for both crack tips are overlapped, the calculation of *J* value might be overestimated and the stress behavior is equal to behavior of single edge crack in finite body. The path integral line should be always apart and controlled in individual condition.

Based on the FFS codes, the multiple cracks are assumed to be independent as single cracks or combined cracks, until or unless certain conditions are satisfied. The intersection point that lies in the present $\gamma_{I,in,D}$ prediction trend curve, which intersects with the single crack prediction, shows good agreement with the outlined FFS codes. It is seen that at range of $0.05 \le a/W \le 0.15$, the $\gamma_{I,in,D}$ value is about to level at 1.072 -1.085. The small changes in these range indicate that shielding effect is very small and promote the unification in interaction at the point of $a/W = 0.15$. The smaller the b/a, the faster unification process starts.

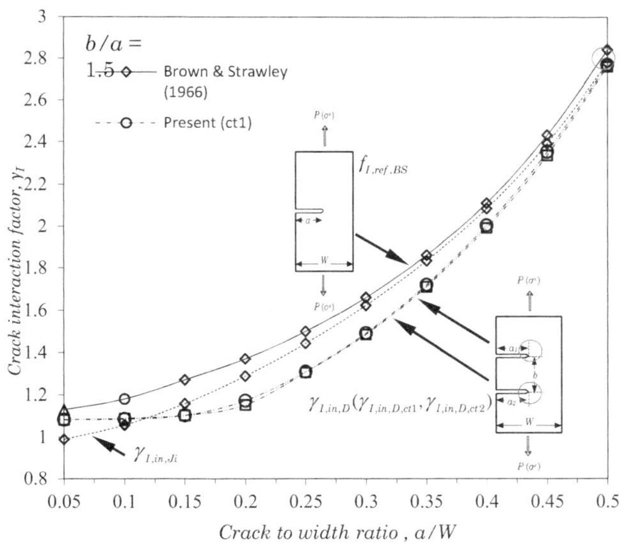

Figure 6. Variation of $\gamma_{I,in,D}$ against a/W for $b/a = 1.5$

4.2. Mode II fracture behavior

Fig. 7 depicts the trend of $K_{I,in,D}$ and $K_{II,in,D}$ against b/a for $a/W = 0.25 - 0.5$. It can be seen that the increase in b/a results in an increase of $K_{I,in,D}$ and a decrease of $K_{II,in,D}$

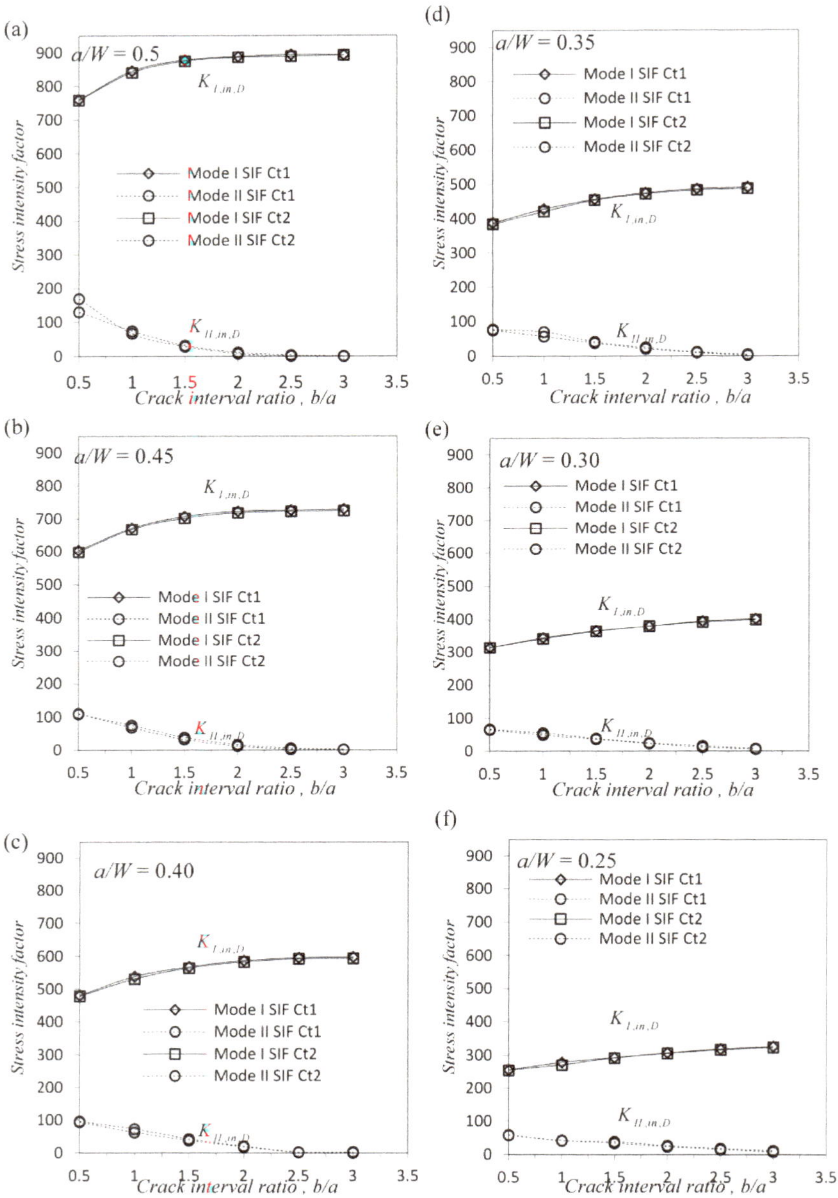

Figure 7. Variation of $K_{I,in,D}$ and $K_{II,in,D}$ for $(a/W = 0.25 - 0.5)$

values. In strong interaction region $0 \le b/a \le 1.0$, the values of $K_{I,in,D}$ increased rapidly for higher crack-to-width ratio $a/W = 0.5, 0.45$ and before grew slowly as the value of a/W decreases to $a/W = 0.25$. Meanwhile, the values of $K_{II,in,D}$ declined significantly at higher crack-to-width ratio $a/W = 0.5, 0.45$ and before decreased slightly as the value of a/W decreases to $a/W = 0.25$. For both crack interaction phase, the value of $K_{I,in,D}$ is always much higher than $K_{II,in,D}$. In weak interaction region $1.5 \le b/a \le 3.0$, it can be seen that the values of $K_{I,in,D}$ increased slightly before growing slowly and then maintaining at the same level to steady state at $b/a = 3.0$. At the same region, the value of $K_{II,in,D}$ declined moderately before decreased slightly and remain stable at the level of $b/a = 3.0$. It means that mode II SIF is less influenced by damage shielding effect than mode I SIF. It also defined that the crack opening is more affected by damage shielding effect than the crack sliding. This has been clearly indicated in Fig. 7 (a)-(f) and Fig. 8(a)-(d).

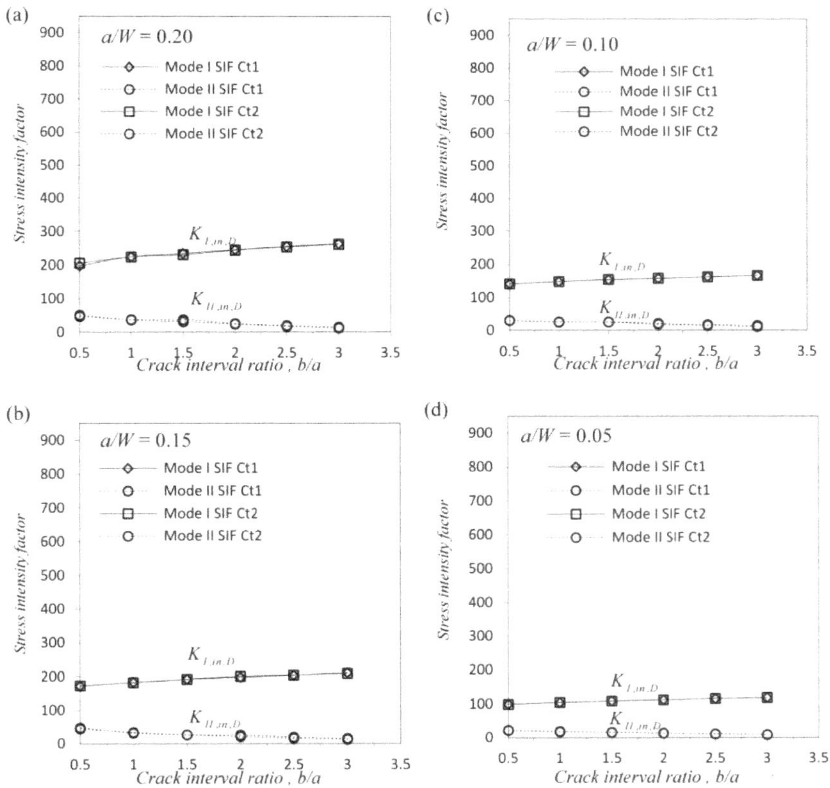

Figure 8. Variation of $K_{I,in,D}$ and $K_{II,in,D}$ for $(a/W = 0.05 - 0.2)$

From these figures, it also observed that the different between mode II SIF and mode I SIF is reduced as the a/W decreases. In the context of crack interaction limit (CIL) based on $K_{I,in,D}$ and $K_{II,in,D}$, by considering the convergence level as the indicator of CIL. It can be seen that the degree and speed of CIL achievement not only depends heavily on the increased of b/a, the reduction of a/W from $a/W = 0.5 - 0.05$ also provides significant impact on CIL determination.

Fig. 8 shows the variation of $K_{I,in,D}$ and $K_{II,in,D}$ for $(a/W = 0.05 - 0.2)$. It can be observed that at smallest $a/W = 0.05$, the value of $K_{I,in,D}$ and $K_{II,in,D}$ is almost hold at constant value and stabilize.

The identification of CIL in this condition is absent because the value of $K_{I,in,D}$ and $K_{II,in,D}$ are about the same value for all b/a. This equalization condition may be referred to the identification of crack unification limit (CIL). Overall, it can be concluded that the higher ratio of b/a and a/W, the more the realization of CIL. Inversely, the lower ratio of b/a and a/W, the more indication to CUL can be realized.

5. Conclusion

The numerical solution based on displacement extrapolation method (DEM) has proved to be more consistent in SIF prediction comparedd to for both crack tips. However, the DEM able to predict the SIF for Mode I and Mode II fracture behaviour. The FE results conclude that the interaction of two cracks is directly influence the reduction of SIF magnitude and γ_I at the crack tips. The parallel cracks have experienced decrease shielding effect as the cracks interval *b/a* decrease. The identification of crack interaction limit (CIL) and crack unification limit (CUL) has been accomplished. The SIF of Mode I is found more significant compared to Mode II in higher or lower b/a and a/W ratio. Mode II SIF can be neglected due to its small effect to the stress shielding effect. The FFS codes rules that define the combination of two cracks as single crack are well translated as CIL and CUL.

Author details

Ruslizam Daud, Ahmad Kamal Ariffin, Shahrum Abdullah and Al Emran Ismail
Universiti Kebangsaan Malaysia, Malaysia

6. References

Andersson, B., Blom, A. F., Falk, U., Wang.G.S, Koski, K., Siljander, A., et al. (2009). *A case study of multiple site fatigue crack growth in the F-18 Hornet bulkhead*. Paper presented at the 25th ICAF Symposium.

ASME. (1998). ASME Boiler and Pressure Vessel Code, Section XI. *New York, USA*.

ASME. (2004). ASME Boiler and Pressure Vessel Code, Section XI. *New York, USA.*

ASME. (2007). 579-1 / FFS-1 Fitness-for-service, Section 9, American Society of Mechanical Engineers. *New York, USA.*

Banks-Sills, L. (1991). Application of the Finite Element Method to Linear Elastic Fracture Mechanics. *Appl. Mech. Rev, 44*, 447-461.

Banks-Sills, L. (2010). Update: Application of the Finite Element Method to Linear Elastic Fracture Mechanics. *Appl. Mech. Rev, 63*, 1-17.

Barsoum, R. S. (1974). Application of Quadratic Isoparametric Finite Elements in Linear Fracture Mechanics. *International Journal of Fracture, 10*, 603-605.

Barsoum, R. S. (1975). Further application of quadratic isoparametric finite element to linear fracture mechanics of plate bending and general rules. *International Journal of Fracture, 11*, 167-169.

Barsoum, R. S. (1976). Application of Triangular Quarter-Point Elements as Crack Tip Elements of Power Law Hardening Material. *International Journal of Fracture, 12*, 463-466.

Brown, W. F., & Strawley, J. E. (1966). ASTM STP 410.

BSI. (1991). British Standard Institute, PD 6493, Section 8. *Guidance on methods for assessing the acceptability of flaws in fusion welded structures.*

BSI. (1997). British Standard Institute, BS7910, Section 8. *Guidance on methods for assessing the acceptability of flaws in structures.*

BSI. (2005). British Standard Institute, BS7910. *Guidance on methods for assessing the acceptability of flaws in metallic structures.*

Burdekin, F. M. (1982). The role of fracture mechanics in the safety analysis of pressure vessels. *Int. J. Mech. Sci., 24*(4), 197-208.

Chandrupatla, T. R., & Belegundu, A. D. (2002). *Introduction to Finite Elements in Engineering* (3rd ed.): Prentice Hall.

Chang, J. I., & Lin, C.-C. (2006). A study of storage tank accidents. *Journal of Loss Prevention in the Process Industries, 19*, 51-59.

Chen, Y. Z., & Liu, H. Y. (1988). Multiple cracks in pressurized hollow cyclinder. *Theoritical and Applied Fracture Mechanics, 10*, 213-218.

Garwood, S. J. (2001). Investigation of the MV Kurdistan casuality. *Engineering Failure Analysis, 4*(1), 3-24.

Gorbatikh, L., & Kachanov, M. (2000). A simple technique for constructing the full stress and displacement fields in elastic plates with multiple cracks. *Engineering Fracture Mechanics, 66*, 51-63.

Henshell, R. D., & Shaw, K. G. (1975). Crack tip finite elements are unnecessary. *Int J Numer Engng, 9*, 495-507.

Hu, W., Liu, Q., & Barter, S. (2009). *A study of interaction and coalescence of micro surface fatigue cracks in aluminium 7050.* Paper presented at the 25th ICAF Symposium.

Ikeda, T., Nagai, M., Yamanaga, K., & Miyazaki, N. (2006). Stress intensity factor analyses of interface cracks between dissimilar anisotropic materials using the finite element method. *Engineering Fracture Mechanics, 73*, 2067-2079.

Isida, M. (1973). Analysis of stress intensity factors for the tension of a centrally cracked strip with stiffened edges. *Engineering Fracture Mechanics, 5*, 647-665.

Jeong, D. Y., & Brewer, J. C. (1995). On the linkup of multiple cracks. *Engineering Fracture Mechanics, 51*(2), 233-238.

Jones, R., Peng, D., & Pitt, S. (2002). Assessment of multiple flat elliptical cracks with interactions. *Theoretical and Applied Fracture Mechanics, 38*, 281-291ssss.

JSME. (2000). *JSME Fitness-for-Service Code S NA1-2000.*

JSME. (2008). *JSME Fitness-for-Service Code S NA1-2008.*

Kamaya, M. (2003). A crack growth evaluation method for interacting multiple cracks. *JSME International Journal, 46*(1), 15-23.

Kamaya, M. (2008a). Growth evaluation of multiple interacting surface cracks. Part II: Growth evaluation of parallel cracks. *Engineering Fracture Mechanics, 75*, 1350-1366.

Kamaya, M. (2008b). Growth evaluation of multiple interacting surface cracks.Part I: Experiments and simulation of coalesced crack. *Engineering Fracture Mechanics, 75*, 1336-1349.

Kamaya, M., Miyokawa, E., & Kikuchi, M. (2010). Growth prediction of two interacting surface cracks of dissimilar sizes. *Engineering Fracture Mechanics, 77*, 3120-3131.

Kirkhope, K. J., Bell, R., & Kirkhope, J. (1991). Stress intensity factors for single and multiple semi-elliptical surface cracks in pressurized thick-walled cylinders. *International Journal of Pressure Vessel and Piping, 47*, 247-257.

Kobayashi, H., & Kashima, K. (2000). Overview of JSME flaw evaluation code for nuclear power plants. *International Journal of Pressure Vessels and Piping, 77*, 937-944.

Lakes, R. S., Nakamura, S., Behiri, J. C., & Bonfield, W. (1990). Fracture mechanics of bone with short cracks. *J. Biomechanics, 23*(10), 967-975.

Lam, K. Y., & Phua, S. P. (1991). Multiple crack interaction and its effect on stress intensity factor. *Engineering Fracture Mechanics, 40*(3), 585-592.

Leek, T. H., & Howard, I. C. (1994). Estimating the elastic interaction factor of two coplanar surface cracks under Mode I load. *International Journal of Pressure Vessels and Piping, 60*, 307-321.

Leek, T. H., & Howard, I. C. (1996). An examination of methods of assessing interacting surface cracks by comparison with experimental data. *International Journal of Pressure Vessels and Piping, 68*, 181-201.

Lewis, P. R., & Weidmann, G. W. (2001). Catastropic failure of a polypropylene tank Part I: primary investigation. *Engineering Failure Analysis, 6*(4), 197-214.

Madenci, E., & Guven, I. (2006). *The finite element method and application in engineering using ANSYS* (1 ed.): Springer Science-Business.

Matake, T., & Imai, Y. (1977). Pop-in behaviour induced by interaction of cracks. *Engineering Fracture Mechanics, 9*, 17-24.

Meizoso, A. M., Esnaola, J. M. M., & Perez, M. F. (1995). Interaction effect of multiple crack growth on fatigue. *Theoretical and Applied Fracture Mechanics, 23*, 219-233.

Milwater, H. R. (2010). A simple and accurate method for computing stress intensity factors of collinear interacting cracks. *Aerospace Science and Technology.*

Mischinski, S., & Mural, A. (2011). Finite Element Modeling of Microcrack Growth in Cortical Bone. *Journal of Applied Mechanics, 78*, 1-9.

Moussa, W. A., Bell, R., & Tan, C. L. (1999). The interaction of two parallel non-coplanar identical surface cracks under tension and bending. *International Journal of Pressure Vessels and Piping, 76*, 135-145.

Murakami, Y. (1976). A simple procedure for the accurate determination of stress intensity factors by finite element method. *Engineering Fracture Mechanics, 8*, 643-655.

Murakami, Y., & Nasser, S. N. (1982). Interacting dissimilar semi-elliptical surface flaws under tension and bending. *Engineering Fracture Mechanics, 16*(3), 373-386.

Newman, J. A., Baughman, J. M., & Wallace, T. A. (2010). Investigation of cracks found in helicopter longerons. *Engineering Failure Analysis, 17*, 416-430.

Newman, J. C., & Raju, I. S. (1981). An empirical stress intensity factor equation for the surface cracks. *Engineering Fracture Mechanics, 15*, 185-192.

O'donoghue, P. E., Nishioka, T., & Atluri, S. N. (1984). Multiple surface cracks in pressure vessel. *Engineering Fracture Mechanics, 20*(3), 545-560.

Pant, M., Singh, I. V., & Mishra, B. K. (2011). A numerical study of crack interactions under thermo-mechanical load using EFGM. *Journal of Mechanical Science and Technology 25*(2), 403-413.

Park, C. H., & Bobet, A. (2010). Crack initiation, propagation and coalescence from frictional flaws in uniaxial compression. *Engineering Fracture Mechanics, 77*, 2727-2748.

Parker, A. P. (1999). Stability of arrays of multiple edge cracks. *Engineering Fracture Mechanics, 62*, 577-591.

Pitt, S., Jones, R., & Atluri, S. N. (1999). Further studies into interacting 3D cracks. *Computers and Structures, 70*, 583-597.

R6. (2006). Nuclear Electric : Assessment of the integrity of structures containing defects, Revision 4. *Gloucester, British Energy Generation Ltd.*

Ratwani, M., & Gupta, G. D. (1974). Interaction between parallel cracks in layered composites. *International Journal of Solids and Structures, 10*, 701-708.

Rutti, T. F., & Wentzel, E. J. (1997). Investigation of failed actuator piston rods. *Engineering Failure Analysis, 5*(2), 91-98.

Saha, T. K., & Ganguly, S. (2005). Interaction of penny-shaped cracks with an eliptic crack under shear loading. *International Journal of Fracture, 131*, 267-287.

Sankar, R., & Lesser, A. J. (2006). Generic Overlapping Cracks in Polymers: Modeling of Interaction. *International Journal of Fracture, 142*, 277-287.

Sekhtar, A. S. (2008). Multiple cracks effects and identification. *Mechanical Systems and Signal Processing, 22*, 845-878.

Shields, E. B., Srivatsan, T. S., & Padovan, J. (1992). Analytical methods for evaluation of stress intensity factors and fatigue crack growth. *Engineering Fracture Mechanics, 42*(1), 1-26.

Shivakumar, K. N., & Raju, I. S. (1992). An equivalent domain integral method for three-dimensional mixed-mode fracture problems. *Engineering Fracture Mechanics, 42*, 935-959.

Ural, A., Zioupos, P., Buchanan, D., & Vashishth, D. (2011). The effect of strain rate on fracture toughness of human cortical bone: A finite element study. *Journal of the Mechanical Behaviour of Biomedical Materials.*

Weidmann, G. W., & Lewis, P. R. (2001). Catastropic failure of a polypropylene tank Part II: comparison of the DVS 2205 code of practice and the design of the failed tank. *Engineering Failure Analysis, 6*(4), 215-232.

Xuan, F.-Z., Si, J., & Tu, S. T. (2009). Evaluation of C* integral for interacting cracks in plates under tension. *Engineering Fracture Mechanics, 76*, 2192-2201.

Yang, Y. H. (2009). Multiple parallel symmetric mode III cracks in a functionally graded material plane. *Journal of Solid Mechanics and Materials Engineering, 3*(5), 819-830.

Z.D.Jiang, A.Zeghloul, G.Bezine, & J.Petit. (1990). Stress intensity factor of parallel cracks in a finite width sheet. *Engineering Fracture Mechanics, 35*(6), 1073-1079.

Z.D.Jiang, J.Petit, & G.Bezine. (1990). Fatigue propagation of two parallel cracks. *Engineering Fracture Mechanics, 37*(5), 1139-1144.

Permissions

The contributors of this book come from diverse backgrounds, making this book a truly international effort. This book will bring forth new frontiers with its revolutionizing research information and detailed analysis of the nascent developments around the world.

We would like to thank Alexander Belov, for lending his expertise to make the book truly unique. He has played a crucial role in the development of this book. Without his invaluable contribution this book wouldn't have been possible. He has made vital efforts to compile up to date information on the varied aspects of this subject to make this book a valuable addition to the collection of many professionals and students.

This book was conceptualized with the vision of imparting up-to-date information and advanced data in this field. To ensure the same, a matchless editorial board was set up. Every individual on the board went through rigorous rounds of assessment to prove their worth. After which they invested a large part of their time researching and compiling the most relevant data for our readers. Conferences and sessions were held from time to time between the editorial board and the contributing authors to present the data in the most comprehensible form. The editorial team has worked tirelessly to provide valuable and valid information to help people across the globe.

Every chapter published in this book has been scrutinized by our experts. Their significance has been extensively debated. The topics covered herein carry significant findings which will fuel the growth of the discipline. They may even be implemented as practical applications or may be referred to as a beginning point for another development. Chapters in this book were first published by InTech; hereby published with permission under the Creative Commons Attribution License or equivalent.

The editorial board has been involved in producing this book since its inception. They have spent rigorous hours researching and exploring the diverse topics which have resulted in the successful publishing of this book. They have passed on their knowledge of decades through this book. To expedite this challenging task, the publisher supported the team at every step. A small team of assistant editors was also appointed to further simplify the editing procedure and attain best results for the readers.

Our editorial team has been hand-picked from every corner of the world. Their multi-ethnicity adds dynamic inputs to the discussions which result in innovative outcomes. These outcomes are then further discussed with the researchers and contributors who give their valuable feedback and opinion regarding the same. The feedback is then collaborated with the researches and they are edited in a comprehensive manner to aid the understanding of the subject.

Apart from the editorial board, the designing team has also invested a significant amount of their time in understanding the subject and creating the most relevant covers. They scrutinized every image to scout for the most suitable representation of the subject and create an appropriate cover for the book.

The publishing team has been involved in this book since its early stages. They were actively engaged in every process, be it collecting the data, connecting with the contributors or procuring relevant information. The team has been an ardent support to the editorial, designing and production team. Their endless efforts to recruit the best for this project, has resulted in the accomplishment of this book. They are a veteran in the field of academics and their pool of knowledge is as vast as their experience in printing. Their expertise and guidance has proved useful at every step. Their uncompromising quality standards have made this book an exceptional effort. Their encouragement from time to time has been an inspiration for everyone.

The publisher and the editorial board hope that this book will prove to be a valuable piece of knowledge for researchers, students, practitioners and scholars across the globe.

List of Contributors

Lucas Máximo Alves
GTEME - Grupo de Termodinâmica, Mecânica e Eletrônica dos Materiais, Departamento de Engenharia de Materiais, Setor de Ciências Agrárias e de Tecnologia, Universidade Estadual de Ponta Grossa, Brazil

Alexander Yu. Belov
Institute of Crystallography RAS, Moscow, Russian Federation

Luiz Alkimin de Lacerda
LACTEC - Instituto de Tecnologia para o Desenvolvimento, Departamento de Estruturas Civis, Centro Politécnico da Universidade Federal do Paraná, Curitiba - PR, Brazil

Karl-Johan Söderholm
College of Dentistry, University of Florida, Gainesville, Florida, USA

Sylwester Kłysz
Air Force Institute of Technology, Warsaw, Poland
University of Warmia and Mazury in Olsztyn, Poland

Andrzej Leski
Air Force Institute of Technology, Warsaw, Poland

Ilya I. Kudish

Kunio Asai
Hitachi Research Laboratory/Hitachi, Ltd., Japan

Narciso Acuña-González
Academic Vice-Chancellor Office, Anahuac-Mayab University, Merida, Mexico

Jorge A. González-Sánchez and Luis R. Dzib-Pérez
Centre for Corrosion Research, Autonomous University of Campeche, Campeche, Mexico

Aarón Rivas-Menchi
Faculty of Engineering, Anahuac-Mayab University, Merida, Mexico

Dino A. Araneo and Francesco D'Auria
GRNSPG, San Piero a Grado Nuclear Research Group, San Piero a Grado, Pisa, Italy

Mahmood Sameezadeh and Hassan Farhangi
University of Tehran, College of Engineering, Iran

Ľubomír Gajdoš and Martin Šperl
Institute of Theoretical and Applied Mechanics, Academy of Sciences of the Czech Republic, Czech Republic

A. Guedri
InfraRes Laboratory; University of Souk Ahras,Algeria
LEM3 Laboratory, UMR CNRS 7239, University of Metz, France

Y. Djebbar
InfraRes Laboratory; University of Souk Ahras, Algeria

Moe. Khaleel
Pacific Northwest National Laboratory, USA

A. Zeghloul
LEM3 Laboratory, UMR CNRS 7239, University of Metz, France

Ruslizam Daud, Ahmad Kamal Ariffin, Shahrum Abdullah and Al Emran Ismail
Universiti Kebangsaan Malaysia, Malaysia